CPS PUBLICATIONS IN PHILOSOPHY OF SCIENCE

edited on behalf of the
Center for Philosophy of Science
University of Pittsburgh

by

Adolf Grünbaum
Larry Laudan
Nicholas Rescher
Wesley Salmon

BEYOND THE EDGE OF CERTAINTY

Essays in Contemporary Science and Philosophy

Edited by

Robert G. Colodny

UNIVERSITY
PRESS OF
AMERICA

LANHAM • NEW YORK • LONDON

University Press of America,™ Inc.
4720 Boston Way
Lanham, MD 20706

3 Henrietta Street
London WC2E 8LU England

Library of Congress Cataloging in Publication Data
Main entry under title:

Beyond the edge of certainty.

    (CPS publications in philosophy of science)
    Reprint. Originally published: Englewood Cliffs,
N.J.: Prentice-Hall, 1965. (University of Pittsburgh
series in the philosophy of science; v. 2)
    Includes bibliographical references and indexes.
    Contents: Newton's first law / Norwood Russell
Hanson — The origin and nature of Newton's laws of
motion / Brian Ellis — A response to Ellis's conception
of Newton's first law/Norwood Russell Hanson—
[etc.]
    1. Physics—Addresses, essays, lectures.
    2. Science—Philosophy—Addresses, essays, lectures.
    I. Colodny, Robert Garland. II. Series. III. Series:
University of Pittsburgh series in the philosophy of
science; v. 2.
[QC71.B46 1983]   530   83-1162
ISBN 0-8191-3057-5
ISBN 0-8191-3058-3

CPS PUBLICATIONS IN PHILOSOPHY OF SCIENCE
Co-published by arrangement with the
Center for Philosophy of Science
University of Pittsburgh

Preface

Since 1960, the Center for Philosophy of Science at the University of Pittsburgh has annually presented a number of public lectures on various current topics in the philosophy of the physical, biological, and social sciences as a permanent part of its program of instruction and research. The lectures delivered during 1960-1961 by scholars from diverse institutions were published in 1962 by the University of Pittsburgh Press under the editorship of Professor Robert G. Colodny in *Frontiers of Science and Philosophy*. The present book of lectures is the second in the series of volumes published under the Center's auspices.

ADOLF GRÜNBAUM
Andrew Mellon Professor of Philosophy
and
Director of the Center for Philosophy of Science
University of Pittsburgh

Contents

Introduction
ROBERT G. COLODNY 1

*Newton's First Law: A Philosopher's Door into Natural
Philosophy* NORWOOD RUSSELL HANSON 6

The Origin and Nature of Newton's Laws of Motion
BRIAN ELLIS 29

A Response to Ellis's Conception of Newton's First Law
NORWOOD RUSSELL HANSON 69

A Philosopher Looks at Quantum Mechanics
HILARY PUTNAM 75

The Thermodynamics of Purpose
DAVID HAWKINS 102

The Physics of the Large
PHILIP MORRISON 118

Problems of Empiricism
PAUL K. FEYERABEND 145

The Ethical Dimension of Scientific Research
NICHOLAS RESCHER 261

Index of Names 279

Index of Topics 283

BEYOND THE EDGE OF CERTAINTY:
Essays in Contemporary Science and Philosophy

ROBERT G. COLODNY
University of Pittsburgh

Introduction

Our age is possessed by a strong urge towards the criticism of traditional customs and opinions. A new spirit is arising which is unwilling to accept anything on authority, which does not so much permit as demand independent, rational thought on every subject, and which refrains from hampering any attack based upon such thought, even though it be directed against things which formerly were considered to be as sacrosanct as you please. In my opinion this spirit is the common cause underlying the crisis of every science today. Its results can only be advantageous: no scientific structure falls entirely into ruin: what is worth preserving preserves itself and requires no protection.

—ERWIN SCHRÖDINGER, *Science and the Human Temperament*

In an age of astounding scientific creativity, it is extremely difficult to obtain a synoptic view of the entire advancing wave of knowledge. The difficulty arises not only from the sheer quantity of published material and the intrinsic complexity of experimental and theoretical findings, but also from the necessity of mapping the "edge of certainty" against some body of knowledge assumed to be established, orthodox, part of a consensus of the international community of scholars. Despite these intractable problems, the modest claim may be made that certain general properties of the process of scientific innovation may be discerned. Without asserting any exhaustiveness or hierarchical ranking, these include:

An increasing tendency of hitherto separate disciplines to flow together, producing conceptual aggregates which are more than the sum of the parts.

As a consequence of the above, as witnessed by such specialties as biophysics, astrophysics, space biology, geochemistry, etc., a comprehensive, unified vision of the cosmos begins to exhibit clearer features.

1

Unsuspected links of unity between separate theories are uncovered; the implications of older, sometimes half-forgotten, hypotheses emerge, and both pioneering scientists and philosophers of science become conscious of how elements of "classical" scientific thought are embedded in contemporary theoretical constructs.

It is at junctures such as these that a daring innovator such as Schrödinger will write: "The old links between philosophy and physical science, after having been frayed in many places, are being more closely renewed. The further physical science progresses, the less it can dispense with philosophical criticism. But at the same time, philosophers are increasingly obliged to become intimately acquainted with the sphere of research, to which they undertake to prescribe the governing laws of knowledge."[1]

Assuming the fulfillment of Schrödinger's rigorous conditions, the world picture presented by the cosmologically oriented scientists and by philosophers of science will tend to differ mainly with respect to the "fine grain." Philosophy of science and "science" in the traditional sense become complementary enterprises.

To the extent that an ensemble of problems is analyzed in depth by a group of philosopher-scientists, the emergent findings will comprise parts of a unified vision, a characteristic of all great ages of scientific achievement.

The reader of these pages will therefore expect to find that the Hanson-Ellis discussions of Newtonian mechanics, Putnam's analysis of the status of quantum mechanics, Hawkins's analysis of thermodynamics and biological teleology, and Morrison's account of astrophysics not only share common historical backgrounds, but also reflect aspects of that unity of physical nature that the various disciplines have been articulating for the last three centuries. It is also of interest to note that the papers of Hawkins and Morrison are excellent examples of that fruitful stimulus to conceptual novelty occasioned by the confluence of several distinct disciplines.

Feyerabend's comprehensive critique of empiricism will be found to illuminate the procedures of all of the foregoing essayists inasmuch as it draws upon similar scientific content and hypotheses for elucidating the nature of scientific progress from the point of view both of discovery and of validation. These extend in time from Newtonian mechanics to wave mechanics, thus clarifying many of the essential problems of the scientific and philosophical dialogue. This does not, however, entail philosophical agreement with the other writers, but rather clarification of the problems posed and the methodology of discourse.

Finally, Rescher's probe of a long-neglected aspect of the scientific enterprise—its ethical dimensions—anchors this volume in its most humane context: the moral dilemmas inherent in that range of activities which culminate in knowledge and which arise from the necessary social-political setting in which scientists are, like other beings, inextricably enmeshed. It is also of interest to note that most of the essays in this work locate particular scientific and philosophical problems in an historical context. This decision arises not out of any antiquarian interest in the past, but from a conviction that many of the most agonizing issues confronting philosopher-scientists arise from the accelerating development of distinct branches of sciences which experienced *contingent* junctures and then proceeded in more or less incomplete fusion, carrying with them unresolved and partly concealed epistemological and metaphysical problems ingredient in the earlier epochs.

The relationship between classical mechanics and many divisions of contemporary science (often referred to in the pages that follow) will illustrate this point. As is well known, Newtonian mechanics made irreversible the process of mathematization of the physical sciences begun by Athenian and Alexandrian scholars of antiquity. The varied elements of the universe were analyzed in terms of the basic concepts of space, time, matter, force, and energy. The comprehensive nature of these elements of dynamical discourse and the elegant mathematical formalisms created in the formulation of their mutual relations endowed the early dynamical laws with the characteristics of perfect examples of a law of nature, and the general method of analysis set forth in the *Principia* and elaborated further by Laplace set a *style* and a tactic for scientific research that advanced from success to success in an ever-widening range of phenomena, until thermodynamics and electrodynamics bore the unmistakable imprint of a particular scientific and philosophic vision.

Here it may be of value to note that in the progression from the Euler-Lagrange formalism to the Hamilton-Jacobi, the domain of the dynamical ideas had increased and embodied extremal principles based upon the work of Fermat and Maupertuis and fully exploiting various conservation theories. There thus emerged a more powerful synthesis of physical knowledge whose supreme instrument was the variational calculus.[2] This led to a renewed philosophical interest in the *foundations* of physical science, an interest associated with the critiques of Hertz, Mach, and Poincaré, and which has been revitalized in the epoch of quantum and relativistic mechanics.

That the transfer of elements from the Newtonian-Hamiltonian world to the Einsteinian-Minkowskian-Weylian has been accomplished only through the closest cooperation of mathematicians, physicists, and philosophers is one of the enduring monuments of modern intellectual history. Here again one should recall that a rather common feature in the history of science is the preparation by one age of the mathematical and technological instruments to be exploited fully by later generations of scientists who will work in a philosophical climate that may be completely different. Thus magnificent scientific constructs repeatedly reveal unsuspected flaws when subjected to a different mode of philosophical scrutiny. What a purely formal logical analysis of theories sometimes fails to reveal is the fact that a complex physical theory is often compounded of elements from quite distinct historical epochs, each of which may have arisen under radically different canons of rigor and different sets of submerged metaphysical presuppositions. Awareness that elements of the scientific world picture reveal in their lack of perfect fit the very disjointed historical processes that created them in the first place may well be one of the reasons for a sense of perpetual crisis in those advancing disciplines that have crossed the "edge of certainty."

Finally, considering the experimental and tentative nature of the formulations of both science and philosophy of science, one is driven to inquire into the climate of opinion that permits the unhampered development of scientific and philosophic ideas. *Here all the contributors to this volume express a consciousness of and a revulsion to any form of authoritarian orthodoxy,* be it grounded in the scientific community or arising external to it. That this is not an issue arising only in alien and enemy lands is well exemplified by Polanyi's assertion that

> Scientific method is, and must be, disciplined by an orthodoxy which can permit only a limited degree of dissent, and . . . such dissent is fraught with grave risk to the dissenter. . . . The authority of current scientific opinion is indispensable to the discipline of scientific institutions; . . . its functions are invaluable even though its dangers are an unceasing menace to scientific progress.[3]

That this represents a minority opinion concerning the relationship of scientific creativity and orthodoxy is epitomized by the recent remarks of that magnificent Pythagorean, Paul Dirac. In his article "The Evolution of the Physicist's Picture of Nature," Dirac, after discussing the circumstances that were prologue to the discovery of Schrödinger's wave equation, states: "I think there is a moral to this story, namely that it is more important to have beauty in one's equations than to have them fit experiment."[4] After an analysis of the later developments of quantum theory, Dirac adds:

> I should like to suggest that one not worry too much about this controversy. . . . The present stage of physical theory is merely a steppingstone towards the better stages that we will have in the future. One can be quite sure that there will be better stages simply *because of the difficulties that occur in the physics of today.*[5]

In a scientific community that has as tools such onetime nonorthodox inventions as Riemannian geometry, Hamilton's noncommutative quaternions, the interconvertability of mass and energy, non-Aristotelian logics, etc., etc., the Diracian point of view as to the audacity of conceptual innovation and a fearless research strategy needs no amens.

The last words belong to Max Born:

> The subordination of fundamental research to political and military authorities is detrimental. The scientists themselves have learned by now that the period of unrestricted individualism in research has come to an end.

They know that even the most abstract and remote ideas may one day become of great practical importance—like Einstein's law of equivalence of mass and energy. They have begun to organize themselves and to discuss the problem of their responsibility to human society. It should be left to these organizations to find a way to harmonize the security of the nations with the freedom of research and publication without which science must stagnate.[6]

Notes

1. Erwin Schrödinger, *Science and the Human Temperament*, James Murphy, trans. (London: George Allen & Unwin, Ltd., 1935).

2. For an elementary, but beautifully lucid account of this, see A. d'Abro, *The Evolution of Scientific Thought*, Chap. 34 (New York: Dover Publications, Inc., 1950).

3. *Science*, Vol. CXLI (Sept. 13, 1963), 1017. Cf. T. S. Kuhn, "The Function of Dogma in Scientific Research," in A. C. Crombie, ed., *Scientific Change* (New York: Basic Books, Inc., 1963).

4. *Scientific American*, Vol. CCVIII (May, 1963), 47.

5. *Ibid.*, 48. Emphasis added. Dirac adds:

It seems to me one of the fundamental features of nature that fundamental physical laws are described in terms of mathematical theory of great beauty and power, needing quite a high standard of mathematics for one to understand it. You may wonder: Why is nature constructed on these lines? One can only answer that our present knowledge seems to show that nature is so constructed. . . . We simply have to accept it. One could perhaps describe the situation by saying that God is a mathematician of a very high order and He used very advanced mathematics in constructing the Universe. . . .

A good many people are working on the mathematical basis of quantum theory, trying to understand the theory better and to make it more powerful and more beautiful. If someone can hit on the right lines along which to make this development, it may lead to a future advance in which people will first discover the equations and then, after examining them, gradually learn how to apply them. To some extent that corresponds with the line of development that occurred with Schrödinger's discovery of his wave equation. Schrödinger discovered the equation simply by looking for an equation with mathematical beauty.

6. Max Born, *The Restless Universe*, W. M. Deans, trans. (New York: Dover Publications, Inc., 1951), 308-309.

NORWOOD RUSSELL HANSON
Yale University

Newton's First Law: A Philosopher's Door into Natural Philosophy

Not the state of rest, but the states of uniform translation form an objectively distinguished class of motions, and this puts an end to the substantial ether. Finally, and fourthly, the general relativity theory re-endows this metric world structure with the capacity of reacting to the forces of matter. Thus, in a sense, the circle is closed.

—HERMANN WEYL, *Philosophy of Mathematics and Natural Science*

Too often the intellectual excitement of science seems geared only to research in *contemporary* physics. Philosophers continually discuss cosmology, relativity, and microphysics. In such areas one's ideas are stretched and strained beyond what our ancestors might have anticipated. Historians of science also have focused attention on past events only by remarking their analogies and similarities with perplexities in physics today. They do not always do this of course, but it happens enough to warrant comment.

However, there are statements, hypotheses, and theories of "classical" science that are rewarding in themselves—without having to be referred to the agonies that now confound quantum theory and cosmology. Specifically, Newton's first law of motion—the "law of inertia"—which has everything a logician of science could desire. Understanding the complexities and perplexities of this fundamental mechanical statement is in itself to gain insight into what theoretical physics in general really is.

Section I

Newton's first law of motion reads:

EVERY BODY FREE OF IMPRESSED FORCES EITHER PER-
SEVERES IN A STATE OF REST OR IN UNIFORM RECTI-
LINEAR MOTION *AD INFINITUM*.[1]

This is not axiomatic in the ancient sense of being "self-evident."
Giants from Aristotle through Archimedes, Ptolemy, Copernicus, and on
to Galileo stood against all or part of this claim. Thus Aristotle:

> All movement is either natural or enforced, and force accelerates natural
> motion (e.g., that of stone downwards), and is the sole cause of unnatural.[2]

Again:

> It is clear, then, that in all cases of local movement there will be noth-
> ing between the mover and the moved, if it can be shown that the pushing
> or pulling agent must be in direct contact with the load. But this follows
> directly from our definitions, for pushing moves things away (either from
> the agent or from something else) to some other place, and pulling moves
> things from some other place either to the agent or to something else.[3]

And again:

> For it is only when something external moves a thing, or brings it to
> rest against its own internal tendency, that we say this happens by force;
> otherwise we do not say that it happens by force.[4]

The only counterthesis to Aristotle's dicta is that of Philoponus (6th
century A.D.), who attacked the idea of *antiperistasis* by insisting that, so
far from being essential to the motion of a projectile, the medium acts
merely to resist such motion. The arrow does not continue its flight because of the parted air burbling around behind the tail of the shaft;
rather, the arrow would go much further were it not for the resistance
of this very burbling air. Thus:

> But a certain *additional time* is required because of the interference of
> the medium. For the pressure of the medium and the necessity of cutting
> through it make motion through it more difficult.[5]

And again:

> *Rather it is necessary to assume that some incorporeal motive force is
> imparted by the projector to the projectile,* and that the air set in motion
> contributes either nothing at all or else very little to this motion of the
> projectile.[6]

However, before Buridan, the Aristotelian outlook prevailed—and
was fairly reasonable, given the available observations. The tugging and
pushing of carts seemed to indicate clearly that a steady force was con-

tinually necessary to keep a cart moving.[7] This law of Newton's, however, pronounces that, once moving, the cart (ideally) will continue to move *ad indefinitum;* the tugging horse functions only to overcome frictional resistance, wind resistance, and the like. But nothing observed by the ancients could have substantiated such a "law." That is why Philoponus's "anti-Aristotelian" views about motion did not carry the day.

> If one wishes to draw a line of separation between the realm of ancient and modern science, it must be drawn at the instant when Jean Buridan conceived his theory of momentum, when he gave up the idea that stars are kept in motion by certain divine intelligences, and when he proclaimed that both celestial and earthly motions are subject to the same mechanical laws.[8]

Thus Buridan:

> With respect to the heavens as a whole, one should envisage one continual influence penetrating all the way to the center: nevertheless that influence has another property and power near the heavens and far off. And heavy and light bodies arrange themselves in this lower world because of this influence which is diversified in power [force?] above and below. And this should not be denied [simply] because we fail to perceive this influence —since we also do not perceive that which is diffused from a magnet to iron through a medium which, nevertheless, is of great force. (*Debemus imaginari a toto caelo unam influentiam continuam usque ad centrum; tamen illa influentia prope caelum et remote habet aliam proprietatem et virtutem, et propter illam influentiam sic virtualiter diversificatam superius et inferius ordinant se gravia et levia in hoc mundo inferiori. Et non debet hoc negari ex eo quod illam influentiam non percipimus sensibiliter, quia etiam non percipimus illam quae de magnete multiplicatur per medium usque ad ferrum, quae tamen est magnae virtutis.*)[9]

Buridan is here generating nothing less than the impetus theory, which is itself genetically connected with Newton's first law:

> One can say that God, when creating the world, has moved, as He pleased, each of the celestial orbits; He has given to each of them an impetus which kept them moving since then. . . . Thus He could rest on the seventh day from the work He had done.[10]

Albert of Saxony reiterates this theory of circular impetus:

> When God created the celestial spheres, He put each of them in motion as He pleased; and they continue in their motion still today by virtue of the impetus which He impressed on them; this impetus is not subjected to any diminution, since the mobile has no inclination which could oppose the impetus, as no corruption there exists.[11]

Nicholas of Cusa comments:

> . . . It is not you nor your spirit who move immediately the globe which is now rotating in front of you. It is, however, you who initiate this motion, since the impulsion of your hand, following your will, produced an impetus and as long as this impetus endures the globe continues to move.[12]

Again:

The spirit of motion, evoked by the child, exists invisibly in the top; it stays in the top for a longer or shorter time according to the strength of the impression by which this virtue has been communicated; as soon as the spirit ceases to enliven the top, the top falls.[13]

But the progress toward Newton's first law is not unbroken; even the great Kepler regards continued force as being necessary merely to maintain the planetary motions:

The body of the sun is circular and magnetic and is turned in its space [path?] carrying the sphere of its influence which is not attractive but [rather] propulsive. (*Solis corpus est circulariter magneticum et convertitur in suo spacio, transferens orbem virtutis suae, quae non est attractoria sed promotoria.*)[14]

Galileo's teacher, Buonamici, writes:

Still, whenever a moving object does not itself generate the force in question, it is then said to be moved by force. That is, while acquiring its ultimate position, it does not itself possess a "moving propensity" since it does not achieve motion from itself. (*Vi autem moveri illa dicuntur quandocunque id quod movetur non confert vim, hoc est non habet illo propensionem, quo movetur, quia. s. non perficiatur ex eo motu, locum illum adipiscens in quo conservetur.*)[15]

Galileo himself is far from clear about precisely how to express his brilliant insight:

. . . Therefore the impetus, ability, energy, or, as one might say, the momentus of descent of the moving body is diminished by the plane upon which it is supported and along which it rolls. (*L'impeto, il talento, l'energia, o vogliamo dire il momento del discendere. . . .*)[16]

Benedetti had already formulated the law of inertia as early as 1585:

Once moving they are never at rest unless impeded. . . . Once in motion only an outside force can restrain them. (*Mota semel nunquam quiescunt, nisi impediantur. . . . Quod semel movetur semper . . . movetur dum ab extrinseco impediatur.*)[17]

Indeed, several natural philosophers—e.g., Descartes, Gassendi, Baliani—were actively exploring along these "inertial" lines. But it was the considerable thoroughness of Galileo's inquiries which finally succeeded in reversing the philosopher's verdict.

Section II

Galileo argued that a sphere, after rolling down a plane on one side of a room, would cross the floor and ascend an inclined plane on the opposite side. Moreover (ignoring friction), the sphere will ascend the second plane to precisely that height above the floor from which it had been released onto the first inclined plane. What the sphere acquires in its descent is thus exactly equal to driving it back up to its original height on the second slope. This is not self-evident either. But intuitively it seems somehow more plausible than any bald statement of the law of inertia which itself, apparently, is now easily generated.

Suppose we incline the second plane less and less steeply. The observation still holds; the sphere will still seek a height on that second plane equal to that from which it started down the first. The second plane is lowered and gets closer to the floor; the sphere travels further and further along the plane "in order to" attain again its original height. (Think of a soapbox derby racer crossing the finish line; it will either come to a stop in a short distance by climbing a steep hill or else travel further by ascending a shallow hill.) Now, as the angle between the floor and the second plane gets closer to zero, the distance the sphere will travel along it and up it will increase; it will, indeed, proceed toward an infinite length of travel as the angle inclination proceeds toward zero. (Thus the derby racer would proceed to infinity were its path unobstructed with hills, steep or shallow, or with friction, wind resistance, or gravity.)[18]

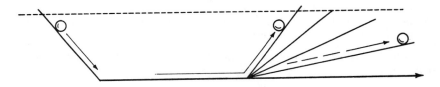

This much alone indicates that a perfect sphere in motion on an ideal (frictionless) floor will move along a straight line to infinity. Only something like our second inclined plane would prevent this by taking from the rolling sphere just that which distinguished it from a motionless sphere resting on the floor.

Galileo's reflections, and those of most physics texts, stop at this point.[19] Doubtless, this much *does* constitute a convincing suggestion as to the nonterminating, rectilinear character of force free motion. But that such motion will be *uniform* is usually assumed to follow qualitatively from such a "thought experiment"; either that or it seems to follow from the very definition of "force free motion."[20]

But why assume what is itself demonstrable? Why discuss qualitatively what can be proved quantitatively? Think of a hypothetical class of *all* those inclined planes that could be nested within the angle between our secondary plane and the floor. As our second plane is inclined less and less steeply, every possible intervening angle with the floor will ultimately have been traversed by the plane, and (ideally) by the ascending sphere. Now imagine a line parallel with the floor, but lower than the original height from which the sphere descended. The line (*A-B*) could then be drawn through all those possible inclinations to which we just referred.

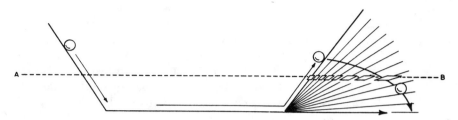

Consider the intersection of *A-B* with each of these inclined planes. At each point of intersection ( ) note the deceleration of the ascending sphere—i.e., the rate at which its velocity is falling off as it climbs this secondary plane. We would expect such instantaneous decelerations to be greater on a steeply inclined plane than at the corresponding point on a plane of shallow inclination. And as this plane is "flattened" from its original inclination down to where it coincides with the floor, so similarly the value of the deceleration variable as it "moves" along the line *A-B* will itself decrease. Ultimately, when the plane *does* join the floor, the deceleration will be absolutely zero (for now the plane and *A-B* will "intersect" only at infinity). A subsidiary argument will apply in the case of acceleration: since there is nowhere in this *Gedankenexperiment* the possibility that the sphere might *gain* in its accelerations as it ascends the second slope, the first argument is decisive. The point here is that the *uniformity* of a body's (force free) motion can be argued for; it need not be merely assumed.[21]

All of which suggests that "rectilinearity," "motion *ad infinitum*," "uniform," and "force free" are interdependent conceptions within classical mechanics. It is possible to treat the idea of uniform, rectilinear motion *ad infinitum* as itself *built into* the notion "force free"—as part of the latter's semantical content. Thinking of a body free of impressed forces, then, is merely thinking of a body either at rest or else in uniform, rectilinear motion *ad infinitum*. But this same semantical game can also be played by packing "force free" into one of the other concepts: *uniform,* or *rectilinear,* or *motion* ad infinitum. And so on.

Thus the first thing one learns about Newton's first law is that its terms are semantically linked. The meaning of some of its constituent terms "unpacks" sometimes from one or two of the others; but then sometimes the meaning of these others unpacks from that of the first. Which are the "contained," and which the semantical "containers," can affect the logical exposition of any mechanical theory built thereupon. In just this way one can distinguish the elegant mechanical theories of Lagrange and Hertz. Archimedes had required an immovable platform away from which to lever the world; so also every physical theory requires a set of stable, primitive conceptions in terms of which all other terms in the theory can be explicated. Although Lagrange and Hertz gave their lives to the *same theory*, classical mechanics, they chose *different semantical platforms* to which to fix their fundamental laws; hence they confronted their theoretical problems in quite different conceptual postures.

"*The* law of inertia," "Newton's first law," is thus really a *family* of schemata. The algorithmic function of the law is contained in this very fact; the theoretician can trace any logical genealogy of mechanical concepts he chooses just from his first decision to invest the law with *this* semantical structure rather than *that* one.

We have considered a typical statement of the law and noted that it is not self-evident. We have also noted what made it seem plausible to Galileo and his successors, as well as the semantical decisions which guarantee that in different formalizations of the theory different meaning relations will obtain between the law's constituent terms: sometimes A, B, and C will be semantically primitive and D will be derived from these, and sometimes D itself will be primitive. One function of Newton's law consists in such interrelating of mechanical concepts. Hence a Mach and a Hertz may interrelate these terms differently; the law does different work within their theories.

These questions about the semantical content, or intratheoretical function, of a physical law are independent of the *truth* of such laws. Newton's first law is not just a linguistic translating device. It is not a formula concerned only with the mutual substitutability of terms within a mechanical game. It began life in factual opposition to an ancient alternative claim, one that was itself amenable to observational tests. The ancients' contention that continued application of force was necessary for motion to continue came simply to the claim that, without it, motion would cease. All terrestrial bodies would therefore come to rest (sooner or later) were no further motive power applied. As a description of what we do observe, as engineers, as physicists, and as travelers, the ancients' claim seems not only substanti*able,* but also substanti*ated in fact.* Hence the negation of that claim ought also to be vulnerable to factual inquiry; since we take that negation as a physical basis for much of the last four hundred years in science, Newton's first law should also be substanti*ated in fact,* at least as directly and plausibly as was the ancient "law" concerning the necessity for continual force behind any moving objects. It was that direct observational plausibility which made that older view the basis for so much of Aristotle's scientific work. Despite so many initial observations to the contrary, Newton's first law should also be referable to certain demonstrable facts, which (however nonobvious) will nonetheless anchor our science of mechanics in an observational foundation of physical truth.

If, however, we restate the law in its logically most transparent form, it will read as follows:

IF THERE WERE A PARTICLE FREE OF UNBALANCED, EX-
TERNAL FORCES, THEN IT WOULD EITHER REMAIN AB-
SOLUTELY AT REST OR WOULD MANIFEST UNIFORM
RECTILINEAR MOTION *AD INFINITUM.*

In this form, the meaning content of the law is clearer. So stated, it is "an unfulfilled hypothetical" or "counterfactual conditional." Although we have no reason for supposing that particles free of unbalanced external forces *do* exist, the law tells us what would obtain if they did. This has the doubly awkward consequence that (1) we cannot experimentally investigate the properties of such bodies, and (2) the law, being a hypothetical claim, cannot be shown to be false. (1) is a factual remark; (2) is a logical consequence of it. The law cannot even be shown to be falsifi*able,* something that many logicians of science have taken to constitute a *necessary condition* for meaningfulness within natural science.[22] Does the counterfactual character of Newton's law of inertia put it into this same class? Not quite. When linked with a network of other physical assumptions, the law has consequences that are testable. This is not true of a "shrunken universe" type of claim—the sort that is entertained in "nocturnal doubling" hypotheses. Given any two moving bodies, one of which is demonstrably freer of external forces than the other, that body will approximate more closely than the other to uniform, rectilinear motion *ad infinitum*—although it can never push that approximation to its formal limit of perfect inertial motion. Semantical difficulties arise. Consider the expression "demonstrably freer." Look at this idea of pushing a physical approximation "to its formal limit." What is meant by saying of one body that it moves "closer" to infinity than another? Indeed, this remark is virtually unintelligible as it stands. But another difficulty lurks beneath this intelligibility issue.

Think not only of the observational fact that no one has ever encountered a force free body, but also of the logical fact that the expressions "uniform" and "rectilinear," if they are to have operational significance and physical meaning, must be linked to possible measuring techniques. How do we establish that a body's motion is rectilinear and uniform? We set up fixed coordinates by reference to which a point's motion from position x, y, z at t to position x', y', z' at t' may be seen to correspond to a rectilinear uniform translation within the space defined by the coordinates. For the physicist or the astronomer this is not an exercise in pure geometry; it is not an exploration of the algorithmic properties of some abstract space. Quite the contrary. It requires setting up *physical* coordinates determined by the locations of actual objects. When these are assumed to be fixed, they will allow the relational, intrageometrical distinctions necessary for describing the point's original motion as traced "within" the resulting reference frame.

Suppose the universe consisted in one and only one punctiform mass.[23] Of its mechanical behavior nothing could be said that had any physical meaning—not even by God. Meaningfully to claim of that mass that it moves uniformly along a rectilinear path it is necessary to fix physical coordinates by assuming other punctiform masses to be securely anchored. (This already wrecks the supposition.) How *many* other masses? As approximate Cartesian coordinates we need one punctiform mass at the "zero point" and three others placed along the three coordinates. Without these and the assumption that they are "absolutely immobile," the motion of our original particle could not be described as uniform and rectilinear; it could not even be said to be in motion at all. But this suggests that to describe the motion of any one particle in the universe as uniform and rectilinear there must be at least five particles in that universe. Any particle one wishes to describe as moving uniformly and rectilinearly *must* then be but one particle in a universe containing at least five: the specimen particle and the four coordinate fixers. But no particle within a five particle universe can be force free—a direct inference from another cornerstone of mechanics, the law of universal gravitation: "Any two particles in the universe are such that they attract each other directly as their masses and inversely as the square of the distance between them."

So the counterfactual character of Newton's law records not merely the fact that no bodies are ever found to be force free: it signals rather the *logical* point that no body whose motion could meaningfully be described as uniform and rectilinear could possibly be force free. Any alternative interpretation would crush the gravitational cornerstone of mechanics, something few physicists would be prepared to undertake just to avoid the counterfactuality of Newton's first law. Any appraisal of the law's logical status is immediately pierced by this point. Newton's law is thus revealed as referring to entities which are *not* such that, although never observed physically, they remain observable, i.e., that we know *what it would be like* to encounter such entities. No. Rather, the law refers to entities that are unobservable in physical principle; either the law conflicts with our conceptions of physical meaning or it conflicts with other laws of mechanics. Either way it is difficult to comprehend.

Concerning the *number* of particles necessary to fix physical coordinates, four is too many. With a zero point particle and but two coordinates determined by two further bodies, the third dimension is easily "mapped away" from the plane so defined. But the same argument can also eradicate one of the remaining three particles; we can always determine from the zero point *two* perpendiculars normal to each other and also to the line connecting the remaining *two* points. But then only one of these last two need remain, for the first coordinate line can be laid out in any arbitrary direction from zero; it is then easy to construct "imaginary" perpendiculars on that first arbitrary coordinate. *But there the reduction halts.* For the zero point at least must, as a matter of physical intelligibility, be construed as a fixed particle, out of which a reference frame may then be constructed with no more difficulty than an exercise in geometrical survey technique. This zero point particle is identical to Neumann's body Alpha, of which more later. The thrust of our earlier argument remains unaffected. In order meaningfully to say of any particle that its motion is uniform and rectilinear, it must be one member of at least a two-particle universe; it thus remains logically impossible both for that particle to be force free and for the rest of mechanics to be true.[24]

Either the theory as a whole is false, or it is physically meaningless to suppose that Newton's first law could be other than a counterfactual-conditional-in-principle. This being so, the function of the law within physical theory is difficult to assess. Is it "true," even though counterfactual? What does it do?

I hope that this much places the "rectilinearity part" of Newton's law of inertia into an analytical box. Concerning the "uniformity part," Professor Grünbaum has so often discussed this with such lucidity that only a mention is necessary. Operationally to insure that the rectilinear motion of a particle is *uniform,* one must introduce a measuring rod into our barren, two-particle universe. Through such a rod a metric may be fixed within the reference frame constructible upon the zero particle, or the body Alpha. For how can we establish that a particle's motion is uniform other than by determining that it traverses equal spaces in equal times—which is what "uniform motion" means?[25] Forget the perplexities involved in understanding "equal times"; the obvious way of insuring that the *spaces* are equal is to lay off a measuring rod first against one translation segment and then against another. If the ends of the segments coincide with the ends of the rod, then the spaces traversed are equal. But saying even this involves assuming that the rod itself suffers no deformation during its transport from one trajectory segment to another. To argue that this is so because of other reasoning generable within classical mechanics—within matter theory, kinetic theory, and elasticity theory—would be circular. These are all derivative subsections within classical mechanics, which itself depends on the meaning and logical roles of Newton's law of inertia. So that law cannot itself be substantiated by considerations which rest on the assumption that the law *is* substantiated, an assumption built into these derivative disciplines. No, that our measuring rod does not expand or contract during translation is itself fixed by "convention," to use a word that has virtually become Professor Grünbaum's very own. This suggests that even to understand the meaning of "uniform" in Newton's first law, a conventional appeal must be made to another cornerstone of mechanics, namely, that *mere transportation does not alter the physical properties of a body.*[26] Of ordinary objects we substantiate this via derivative disciplines within classical mechanics. But the ideal rod introduced to understand the meaning of Newton's law cannot itself be tested by such derivative mechanical disciplines, since the law is fundamental to classical mechanics as a whole, and hence to all its subdisciplines. The "uniformity part" of the law therefore requires for its very comprehension a sophisticated appeal to the conventionality thesis now so familiar to us from the Poincaré-Grünbaum analytic tradition. Such an exploration is not a philosophical frosting on the mechanical cake. It is necessary even to grasp the *meaning* of Newton's first law of motion.

These reflections have exciting consequences for the theoretical development of Newton's first law. Once it is apparent that every physical reference frame is, and must be, itself accelerated, i.e., *not* force free, it follows that the concepts of *rectilinearity* and *infinity* must either be defined physically in terms of such physical frames, or else they, and the law, must unquestionably assume an absolute space. If rectilinearity and infinity are anchored to the properties and behavior of physical objects, then these terms cannot be understood in their original geometric and number theoretic manner. Their entire significance must depend on the local peculiarities of particular physical spaces and processes. Yet it was an objective of the founders of modern mechanics that precision should be brought into studies of spaces and processes by translating geometrical ideas from mathematics into physics. Were that attitude to control our understanding of Newton's first law, however, we would have to take pure geometrical space, i.e., absolute space, as the force free framework within which particles and processes reside. Without this assumption, one cannot comprehend geometrically founded statements about the inertial motion of particles. Either the meanings of "rectilinearity" and "infinity" come through to us from pure geometry and number theory exclusively—in which case absolute space is required as the envelope for all mechanical subject matters—or else these terms are defined through possible physical configurations—in which case Newton's first law is in principle a counterfactual conditional, since it denotes entities that are in principle nonobservable. Geometrical meaning or physical meaning; absolute space or counterfactual conditionality.[27]

Every operational difficulty that attaches to local spaces whose geometrical coordinates are not fixed by particles is magnified when reference is made to absolute space itself.[28] The complex discussions by Leibniz, Euler, Laplace, Gauss, Hertz, and Mach, as well as Neumann's attempt to fuse pure geometry with "impure" physics by his postulated "body Alpha" —all this need not detain us.[29] What must detain us, however, is precisely what detained these thinkers, namely, the conceptual relationship between mathematics and physics. *The* insight of the giants of the Scientific Revolution was that the world, and its constituent processes, is built on geometrical-mathematical lines. But one of the "counterinsights" of contemporary mathematicians has been that progress in mathematics often depends on recognizing that there are always more natural processes *not* amenable to mathematical treatment than vice versa. The perplexities of physicists have always been a mine for mathematical research from which nuggets of mathematical truth have been extracted.

Thus, on the one hand, it seems that physics got off the ground only when it was mathematicized. On the other, it appears that mathematics can continue its ascent only when its research is elevated on the unsolved problems of physics. Philosophical attitudes toward Newton's first law have oscillated accordingly. Either the law's meaning seems to be imported into physics from pure mathematics, or else that meaning is determined by physical considerations and then fed back into mathematics for further formal development.

For argument, let mathematics and physics be construed as but different aspects of one comprehensive discipline, a kind of mathematics-physics —the sixteenth-century ideal. The law of inertia then can come into physics, as historically it was felt to do, with predetermined geometrical meaning. It will serve as a paradigm for physics as a whole, just as geometry itself once served as a heuristic device for understanding nature. The outlooks of Copernicus, Kepler, Galileo, Descartes, and Newton were clearly of this kind. But now stress the logical differences between mathematics and physics. Stress that geometrical statements are descriptively *true* only in that their negations, although possibly consistent, do not describe natural facts. The hunt must then be on for the physical meaning of "uniformity," "rectilinearity," and "infinity," not to mention terms like "equals," "is proportional to," "is commutative," "is of the second order in time," "is divergent," etc. Only by finding physical meanings for such expressions can one use them in making true physical statements, i.e., statements whose negations are not merely inconsistent. Mathematics and physics on this account seem *logically* different disciplines, such that the former can only occasionally solve the latter's problems. Geometry as *Raumphysik* is thus based on a confusion of logical types.

Section III

Neumann's body Alpha is a fascinating halfway house. Yearning for definitions of "uniformity," "rectilinearity," and "infinity" that would not be subject to possible overhaul within each new physical reference frame, Neumann "invented" Alpha: it had, by hypothesis, unquestionable fixity both within geometry and within physics. Through the assumption of this one body, Neumann sought to convey the absolute definitions of Euclidean geometry into observational physics. One result would then have been that in a two-particle universe (one "moving particle" and Alpha) one could recognize the motion as rectilinear or otherwise simply by reference to this physically fixed, but gravitationally inconsequential body. However, physical unintelligibility is the only upshot both for Alpha and for its relationship with the other particle.[30] What physical sense is there in describing Alpha as "gravitationally inconsequential"? So described, Alpha must also be *geometrically* inconsequential for the purposes of physics, just as was its precursor, the Fifth Postulate of Newton's *Principia*. And, after Neumann, physicists have in unison pronounced "Let no man join what nature hath sundered," namely, the *formal creation* of spaces and the *physical description* of bodies.

Once the logical connection between meaning and operations was stressed in the mid-nineteenth century, further perplexities became apparent for Newton's first law. Once the description of motion as rectilinear is seen to be meaningful only via coordinates fixed by material objects, it appears that these coordinate fixers can have greater or less perturbational effect on the motion under study, depending on their mutual separation. If the "fixers" are lead balls spaced one foot apart, their effect on a brass sphere will be greater than would be the corresponding effect of four H-atoms one billion miles apart. These latter will be assumed "to have no effect at all" on a grain of sand equidistant from them all.[31] But the very possibility of extrapolating until differences between a geometrical space and a physically defined space become negligible—this possibility is itself articulable only via a theory of measurement harnessed to a sophisticated perturbation theory. Why should not the perturbational effect of bodies on each other *increase* with their separation? Why should not the *measurement* of perturbation affect the measured motion (à la the indeterminacy principle)—the smaller the effect, the greater the interaction between phenomenon and detector? A fundamental theory of measurement can be unpacked from little more than the perplexities involved in understanding the meaning of Newton's first law. The point here is that one's extrapolation techniques cannot be supposed in advance to be approximating to some ideally geometrical, physically unperturbed space. Rather, the extrapolation techniques themselves must be controlled by physical considerations built into any operationally useful theory of measurement. One might even discover that in certain contexts the reduction of perturbational effects in the reference frame will disclose the specimen object to be traversing not a rectilinear Euclidean path, but a geodesic in some *non-*Euclidian space. Concerning the original notions built into "rectilinearity," the motion of such a specimen particle would not conform to Newton's first law at all.[32]

In such a context, however, the physicist may decide to redefine the relationship between reference frames and inertial motions. He might say that *any* constellation of particles can constitute an inertial framework if the motion of some specimen object is geodesic within *some* space defined by that particle constellation. This procedure would not initially force any Euclidian content into, or out of, the law of inertia. It simply characterizes reference frames and motions as "inertial" when the former determine the latter as *geodesic in some space.* The theoretical possibilities for the extension of this law are thus completely unlimited. For now the actual path of a particle will convey as much about the space it traverses as it used to be taken to convey about the perturbational effect of other particles nearby. When space is no longer a passive envelope *within which* particle interactions occur, but rather in some sense a *property* of interactions as defined by the coordinate fixers, then Newton's first law can become a "blank check" for the theoretician. He is free to do what he wishes with it, to entertain a variety of possible spatial frameworks, rather than simply what was earlier built a priori into the terms "uniform" and "rectilinear."[33] He is not hampered by having to strap these terms to physical objects and processes. Rather, he is now free to strap them to which objects and processes he chooses, the better to facilitate whatever immediate physical inquiry concerns him.

A residual difficulty may still lurk in the question, "Is Newton's first law true?" We have seen that determining the truth of a hypothetical claim whose protasis is unfulfillable raises difficulties. Meaningfully to describe the motion of an object as "rectilinear" one must relate it to a physical reference frame. But in this universe all physical reference frames are themselves accelerated; we do not even have the concept of an alternative to this—not, at least, one in which mechanics (classical or quantum) still holds. Hence the motion of no object can be perturbation free, free of unbalanced, impressed forces. Making claims about such nonexistent objects, such as that they either remain at rest or move with uniform rectilinearity *ad infinitum,* seems somewhat like making claims about centaurs or mermaids. More strongly, we know what it would be like for a creature to have "biological properties" like centaurs and mermaids. The idea of a force free body, however, seems like that of a quadrilateral triangle, or a squared circle. But that may be *too* strong. These last involve formal contradictions within mathematical theory. The idea of a force free body is not formally inconsistent. If it were, Newton would have detected this immediately. At most this idea involves a physical contradiction. The law of universal gravitation might be false and yet still consistent. Assuming that law to be true, however, at least within some more comprehensive cosmological theory, references to a force free body must then be references to what is either physically inconsistent or physically empty. In what sense, then, can Newton's first law be true? In a sense analogous to that in which the law of ideal gases is true, or the laws of perfectly elastic bodies, perfectly rigid levers, and perfect conductors. No such objects as these exist either. Nor could they—not, at least, without upsetting other established empirical relationships. Nonetheless, reference to such hypothetical objects clears up formal calculations within kinetic theory, mechanics, and electrical circuit theory. They do not describe anything actual, so they cannot be false descriptions or vulnerable generalizations. They encapsulate in an algorithmically powerful pattern, however, whole classes of formal relationships physically realized in nature.

The truth of Newton's first law is thus also derivative. It accrues to the law because an immense number of observational consequences derivable in part therefrom turn out to be physically true. In short, this law, like many others in mechanics, patterns and organizes a host of formally expressed relationships (physical equations) each of which, at a derivative level, can be interpreted so as to generate observation statements confirmed in fact. In some contexts the law may express a priori relationships within mechanical theory, e.g., the semantical connections between "uniform," "rectilinear," "ad infinitum," "acceleration," etc. On other occasions, the law may reflect, a posteriori, a range of facts that support mechanics as a set of empirical descriptions and pragmatically useful calculational techniques. In this last sense, its "truth" as a physical law is not difficult to fathom. It is rather like Boyle's law in 1662, a first-order description of observed phenomena. In the former sense, however, asking after the physical truth of what is functioning largely as a formal patterning formula—appropriate more to algebraic manipulations than to descriptions of phenomena—seems out of place. Still, the question has a philosophical moral: orthodox statements of Newton's first law, when advanced without reference to any physical context or theoretical problem, are really references to a "law function," not to a genuine law of nature. Until all the values are specified for the variables, the apparent perplexity is the result only of running together many independent questions and many different functions of this law schema. Poincaré's account of the law as a definition and Mach's as a remote generalization are thus incompatible only if both accounts are taken to be exhaustive and exclusive. In a particular physical context, however, whether the law is functioning as a definition or a derivative generalization usually makes itself clear without perplexity or contradiction with other uses of the law.[34]

Note also how Newton's first law forces attention onto the status of theoretical terms within physical theory. We need not attend only to colorless symbols like ψ in quantum theory or i in statistical electrodynamics. "Inertia," "force," "uniform," etc., also function within an intricate tradition of theoretical constructions. Consider "acceleration" and what this term must have been to Galileo before the era of elevators, Buicks, 707's and seatbelts. The function of such a theoretical construction is easy to characterize. Just as in explanations, one cannot explain everything at once. We explain x, y, z only by leaving a and b unquestioned in that particular context, explaining the former in terms of the latter.

a and b may themselves prove to be explicable in terms of further notions like α and β. But at any one time, some conceptions *must* function, within science, as *explained nonexplainers* (e.g., a semipermeable membrane separating a solution in the expected way). Others will function as *explained explainers* (the atoms of the membrane and of the solution, as dealt with in microphysics). And there will always be some *nonexplained explainers* (ψ, atomic diffraction, pair creation).[35] Thus reference to the properties of atoms goes a long way toward explaining the characteristics of semipermeable membranes. But the characteristics of these atoms will themselves be explicable via electron theory and quantum electrodynamics, the latter disciplines resting on δ, ψ, etc.—unexplained explainers. In much this way the concept of a perturbation free object, moving with rectilinear uniformity *ad infinitum,* often functions within classical mechanics. It certainly functions thus within Hertz's theory.

If all this can be argued for the first law, it follows immediately for the second law. As is well known, once one relates the ideas of *force* and *acceleration* as the second law does, Newton's first law tumbles out as a special case. Where the *sum of all external, unbalanced* forces equals zero, there will be no accelerations evident in a body:

$$\left\{ \Sigma F = 0 \right\} \rightarrow a = 0$$

Hence every body in motion will continue thus without acceleration, and, if at rest, it will remain thus.[36]

Corresponding (but distinguishable) analyses apply to the third law— that concerned with "action" and "reaction." But we have done enough here to suggest that every law within physics is a cornucopia of philosophical perplexities and conceptual excitement. Every such law functions in organizing part of a science's subject matter, in patterning the structure of its arguments and its permissible intellectual moves. And if the discipline which embodies it effectively describes nature, such a law may be said to tell the truth. The fundamental laws of statics and kinematics, of optics and dynamics, of celestial perturbations and microphysical interactions, these contain the most profound challenges to the human understanding to be confronted in our time. And Newton's first law, sometimes characterized as the simplest of them all, turns out to embody challenges as profound as any.

Epilogue

The apparently simplest mechanical principles are of a very complicated character, . . . these principles are founded on uncompleted experiences, nay on experiences that never can be fully completed, . . . practically, indeed, they are sufficiently secured, in our view of the tolerable stability of our environment, to serve as the foundation of mathematical deduction, but . . . they can by no means themselves be regarded as mathematically established truths but only as principles that not only admit of constant control by experience but actually require it.

—MACH

Notes

1. *Lex I* (editions of 1687 and 1713): *Corpus omne perseverare in statu suo quiescendi vel movendi uniformiter in directum, nisi quantenus a viribus impressis cogitue statum illum mutare.* *Lex I* (edition of 1726): *Corpus omne perseverare in statu suo quiescendi vel movendi uniformiter in directum, nisi quatenus illud a viribus impressis cogitue statum suum mutare.* [Philosophiae Naturalis Principia Mathematica (Royal Society, London, 1687; Cambridge, 1713; London, 1726).]

2. *De Caelo,* 301b 18, W. K. C. Guthrie, trans. (Cambridge: Harvard University Press, Loeb Classical Library, 1939), p. 279.

3. *Physica,* Lib. VII, Cap. I, 243b, Vol. II, P. H. Wicksteed and F. M. Cornford, trans. (Cambridge: Harvard University Press, Loeb Classical Library, 1934), p. 207.

4. *Ethica Eudemia de Virtutibus et Vitiis,* 1224b 6, J. Soloman, trans., in W. D. Ross, ed., *The Works of Aristotle* (London: Oxford University Press, 1925).

5. Joannes Philoponus, *Commentary on Aristotle's Physics,* Vitelli, ed. (Berolini: Reimeri, 1887-1888), pp. 678.24-684.10.

6. *Ibid.,* pp. 639.3-642.9.

7. Before Galileo *force* was known solely as *pressure,* by the principle *cessante causa cessat et effectus.*

8. Pierre Duhem, *Etudes sur Léonard de Vinci,* Vol. III (Paris: De Nobele, 1955), p. ix.

9. John Buridan, *Quaestiones super Libris Quattuor de Caelo et Mundo,* Lib. IV, Qu. 2, E. A. Moody, ed., Mediaeval Academy of America Publication No. 40, Studies and Documents No. 6 (1942), p. 250.

10. John Buridan, *Quaestiones Octavi Libri Physicorum,* Bibl. Nat. Fonds Lat., Ms. 14723, Fol. 95, Col. b, Qu. 12: *Utrum projectum post exitum a manu projicientis moveatur ab aere, vel a quo moveatur.*

11. Albert of Saxony, *Sublissimae Quaestiones in Libros de Caelo et Mundo,* Lib. II, Qu. 14. Cf. also his *Quaestiones in Libros de Physica Auscultatione,* Lib. VIII, Qu. 13.

12. Nicolas of Cusa, *Dialogorum de Ludo Globi,* Lib. I, Maurice de Gandillac, trans., in *Oeuvres de Nicolas de Cues* (Paris: Aubier, 1942), p. 527.

13. *Dialogus Trilocutorius de Possest.*

14. *Gesammelte Werke,* Max Caspar, ed., Vol. XV (Munich: C. H. Beck, 1937-1960), p. 172.

15. My translation of *De Motu* (1591), Lib. V, Cap. 35. Cf.:

> . . . Impetus is a thing of permanent nature, distinct from the local motion in which the projectile is moved . . . impetus is a quality naturally present and predisposed for moving a body in which it is impressed, just as it is said that a quality impressed in iron by a magnet moves the iron to the magnet . . . [impetus] is remitted, corrupted, or impeded by resistance or a contrary inclination.

Buridan, *Questions on the Physics* in Marshall Clagett, *The Science of Mechanics in The Middle Ages* (Madison: University of Wisconsin Press, 1959), p. 537.

16. *Dialogues Concerning Two New Sciences,* Henry Crew and Alphonso de Salvio, trans. (New York: The Macmillan Company, 1933), p. 174.

17. *Oeuvres de Descartes,* Charles Adam and Paul Tannery, eds., Vol. X (Paris: Cerf, 1908), pp. 60, 225.

18. ". . . When this friction is drastically reduced, as on an ice rink, their velocity is maintained for a long time. This makes Galileo's discovery seem plausible." Dennis Sciama, *The Unity of the Universe* (New York: Doubleday & Company, Inc., 1959), p. 85. Cf. Ernst Mach, *Science of Mechanics,* T. J. McCormack, trans. (London: Routledge & Kegan Paul, Ltd., 1907), pp. 330 ff.

19. *Two New Sciences, op. cit.,* Third Day, pp. 242-246.

20. As whenever the first law is characterized as being but a limiting case of the second law, i.e., when $\Sigma F = 0$, it follows that $a = 0$ (whether $m \lessgtr 0$). That is, $d^2x/dt^2 = 0$; $d^2y/dt^2 = 0$; $d^2z/dt^2 = 0$: the second time derivations of coordinates x, y, z vanish when a body is force free. From this much, however, one can infer the rectilinearity of an inertial path only by building that concept into "force free."

> But along a horizontal plane the motion is uniform since here it experiences neither acceleration nor retardation. . . .
>
> Furthermore we may remark that any velocity once imparted to a moving body will be rigidly maintained as long as the external causes of acceleration or retardation are removed, a condition which is found only on horizontal planes; for in the case of planes which slope downwards there is already present a cause of acceleration, while on planes sloping upwards there is retardation; from this it follows that motion along a horizontal plane is perpetual; for, if the velocity be uniform, it cannot be diminished or slackened, much less destroyed. Further, although any velocity which a body may have acquired through natural fall is permanently maintained so far as its own nature (*suapte natura*) is concerned, yet it must be remembered that if, after descent along a plane inclined downwards the body is deflected to a plane inclined upwards, there is already existing in this latter plane a cause of retardation; for in any such plane this same body is subject to a natural acceleration downwards. Accordingly here we have the supposition of two different states, namely, the velocity acquired during the preceding which if acting alone would carry the body at a uniform rate to infinity, and the velocity which results from a natural acceleration downwards common to all bodies. It seems altogether reasonable, therefore, if we wish to trace the future history of a body which has descended along some inclined plane and has been deflected along some plane inclined upwards, for us to assume that the maximum speed acquired during descent is permanently maintained during the ascent. In the ascent, however, there supervenes a natural inclination downwards, namely, a motion which, starting from rest, is accelerated at the usual rate. (*Ibid.,* pp. 215-216.)

21. Cf. Mach, *op. cit.,* pp. 168-169.

22. An *un*falsifiable claim—one that is compatible with anything whatever—is sometimes termed "insignificant." Thus the assertion that the universe shrank last night, being now one billionth part smaller than yesterday—including ourselves, all measuring instruments, theodolites, micrometers, diffraction gratings, the wave lengths of standard radiation, elementary particles, etc.—this claim, since it cannot be falsified *ex hypothesi,* and makes no difference to anything we do observe, is physically meaningless.

23. This very supposition is internally inconsistent for anyone who accepts

Mach's kinetic definition of mass. Since *ex hypothesi* a single particle cannot interact with other particles, it is idle to discuss its mass in Machian terms. But are there really any other meaningful terms?

24. As Berkeley wrote:

> If every place is relative then every motion is relative and as motion cannot be understood without a determination of its direction, which in its turn cannot be understood except in relation to our or some other body. Up, down, right, left, all directions and places are based on some relation and it is necessary to suppose another body distinct from the moving one . . . so that motion is given in relation to which it exists, or generally there cannot be any relation, if there are no terms to be related.
>
> Therefore if we suppose that everything is annihilated except one globe, it would be impossible to imagine any movement of that globe.
>
> Let us imagine two globes and that besides them nothing else material exists, then the motion in a circle of these two globes round their common centre cannot be imagined. But suppose that the heaven of fixed stars was suddenly created and we shall be in a position to imagine the motion of the globes by their relative position to the different parts of heaven. (Berkeley, *De Motu, op. cit.*, written thirty years after the publication of the *Principia* of Newton.)

25. "Definition: By steady or uniform motion, I mean one in which the distances traversed by the moving particle during any equal intervals of time, are themselves equal." (Galileo, *Two New Sciences, op. cit.*, Third Day.)

26. Compare a different, but somewhat analogous claim by Keynes:

> The Law of the Uniformity of Nature appears to me to amount to an assertion that an analogy which is perfect, except that mere differences of position in time and space are treated as irrelevant, is a valid basis for a generalisation, two total causes being regarded as the same if they only differ in their positions in time or space. This, I think, is the whole of the importance which this law has for the theory of inductive argument. It involves the assertion of a generalised judgment of irrelevance, namely, of the irrelevance of mere position in time and space to generalisations which have no reference to particular position in time and space. It is in respect of such position in time or space that "nature" is supposed "uniform." [*A Treatise on Probability* (London: Macmillan & Co., Ltd., 1921), p. 226.]

27. A summary of Newton's interpretation of his "bucket experiment" (*Principia*, "Scholium on Space and Time") is that absolute rotation has nothing whatever to do with the relative rotations which are directly observed (e.g., the spinning bucket), but that nevertheless we can determine experimentally the amount of absolute rotation possessed by a body. All we need do is measure the curvature of a water surface rotating with the body. This determination of absolute rotation consists in *measuring centrifugal force*. Foucault's pendulum experiment reveals the motion of the pendulum plane to be acted upon by a Coriolis force. Cf. Mach:

> For me only relative motions exist. . . . When a body rotates relatively to the fixed stars, centrifugal forces are produced; when it rotates relatively to some different body and not relative to the fixed stars, no centrifugal forces are produced. I have no objection to just calling the first rotation so long as it be remembered that nothing is meant except relative rotation with respect to the fixed stars.
>
> Obviously it does not matter if we think of the earth as turning round on its axis, or at rest while the fixed stars revolve round it. Geometrically

these are exactly the same case of a relative rotation of the earth and the fixed stars with respect to one another. But if we think of the earth at rest and the fixed stars revolving round it, there is no flattening of the earth, no Foucault's experiment, and so on—at least according to our usual conception of the law of inertia. Now one can solve the difficulty in two ways. Either all motion is absolute, or our law of inertia is wrongly expressed. I prefer the second way. The law of inertia must be so conceived that exactly the same thing results from the second supposition as from the first. By this it will be evident that in its expression, regard must be paid to the masses of the universe. (Mach, *op. cit.*)

Cf. also Newton's *Principia Mathematica Philosophiae Naturalis,* Fifth Corollary.

28. Consider the conceptual difficulties Hertz encountered when he laid it down that:

> Every unfree system we conceive to be a portion of a more extended free system; from our point of view there are no unfree systems for which this assumption does not obtain. If, however, we wish to emphasise this relation, we shall denote the unfree system as a partial system, and the free system of which it forms a part, as the complete system. [*The Principles of Mechanics* (London: Macmillan & Co., Ltd., 1899), p. 144.]

This statement is continuous, of course, with Hertz's own "fundamental law"—namely, the law of inertia:

> FUNDAMENTAL LAW: Every free system persists in its state of rest or of uniform motion in a straightest path. (*Systema omne liberum perseverare in statu suo quiescendi vel novendi uniformiter in directissiman.*) (*Ibid.,* p. 178.)

29. These discussions all turn on the fact that, according to Newton's theory, the only way rotation relative to absolute space can be detected is from the existence of centrifugal forces (the "bucket") and Coriolis forces (the Foucault pendulum). But absolute space in the *Principia* was "invented" precisely to account for these forces—so it adds nothing to what we knew before it was "invented." Whenever an "explanatory" hypothesis has as its consequences *only* those anomalous phenomena for which we originally sought an explanation, that hypothesis may quite rightly be regarded as being but a "quasi-explanatory" *restatement* of the very descriptions with which inquiry began. As such it is not an explanation at all—or, at best, an *ad hoc* explanation. Leibniz, Huyghens, Bernoulli, and Poleni originally reacted thus to $F \propto \gamma \ (Mm/r^2)$, noting that its only consequences were descriptions of planetary perturbations, which latter set the initial problem. Further consequences of the law were soon drawn, making its claims to be a physical explanation thereby more plausible.

30: Cf. Bertrand Russell: "It seems evident that the question whether one body is at rest or in motion must have as good a meaning as the same question concerning any other body; and this seems sufficient to condemn Neumann's suggested escape from absolute motion." [*The Principles of Mathematics,* 2nd ed. (New York: W. W. Norton & Company, Inc.), p. 464.]

31. Notice that this counters Mach's Principle, to wit, that other matter in the universe will make a small contribution to a body's total inertia, rather than no contribution at all—*because,* according to Mach, a body has inertia only because it interacts (in some way) with *all* the matter in the universe. This directly opposes the views of Galileo and Newton, for whom inertia was an *intrinsic* property of matter. Einstein, who was greatly influenced by this position, is responsible for naming it "Mach's Principle." The view was held also by Stallo, J. J. Thomson, Lange, Kleinpeter, J. G. MacGregor, and Pearson.

32. Thus Einstein noted the similarities between gravitational and inertial forces—both are proportional to the mass of the body acted on—and the dissimilarity of both vis-à-vis electric and magnetic forces, which do not induce the same accelerations in all bodies (consider neutral bodies). This similarity so impressed Einstein that he came to claim that it was impossible to distinguish gravitational from inertial forces. If challenged to discover whether a field of force is gravitational or electrical, we just measure the accelerations of a neutral and a charged body. The acceleration of the former discloses the strength of the gravitational part of the field. The additional acceleration of the latter gives the strength of the electrical force. The same experiment with "inertial force" substituted for "electrical force" will permit of no comparable distinction in measurement: the *total* strength of the field is determinable, but not the individual contributions of the gravitational and inertial components. A man's trousers may fall either when he hits the ground after jumping from a ledge or when he is lifted quickly from the ground in a rapidly accelerating rocket. From the man's point of view, the situations are mechanically indistinguishable. There is thus no *general* criterion for distinguishing inertial from gravitational forces; this is Einstein's Principle of Equivalence. It is "resolved" by treating all inertial forces as being fundamentally gravitational. Our observed inertial forces result from stellar accelerations. Without the latter, the former would not exist. Cf. P. G. Bergmann, *Theory of Relativity* (Englewood Cliffs, N.J.: Prentice-Hall, Inc., 1942), pp. 153 ff.

It is interesting to note that Euler defended the necessity of absolute space by pointing out that since (according to the law of universal gravitation) all material reference frames are accelerated, inertial motion could exist only relatively to a nonmaterial reference frame, i.e., *absolute space*. This constitutes a preservation of the Euclidean geometrical meaning of "rectilinearity" at the complete sacrifice of its physical meaning.

33. Another way of making this point is to say that not all frames of reference are equivalent for the formulation of the laws of motion. There does exist in nature a *special* set of reference frames in which Newton's laws hold without additional inertial forces' (e.g., Coriolis forces') having to be introduced. But not all phenomena in fact occur within that special set.

Mach argued (*op. cit.*, pp. 286 ff.) that inertial reference frames were just those which were unaccelerated relative to the "fixed stars"—i.e., some suitably defined average of all the matter in the universe. For him matter has inertia only because there is other matter in the universe. Redistribute that other matter, and our conception of inertia might have to be changed at once. Bertrand Russell denies this:

> Mach has a very curious argument by which he attempts to refute the grounds in favour of absolute rotation. He remarks that, in the actual world, the earth rotates relative to the fixed stars, and that the universe is not given twice over in different shapes, but only once, and as we find it. Hence any argument that the rotation of earth could be inferred if there were no heavenly bodies is futile. This argument contains the very essence of empiricism, in a sense in which empiricism is radically opposed to the philosophy advocated in the present work. The logical basis of the argument is that all propositions are essentially concerned with actual existents, not with entities which may or may not exist. For if, as has been held throughout our previous discussion, the whole dynamical world with its laws can be considered without regard to existence, then it can be no part of the meaning of these laws to assert that the matter to which they apply exists, and therefore they can be applied to universes which do not exist. Apart from general argu-

ments, it is evident that the laws are so applied throughout rational Dynamics, and that, in all exact calculations, the distribution of matter which is assumed is not that of the actual world. It seems impossible to deny significance to such calculations; and yet, if true or false, then it can be of no necessary part of their meaning to assert the existence of the matter to which they are applied. This being so, the universe is given, as an entity, not only twice, but as many times as there are possible distributions of matter, and Mach's argument falls to the ground. (*Op. cit.*, pp. 492-493.)

Cf. Cassirer:

We need only apply these considerations to the discovery and expression of the principle of inertia in order to recognize that the real validity of this principle of inertia is not bound to any definite material system of reference. Even if we have found the law at first verified with respect to the fixed stars, there would be nothing to hinder us from freeing it from this condition by calling to mind that we can allow the original substratum to vary arbitrarily without the meaning and content of the law itself being thereby affected. (Quoted in Sciama, *op. cit.*, p. 98.)

34. Cf. N. R. Hanson, *Patterns of Discovery* (London: Cambridge University Press, 1958), Chap. 5 with a much less flexible approach, to wit:

What is the status of this Law? One possible approach to the subject would be to regard Newton's theory as axiomatic, and to work out as an exercise in abstract mathematics the consequences of the theory, without claiming any relationship between the results of the theory and the events of the actual world. But this would be a sadly mistaken policy. It would rob the theory of the greater part of its interest, and it would deprive us of the immensely valuable help provided by our knowledge of what does in fact happen in the real world. It is therefore altogether preferable to regard Newtonian mechanics as a science based upon experiment. [L. S. Pars, *Introduction to Dynamics* (London: Cambridge University Press, 1953), p. 33.]

This is reminiscent of Mach, *op. cit.* and C. D. Broad, *Perception, Physics, and Reality* (1913), both of whom characterize the second law of motion as being an abstract induction from experience.

35. I owe this terminology to Wilfrid Sellars.

36. In contemporary physics the second law is usually amended to read: force is mass times *absolute* acceleration. Were the earth at rest and the sun moving around it, $F - Ma$ would not be satisfied since *ex hypothesi* the sun's gravitational force would produce no acceleration on the earth at all. Only if the accelerations of bodies are measured in a special way, therefore, will the second law hold: the accelerations must be measured relative to an inertial frame of reference, i.e., a collection of bodies on which no forces act and which have no *absolute* acceleration. Plenty of objects on earth are not accelerating *for us,* although they all have an absolute acceleration. Additional inertial forces—like the Coriolis force—must sometimes be introduced in order to cope with unusual reference

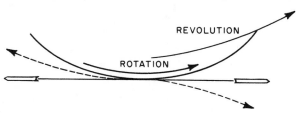

frames, e.g., that in which two rockets are launched at exactly the earth's rotational $v+$ revolutionary velocity; one rocket is shot "forward" with the earth's velocities, the other opposing them.

Cf. Russell:

> Nevertheless Newtonian dynamics will explain Foucault's pendulum and the flattening of the earth at the poles if the earth rotates, but not if the heavens revolve. This shows a defect in Newtonian dynamics, since the empirical science ought not to contain a metaphysical assumption which can never be proved or disproved by observation—and no observation can distinguish the rotation of the earth from the revolution of the heavens. This philosophical principle, that distinctions which make no difference to observable phenomena must play no part in physics, has inspired a good deal of the work on relativity, and is advocated by many writers. (Quoted in Sciama, *op. cit.,* p. 100.)

BRIAN ELLIS
University of Melbourne

The Origin and Nature
of Newton's Laws of Motion

Are the laws of acceleration and of the composition of forces only arbitrary conventions? Conventions, yes; arbitrary, no—they would be so if we lost sight of the experiments which led the founders of the science to adopt them, and which, imperfect as they were, were sufficient to justify their adoption. It is well from time to time to let our attention dwell on the experimental origin of these conventions.

—HENRI POINCARÉ, *Science and Hypothesis*

When we speak of a law of motion, we do not speak of the way in which bodies actually move. Rather we speak of the way in which they *would* move, given that they are subject to the action of certain forces. Laws of motion, therefore, whether they be Newtonian, Aristotelian, or relativistic are laws of *dynamics* rather than kinematics.

The study of the nature of a law of motion is the study of the role that it has within the body of science, of the reasons for its acceptance, and of the conditions under which it might be rejected. In making such a study, we should first want to know how the law in question initially came to be accepted—not because its present role in science is thereby determined, or because the reasons why it is now accepted are necessarily the same as the original ones, but rather because doubts and misconceptions about the origin of a law may spill over into doubts and misconceptions as to its present status. If, for example, we read that Newton claimed that his laws were directly confirmed experimentally, then all arguments as to the impossibility of this may fail to carry conviction, for in the absence of an historical investigation, doubt may remain that Newton had in fact done what our arguments seemed to show was impossible. The historical investigation should clear the air of possible doubts and misconceptions arising from such sources.

Next, we should want to know on what grounds or what conditions a law of motion *might* be accepted or rejected. And this is now a philosophical question rather than an historical one, for in answering it, we are free to imagine whatever sort of world and whatever sorts of experimental outcomes we please. We are not constrained by the feasibility of there being such a world, of conducting such experiments, or of their having the outcomes that we suppose. The problem here is to describe *generally* the sorts of grounds on which laws of motion might be accepted or rejected.

The study of this problem will naturally involve a study of the concept of force, for the laws of motion are laws of dynamics, and force is obviously the central concept. We should want to know what sort of thing a force is, under what conditions a force may be said to be acting upon a body, and under what conditions a body may be said to be free from the action of the forces. If possible, we should try to state these conditions generally. Our discussion should not be limited to the criteria for the existence or nonexistence of *particular kinds* of forces, for the problem is a very general one. Laws of motion do not distinguish between different kinds of forces.

Although Newton's laws of motion are the immediate subject of this essay, the arguments that will be produced are not intended to apply only to Newton's laws. Rather they are intended to apply to *any* law of motion, i.e., to any law that says how bodies would move given that they are subject to the action of certain forces (or given that they are not subject to the action of any forces). Newton's laws feature only by way of example. For the purposes of this essay, therefore, it is not necessary to consider the relativistic corrections that must be made to Newton's laws, for this would serve no purpose other than to complicate the example. In the following sections of this essay, it will simply be assumed that no such corrections are necessary.

Section I will be concerned with the historical origins of Newton's laws of motion. It will be argued that no detailed experimental evidence existed that would have warranted the acceptance of Newton's laws of motion in the seventeenth century. It is often supposed that the experimental foundation for Newton's laws is to be found in the kinematics of Galileo. But quite apart from the question of whether any study of kinematics could serve as the experimental foundation for a system of dynamics, it will be shown that it is extremely doubtful whether there is even any close *historical* connection between the kinematics of Galileo and Newton's laws of motion. Rather it is much more plausible to suppose that Newton's laws of motion were derived directly from Cartesian physics, and that the only experimental evidence that was in any way directly relevant to the truth of Newton's laws was the evidence upon which

Descartes and Huyghens supported their law of conservation of momentum.

The problem of assessing the weight of the evidence upon which Newton's laws were historically founded thus reduces to that of assessing the historical evidence for the law of conservation of momentum. The historical origin of this law beginning with Descartes' crude formulation of it (c. 1629) will therefore be discussed.

Section II will deal with the law of inertia and the concept of force, and the attempt will be made to formulate general criteria for arguing that a body is or is not subject to the action of a force. It will be seen that there is a general class of concepts in science that might legitimately be called force concepts, and that our concept of motive or dynamical force is only one member of this more general class. The distinguishing feature of forces generally is that in some sense their existence *entails and is entailed by* the existence of the effects they are supposed to produce. This being the case, the ontological status of forces within science is a very peculiar one, since, it will be argued, there is always an element of convention in deciding what we should regard as an effect, and hence, to the extent to which there is this element of convention, the existence of forces is also conventional.

But to this very extent, the laws of dynamics must also be conventional. And to illustrate this, a system of dynamics, at least as powerful as Newton's but which contains a different law of inertia, will be constructed.

Section III will be a discussion for the logical status of Newton's second law of motion—with particular emphasis on the relationship between the concepts of force and mass. Section IV will deal with the principle of action and reaction.

I. The Origin of Newton's Laws of Motion

The argument in this section will be mainly concerned with the relationship between Galileian kinematics, Cartesian physics, and Newtonian dynamics. It will be shown that, contrary to popular belief, all three of Newton's laws were probably derived from Cartesian physics, and not, as is often supposed, from Galileo's kinematics. But the main object will be to reinforce the view of Butterfield[1] that Newton's laws were primarily *conceptual* in origin, in that they represented a new way of conceiving dynamical problems. It will be shown that they were not originally supported by any evidence which need in any way have upset the views of anyone who accepted certain medieval postulates concerning force and motion. If any one factor can be said to be primarily responsible for bringing about the conceptual change, it was the anachronism of a homocentric system of dynamics in a world that was no longer believed to be earth-centered.

Galileian Kinematics. The *Two New Sciences*[2] of Galileo is often said to have laid the foundations of strength of materials and dynamics. But as I understand the term, it is inappropriate to describe the science discussed in the third and fourth "Days" of this great dialogue as dynamics, for dynamics is the science that deals specifically with causes of motion or changes of state of motion. It attempts to lay down the principles from which the motion of a body can be inferred from a knowledge of the forces acting upon it, or conversely, principles from which the forces acting upon a body can be inferred from a knowledge of its motion. Kinematics, on the other hand, is the study of the motions that bodies actually have in certain circumstances, and it seeks merely to represent these motions and to describe them without inquiring into the causes of their production. And since Galileo specifically rejected this latter inquiry as lying outside the scope of his essay,[3] his system can hardly be described as a system of dynamics.

The relevant parts of Galileo's *Two New Sciences* dealt mainly with the problem of describing free fall and projectile motion, and of stating the principles that govern these phenomena. His achievement in dealing successfully with this problem is certainly of first-rank importance in the history of science. Nevertheless, it is apt to be overestimated. Galileo did not state the law of inertia, as is sometimes supposed, and his *Two New Sciences* makes no commitment to any particular system of dynamics. He did, it is true, say that if a body is set in motion along a perfectly smooth horizontal plane, it will continue to move "with a motion which is uniform and perpetual provided that the plane has no limits."[4] But when (in the dialogue) he was questioned about this plane, and asked whether he meant a plane tangential to the earth's surface, he withdrew and said that his statement was at least as good an approximation to the truth as Archimedes' assumption that the scale pans of a beam balance hang parallel to each other.[5] In any case, Galileo said nothing whatever here about forces. He simply said that in such and such circumstances (described without reference to forces) a body would move in such and such a way. His law is therefore a purely kinematic law.

Moreover, it is extremely doubtful whether Galileo would have considered that a body in the state described would not be subject to the action of any forces, for at the time of writing he was evidently sympathetic to Buridan's impetus theory, according to which every change of *position* requires the action of some force.[6] Added to this, in the one place where Galileo did discuss dynamical questions—in his *Two Chief World Systems*[7] —he clearly and emphatically rejected uniform straight line motion as natural motion, and maintained equally emphatically that uniform circular motion was the only motion suitable to the preservation of order in the universe.[8] Consequently, when Galileo's statement about the motion of an object along a smooth horizontal plane is viewed in the light of his general dynamical views (expressed only a few years previously), it is hard to read it as a statement of the law of inertia.

It is, of course, possible that Newton read Galileo's statement and saw it as a statement of the law of inertia. But one who rejected the law of inertia or accepted another system of dynamics would be unlikely to give it this interpretation—although he might well accept it as true, since it is a purely kinematical law. It seems likely then that Newton already accepted the law of inertia and merely found in Galileo an historical anticipation of it.

Cartesian Physics. Descartes was in many ways the antithesis of Galileo. Philosophically he was a rationalist, that is, he believed that the various principles which govern the behavior of things in the world are derivable from certain a priori truths—propositions whose truth is immediately, clearly, and certainly apprehended by anyone who sufficiently understands the terms in which they are expressed.

Descartes took the geometry of Euclid as his model of science. To him this represented the ultimate in scientific achievement, for as he viewed the matter, it is a system of knowledge every element of which is known with certainty to be true. The axioms and postulates were considered to be a priori truths. The theorems are derived from the axioms by simple logical transitions, and are thus invested with the same order of certainty as the initial axioms themselves. Descartes believed that all scientific knowledge should be similarly derivable, provided only that the correct axioms from which to proceed could be found.[9]

As a rationalist, Descartes believed that he could demonstrate the existence of God—an immutable God—and from this immutability could demonstrate certain propositions concerning matter and motion, for according to Descartes' ontology, the universe consists simply of matter and motion. But God is immutable, hence He must preserve in the universe as much matter and motion as He originally included. Consequently, the total quantity of matter in the universe must be conserved. Changes in the quantity of matter in any individual object may indeed occur, but all such changes must be accompanied by equal and opposite changes in the total quantity of matter in the rest of the universe. Likewise, he argued, the total quantity of *motion* in the universe must be conserved. Changes of motion may occur—indeed all changes in the universe were considered to be mere changes of motion—but every change of motion must be accompanied by an equal and opposite change in the motion of something else, the total being thus conserved.[10]

The principle of *conservation of motion*, which represents the first crude statement of the law of conservation of momentum, was one of the central doctrines of Cartesian physics. No one who is even vaguely familiar with Cartesian natural philosophy could be unfamiliar with this principle. Motion was defined by Descartes as the product of velocity and quantity of matter; hence his statement comes very near to the modern principle. The only essential difference concerns Descartes' concept of velocity. Descartes did not conceive of velocity as a fully vector quantity, for he had no rules for adding or subtracting velocities when those were not similarly directed. It is better, therefore, to call Descartes' principle "conservation of motion" and to reserve the phrase "conservation of momentum" for use in connection with the modern law.

Descartes first expressed these ideas in his *Le Monde, ou traité de la lumière*,[11] which he wrote in 1629-1633, that is, nearly sixty years before the publication of Newton's *Principia*. The manuscript was not published at the time but was revised and published posthumously in *Principles of Philosophy*, Part II (1644).[12] In this original manuscript, Descartes set out the law of conservation of momentum in the crude form in which I have just expressed it. But more than this, he also gave us the law of inertia—and gave it to us in its complete and modern form.

Descartes presented his law of inertia in two parts, dealing separately with speed and direction. (We shall here consider these two parts together.) He argued in the following way. At any given moment a body has a definite state of motion—a definite speed and direction of motion. But every body remains in the state in which it is unless it is caused by some external agency to change that state. Thus at any given moment a body has a definite shape, size, color, texture, and so on, and it will retain these specific properties unless it is acted upon by some external agency. Consequently, Descartes argued, a body must retain its speed and direction of motion unless it is acted upon by some external agency. Or, in more modern language, every body must continue in its state of rest or uniform motion in a straight line unless it is caused by some external force to change that state.

The argument would, of course, be quite unconvincing to anyone who did not already accept the law of inertia. In the first place, what reason had Descartes to describe the speed and direction of a body as a *state* of that body? Could we not equally well argue that at any given moment a body has a definite *position,* and that consequently it must remain in the *place* in which it is unless it is acted upon by some external agency? In other words, could we not use this argument to support a dynamics of rest? In the second place, why should we accept the general premise that every body remains in the state in which it is unless it is caused by some external agency to change that state? After all, we believe that things do change in shape and temperature, for example, without the influence of any external agencies. A vibrating rod changes in shape without any such external influence, and a hot body will cool down unless there is some source of heat to maintain its temperature above that of its surroundings. Finally, why *external* agency? If one accepts the impetus theory of Jean Buridan and Nicolas of Orêsme, then one would surely reject the thesis that every body remains in the state in which it is unless acted upon by some external agency, for this is precisely what the impetus theory denied.

The argument is therefore little more than an absurd piece of soph-istry, and it is hard to believe that it was this argument which led Descartes to accept the law of inertia. It seems far more likely that he arrived at it via his general metaphysical position and his principle of conservation of motion. He does, it is true, offer some empirical support for this law. But in the absence of criteria for the existence (or absence) of forces inde-pendent of the law of inertia, this, too, is little better than a piece of sophistry.

The experiment he describes is that of whirling a stone in a sling, releasing it, and noting the speed and direction of its subsequent motion.[18] He remarks that the speed of the stone will be the same as its orbital speed, and that its direction will be tangential to its orbital path at the point of release. And we are invited to conclude that when the constraining force is removed, the body continues in the state of motion which it had at that instant. Neither of these contentions could be said to be experimentally established by Descartes, but even if we grant that they were, the law of inertia most certainly does not follow. It lends not even the slightest in-ductive support for the law of inertia.

The law of inertia states that every body not subject to the action of forces continues in its state of rest or uniform motion in a straight line. But a follower of Buridan and Nicolas of Orêsme would immediately ob-ject that the stone is subject to a force after it leaves the sling—an *internal* force, or impetus. He would therefore reject Descartes' experiment as quite irrelevant to the issue of whether or not a body would continue to move with uniform motion when it was not subject to the action of a force.

But more than this, even if we suppose that Descartes had criteria for the absence of internal forces such as Buridan postulated, the experiment would still be irrelevant on other grounds, for it is patently false that the stone continues to move uniformly in a straight line after it leaves the sling. It describes a roughly parabolic path. True, we should say that this is due to gravity. But in the first place this is simply an admission that the experiment is irrelevant. One cannot prove one's ability to run a four-minute mile before lunch by demonstrating one's inability to run a four-minute mile after lunch. Likewise, Descartes cannot prove that a body would continue to move with uniform motion in a straight line without gravity by showing that it does not move in a straight line with gravity.

Finally, what reason has Descartes for saying (if he does) that there is a force of gravity at all apart from the evident fact that projectiles do not move in straight lines? What independent criteria does he have for the existence of such a force? He might point to the fact that bodies have weight. But at best this only shows that a force is required to retain a body in a state of rest above the surface of the earth—a conclusion that *prima facie* at least *contradicts* the law of inertia.

The last point raises a very important issue, namely, what are the criteria for the existence of forces? The whole question of the logical status of Newton's laws of motion hinges upon it. But it is mentioned here only in passing; it will be considered in detail in Part II.

Following Descartes, Huyghens continued with investigations relevant to the law of conservation of momentum. Descartes had merely stated that the total quantity of motion in the universe must be conserved. By "motion" he meant "product of quantity of matter and velocity." Hence "motion" may be roughly translated by our modern term "momentum." But Descartes was not clear about momentum as a *vector* quantity, and hence cannot be said to have stated the law in its complete and modern form. As a result of a number of collision experiments, the necessary corrections were made by Huyghens in the 1650's, and his results were eventually published in the *Journal des Savants* in March, 1669.

Newtonian Dynamics. The key to the origin of Newton's laws of motion is to be found in the wording of the second law. The original Latin reads:

> MUTATIONEM MOTUS *PROPORTIONALEM ESSE VI MOTRICI IMPRESSAE, ET FIERI SECUNDUM LINEAM RECTAM QUA VIS ILLA IMPRIMITUR.*[14]

This is correctly translated by Cajori to read:

> *THE CHANGE OF MOTION* IS PROPORTIONAL TO THE MOTIVE FORCE IMPRESSED; AND IS MADE IN THE DIRECTION OF THE RIGHT LINE IN WHICH THAT FORCE IS IMPRESSED.[15]

The important words are *mutationem motus*.[15] Nearly everyone has assumed that when Newton said "change of motion" he really meant "rate of change of motion." Jammer, for example, remarks that ". . . what Newton meant by saying *mutatio motus* should be rendered in modern English as "rate of change of momentum."[16] There is, of course, no reason to quarrel with the translation of *motus* as "momentum," for Newton himself explicitly defines *motus* as the product of mass and velocity. But why translate *mutatio* as "rate of change"? What justification is there, linguistic or contextual, for this translation? It will be argued here that there is no justification for this translation—*either* linguistic *or* contextual. Moreover, it will be shown that this translation is inconsistent with several passages in the immediate context.

This being the case, several important consequences follow. First, Newton's concept of motive force is very different from our modern concept. Second, the concept of motive force that Newton actually employed is not the kind that would naturally be used to explain the phenomena of free fall or projectile motion. Rather it is the kind of concept that is naturally suited to the explanation of impact phenomena. Consequently, it is reasonable to suppose that there is a close historical connection between the laws of motion and the laws of impact (e.g., conservation of momentum), a supposition that becomes greatly reinforced when it is seen that Newton's laws, correctly interpreted, can be derived very easily from the law of conservation of momentum. Thus there is good reason to believe that all three of Newton's laws of motion were derived directly from Cartesian physics.

There is, to begin with, no justification in classical or medieval Latin for translating *mutatio* as "rate of change," and no one has ever argued that there is such a justification. If the translation is to be justified, therefore, it must be on contextual grounds.

What, then, is the immediate context? The sentence that expresses Newton's second law of motion is immediately followed by the words: "If any force generates a motion, a double force will generate double the motion, a triple force, triple the motion, *whether that force be impressed altogether and at once or gradually and successively.*"[17] Now clearly, this does not provide any positive support for the translation of *mutatio* as "rate of change." On the contrary, to make this translation is to make nonsense of Newton's second law, for according to this translation we must suppose Newton to be saying that the *rate of change* of momentum is proportional to the motive force impressed "whether that force be impressed altogether and at once or gradually and successively." Clearly, Newton meant to say that the change of momentum is proportional to the (total) motive force impressed "whether that force be impressed altogether and at once or gradually and successively," that is, exactly what he did say.

Next, consider the First Corollary, which occurs immediately after the statement of the three laws of motion. It concerns the composition of forces. The proof begins with the words: "If a body in a given time, by the force M impressed apart in the place A, should with an *uniform* motion be carried from A to B . . ."[18] Now if the magnitude of the force M is proportional to the *change* of motion that it produces, then this sentence is intelligible, and M is the measure of what we would call "impulse." But if M is proportional to the *rate of change* of momentum, then the sentence should read: "If a body, initially in the place A, should be subject to a force M which carries it with *uniformly accelerated* motion to the place B." Consequently, the argument of the First Corollary is also inconsistent with the translation of *mutatio* as "rate of change."

Looking further afield, we see that throughout the whole section on "Axioms, or Laws of Motion" in Newton's *Principia* there is not a single phrase that would suggest the usual interpretation of Newton's second law. It is therefore quite incontrovertible that Newton meant what he said. When he said "change of motion," he meant "change of motion." It would, in any case, be rather extraordinary if Newton did not exercise a little more than his usual care in formulating the basic principles upon which his whole system of dynamics depends.

Given, then, that Newton meant "change of motion" when he said "change of motion," it becomes immediately clear that Newton's concept of "motive force impressed" is radically different from our modern one. It corresponds far more closely to the primitive concept of a push or a kick than it does to the strength of a push at any given instant.

Consider once again Corollary I. Quite clearly, Newton thought of the motive force M impressed apart at the place A as a push or a kick given to the object at this point that carries it along with *uniform* speed to the place B in the given time. And by his second law of motion, he considered the magnitude of this kick to be proportional to the *total* change of momentum that the body undergoes. Hence Newton's concept of motive force is much nearer to our present-day concept of impulse—the only essential difference being that, whereas Newton regarded motive force as a *primitive* concept, we define impulse in terms of instantaneous force.

To reinforce the interpretation, consider the following passage from the Scholium at the end of the section on "Axioms, or Laws of Motion":

> When a body is falling, the uniform force of its gravity acting equally, impresses, *in equal intervals of time,* equal forces upon that body, and there-fore generates equal velocities; *and in the whole time, impresses a whole force, and generates a whole velocity proportional to the time.*[19]

What could be more explicit? The force impressed upon a body (by gravity) is a function of the time. The longer a body is allowed to fall, the greater the impressed force that acts upon it. There is simply no other interpretation that can be put upon this statement. It is quite absurd, therefore, to maintain that by "motive force impressed" Newton meant anything like "instantaneous force." And it is quite beyond any reasonable doubt that "motive force impressed" meant something like "impulse."

From the point of view of the mathematical development of Newtonian dynamics, there is little to choose between the two interpretations under discussion. Newton's law is correctly formulated:

$$\triangle I = \triangle (MV) \tag{1}$$

where $\triangle I$ is the impressed force imparted to a body in a time $\triangle t$, and $\triangle (MV)$ is the change of momentum that the body undergoes in the time $\triangle t$, hence:

$$\frac{\triangle I}{\triangle t} = \frac{\triangle (MV)}{\triangle t} \tag{2}$$

Then if f (instantaneous force) is defined by:

$$f = \frac{\lim}{\triangle t \to 0} \frac{\triangle I}{\triangle t} \tag{3}$$

we have at once that:

$$f = \frac{d(MV)}{dt} \tag{4}$$

Nevertheless, there is an important conceptual difference between the two formulations (1) and (4). And if we are interested in finding the origin of Newton's laws, it is important to consider the conceptual scheme that Newton actually used.

If, as has been maintained, impulse was the primitive force concept in Newtonian dynamics, then the first conclusion to be drawn is that Newtonian dynamics was historically more closely linked to the theory of impact than to the theory of free fall or projectile motion, for in the latter connection, impulse is necessarily a derivative notion. Gravitational forces do not come into being or cease to operate. Hence to define the impulse given to a body in free fall, an arbitrary time interval has to be specified. The primitive force concept in this connection is obviously *weight*. However, the situation is reversed when we turn to the phenomena of impact. Here it is the modern concept of instantaneous force that is necessarily derivative and the concept of impulse that is primitive, for we naturally conceive the action of one body on another in a collision as a simple phenomenon—a whole temporally extended event. Thus we have terms in ordinary language like "a push," "a kick," and "a thrust" to refer to such events. It is reasonable to suppose, therefore, that Newton's laws of motion were closely tied to the laws of impact—if not actually derived from them. And certainly the suggestion that they were derived from Galileo's work on free fall seems highly implausible.

Now in fact there was only one law of impact that was widely accepted in the 1660's when Newton was working out his main ideas on dynamics. This was the Cartesian law (as refined by Huyghens) of conservation of momentum. If, therefore, Newton's laws of motion were derived from the laws of impact, they must have been derived from conservation of momentum. And when we examine the matter, we see that this is very plausible indeed, for all three laws of motion (as here interpreted) are direct consequences of the law of conservation of momentum (L.C.M.), provided only that we are prepared to assume: (1) that a force is the cause of a change of motion, and (2) that a cause is proportional to (or equal to) its effect. And since (1) is simply a paraphrase of Newton's own definition of motive force impressed (Definition IV),[20] and since (2) was a universally received doctrine in the seventeenth century,[21] it is not unreasonable to suppose that Newton would have accepted both of these assumptions.

The derivation is very simple. The law of inertia follows immediately from L.C.M. when we apply this law to a system consisting of a single body. The second law of motion follows immediately from assumptions (1) and (2). The change of momentum (the effect) is proportional to the motive force impressed (the cause) and takes place in the direction of the right line in which that force is impressed (additional stipulation). The third law of motion follows in two steps. If momentum is conserved, then every change of momentum must be accompanied by an equal and opposite change of momentum. Hence every cause of change of momentum that comes into being (i.e., every action) must be accompanied by an equal and opposite cause of change of momentum (a reaction). Hence action and reaction must be equal and opposite.

There is no denying the plausibility of this derivation. Of course it is impossible to be certain that it accurately reflects Newton's own reasoning on the matter. Nevertheless, it seems far more plausible than any other alternative, and we may be reasonably confident that it is not too far from the truth.

Conclusion. If the thesis that has been presented here is substantially correct, then there can be no doubt that Newton's laws were primarily conceptual in origin. They were neither derived from nor supported by careful observations or detailed experiments. Rather they were the delayed products of Descartes' wondering what sorts of laws would obtain in a world presided over by an immutable God. They were the product of asking how things *ought* to be, rather than how things actually are. That this was Descartes' procedure there can be no doubt whatever. In *Le Monde* he remarks:

> But even if all that our senses have ever experienced in the actual world would seem manifestly to be contrary to what is contained in these two rules, the reasoning which has led me to them seems so strong that I am quite certain that I would have to make the same supposition in any new world of which I have been given an account.[22]

A more specific rejection of empiricism is hard to find anywhere in the literature of science. That Newton's laws were derived from Cartesian physics in the way suggested may be somewhat more doubtful, but the account is the best that we have. If it is true, then Newton's laws can hardly claim a stronger empirical foundation than Descartes'.

Of course it is true that empirical considerations had an important if not decisive role in bringing this conceptual revolution about. The planetary theory of Kepler, and thus, ultimately, the astronomical observations of Tycho de Brahé, had created the need for a nonhomocentric system of dynamics. When it became no longer possible to accept that the earth was the center of the universe, a system of dynamics that was not earth-centered was clearly needed. The homocentric dynamics of Aristotle were an anachronism in the heliocentric universe of Kepler.

It is also true that considerable research work was needed to convert Descartes' crude law of conservation of motion into Huyghens' law of conservation of momentum. But even when these qualifications are made, it remains a sound historical judgment to say that Newton's laws of motion were not primarily the expression of new empirical findings unknown to the ancient world. Rather, to paraphrase Butterfield, they expressed the result of putting on a new conceptual thinking cap.

II. The Logical Status of Newton's Law of Inertia

It has been argued that Newton's laws of motion were primarily conceptual in origin. The question now arises: What is their status today? Could they in fact, or even in principle, be shown experimentally to be true?

In one way, the answer to this question is obvious. Laws of motion have *in fact* been replaced by other laws of motion—the Aristotelian and medieval laws by the Newtonian, and the Newtonian by the relativistic. But this answer fails to distinguish between conceptual adequacy and empirical truth. The fact that Newton's laws have been replaced by other laws does not tell us much about their logical status. An empirical proposition may be rejected and (in a sense) replaced by some other empirical proposition. But a conceptual schema can also be replaced by a more adequate conceptual schema. Hence the fact that Newton's laws have been superseded does not tell us whether they are empirical propositions or merely conceptual truths. The logical status of these propositions is still an open question.

Let us suppose, then, that Newton's laws had never been superseded, and that no reason had ever been found for thinking Newtonian dynamics to be an inadequate or anachronous system. We could still ask, could we not, whether these laws could in fact or in principle be shown experimentally to be true. It will be argued here that it is in principle impossible to do this, since the existence of forces is always to some extent conventional. And to illustrate this thesis, the conventionality of the law of inertia will be established by the construction of a system of dynamics in every way as adequate as Newton's, but using a different principle of natural motion.

The Existence of Gravity (Preliminary Remarks). In Part I of this essay, Descartes' example of releasing a stone from a sling was discussed. The example was supposed to provide at least some empirical support for the law of inertia. It was argued that in the absence of independent criteria for the existence of forces, it does nothing at all to support this law. The final point raised turned on the question of the existence of gravity. What reason have we, independent of the law of inertia, for saying that gravitational forces exist? It was remarked that to say that we know there is a gravitational field because projectiles do not move with uniform motion in a straight line is obviously to beg the question. It is simply to assume the truth of the law whose truth we wish to establish. It was also remarked that to say that we know there is a gravitational field because bodies have weight is equally absurd. For at best this shows only that a force is required to *maintain* a body in a state of rest or uniform motion in a straight line—a conclusion that *prima facie* contradicts the law of inertia.

Of course it may be objected that a force is required only to maintain a body in a state of rest or uniform motion in a straight line just because of the existence of a gravitational field. But this is clearly circular, for it presupposes the existence of the very entity whose existence is in question.

Again, it may be argued that the existence of a gravitational force equal and opposite to the sustaining force exerted (e.g., by a spring balance) follows from Newton's principle of action and reaction. This raises the question of why we should accept Newton's third law. But for the moment let us accept it. Even then the existence of a gravitational force does not follow. For the principle of action and reaction merely tells us that to every action there is an equal and opposite reaction. And the reaction in this case is located in whatever supports the spring balance. The upward force exerted by the spring balance on the suspended body finds its reaction in the downward force exerted by the spring balance on its support.

Now it may be said: "Surely the *body suspended* must be acted upon by a force equal and opposite to the force exerted by the spring balance." But why? What reason have we, independent of the law of inertia, for saying such a thing? If we accept the law of inertia (and the principle of composition of forces), then we must accept that the body suspended is acted upon by a force equal and opposite to the sustaining force, for otherwise we must say that the body would be accelerated upward. But if we do not accept the law of inertia, then no such conclusion follows. We may, for example, be prepared to say that the suspending force exerted by the spring balance is an *unbalanced* force. The argument seems to have force only because we naturally think of a state of suspension as being a state of equilibrium. But our tendency to think in this way is a *consequence* of our accepting the law of inertia, and hence we cannot use this fact to *support* the law.

Some of the persuasiveness of this argument arises from the fact that the word "equilibrium" is ambiguous. There is one sense in which we may all agree that the suspended body is in equilibrium—whatever dynamic principles we accept. This is the sense of "kinematic equilibrium," where "equilibrium" simply means "does not move." But kinematic equilibrium does not entail *dynamic* equilibrium unless we hold the dynamic principle that a body which does not move is not subject to the action of any unbalanced forces. And since this is part of the law of inertia, it follows that we cannot argue to the existence of a gravitational force equal and opposite to the suspending force without presupposing the relevant part of the law of inertia. The argument from the kinematic equilibrium of suspended bodies to the existence of gravitational forces is therefore invalid.

The Existence of Forces. It is evident that if we are to see our way out of this maze, we must make the attempt to formulate criteria for the existence of forces, for the question of the logical status of the law of inertia obviously depends upon whether we have any criteria that are independent of this law.

What, then, are our criteria for the existence of forces? There are, it seems, two completely general criteria. We may say that a system is subject to the action of a force if and only if (1) the system persists in what we regard as an *unnatural state,* or (2) the system is changing in what we consider to be an *unnatural way.* The concepts of natural state and natural change must, of course, be explicated. But there will be some advantage in clarity gained by postponing this for the present. Here we shall say merely that a system is thought to be an unnatural state if and only if we consider that its persistence in that state *requires causal explanation,* and a system is considered to be changing in an unnatural way if and only if we think that its changing in that way requires causal explanation. The conditions under which we should say that the behavior of a system does or does not require causal explanation will be discussed later. For the moment we shall proceed intuitively, and show the application of these criteria by means of examples.

Example 1. In Aristotelian dynamics it was considered natural for a body to return to its proper place by the shortest path. A freely falling body was therefore not considered to be subject to the action of any forces. A force was required only to *arrest* this natural motion. In mediaeval dynamics the uniform motion of an object along a horizontal plane (e.g., a rolling wheel) required the action of a force (impetus) for its maintenance. Any change of position of an object that did not result from its motion directly to its natural place was regarded as an unnatural change, and hence required a force for its explanation.

Example 2. In Cartesian and later in Newtonian dynamics, the Aristotelian-mediaeval concepts of natural place and natural motion were replaced by new concepts. The concept of natural place was abandoned, and uniform straight line motion was considered to be natural motion. Of course this does not mean that every case of uniform straight line motion would be counted as a case of natural motion, for in many cases the continuance of a body in such a state would require causal explanation. The point is simply that while there are no circumstances in which we should consider nonuniform motion to be natural motion, there may be circumstances in which the continuance of a body in a state of rest or uniform motion would not require causal explanation. This is what is meant by describing uniform straight line motion as natural motion.

Example 3. Every solid object that can be said to be deformed or in a state of *strain* must be conceived as having a natural shape and size against which this deformation can be gauged, i.e., it must be conceived as having a shape and size, continuance in which would not necessarily require causal explanation. The natural shape of a body is that shape which a body has in circumstances in which: (a) its shape is stable (i.e., not a function of time), (b) its shape is independent of its orientation, and (c) the change in shape that the body undergoes when it is given a positive or negative rotation about any given axis is independent of the *sense* of its rotation about that axis. Thus a body sitting on a horizontal table is deformed, since under these circumstances, we believe, its shape is a function of its orientation. Of course in one sense its shape need not depend on its orientation under these circumstances, for it may assume the *same geometrical figure* whatever its orientation. But that is not what is meant here by *sameness of shape*. An object retains its shape (in the required sense) if and only if the distance between every pair of points in its surface remains the same. Thus even a ball sitting on a horizontal table, whose geometrical figure (some kind of oblate spheroid) may be independent of its orientation, may nevertheless be in a state of strain.

The application of (a) is obvious. Every vibrating object is at least sometimes in a state of strain. By (c) it is intended to rule out strains produced by spin, for if an object is spinning about a given axis, criteria (a) and (b) may well be satisfied. Nevertheless, we may wish to say that it is in a state of strain. But if now the object is given an *additional* rotation (positive or negative) about that axis, then the change of shape that the body undergoes will depend upon the sense of this additional rotation. Hence by criterion (c) the body can be concluded to be in a state of strain.

Every solid object that by any of these criteria is deformed requires the existence of some force or forces to explain its deformation.

Example 4. Suppose that a body initially at the same temperature as its surroundings begins to increase in temperature. Such a body would be conceived by us to be changing in an unnatural way, and a causal explanation of its changing in that way would be demanded. We should say that it must have some source of supply of *energy*.

Now it will be said that energy is not a force concept. But as force concepts are here understood, energy is a force, and supplying energy to a body is a way of acting upon that body by a force. There is, in fact, historical precedence for this. Until the mid-nineteenth century, the quantities we now call forms of energy were always described as forces.

Example 5. Suppose that a body undergoes a change of mass. Then such a change may be conceived by us to be an unnatural change, and hence to require the existence of a force to produce it. Such changes do not in fact occur in isolation from other changes. Relativistic changes of mass are regarded as *side effects* of acceleration-producing forces. The reason for this is not obvious. Why a change of state of motion should be regarded as a *central* effect of a force and a change of mass as a *side* effect needs to be explained. The reasons may be partly historical and partly a matter of the obviousness of the changes that occur. But whatever the reasons, if changes of mass were to occur in the absence of changes of state of motion (and other obvious changes), such changes would no doubt be regarded as unnatural (i.e., as requiring causal explanation) and purely mass-changing forces would be needed to account for them.

In general, then, we may say that a system is acted upon by a force (or forces) if and only if we consider that the system persists in an unnatural state or that it is changing in an unnatural way.

We have remarked that a system is considered to be in an unnatural state if and only if we consider that its continuance in that state requires what has been called "causal explanation." But under what conditions should we say that we consider (some aspect of) the behavior of a given system to require causal explanation? In answering this, we do not propose to offer any analysis of causality or to give any positive characterization of causal explanations. For the purposes of this essay, we shall simply *say* that the behavior of a given system is considered to require causal explanation if and only if we feel that this behavior is not sufficiently explained by its subsumption under a law of *succession*.

A law of succession is any law that enables us to predict the future states of any system (or given class of systems) simply from a knowledge of its present state, assuming that the conditions under which it exists do not change. The law of radioactive decay is such a law. From a knowledge of the number of atoms contained in a given sample of a radioactive element, the number of atoms of that element contained in that sample at any future time can be predicted (provided that the decay constant is known). The law of free fall is another such law. From a knowledge of the present position and state of motion of a freely falling body, the future positions and states of motion can be predicted (provided that g is known). Kepler's laws of planetary motion provide yet other examples of laws of succession. From a knowledge of the present position and state of motion of a given planet, the future positions and states of motion of that planet can be predicted.

Now it is maintained that provided that subsumption under such a law is considered to give a *final explanation* of the behavior of a given system, then the system is considered to be in a natural state or to be changing in a natural way. That subsumption under such a law is considered to be a final explanation is shown by the fact that we would reject any request for an explanation of why systems behave in that way or remain in that state as *inappropriate* in the context of the given inquiry.

The law of inertia is a law of succession according to the above criteria. But this law, unlike the others cited, has a special role within our conceptual scheme, for if the behavior of a given object could be explained simply by subsumption under this law, then we should reject the question of why that object remains in that state. Normally, however, matters are more complicated than this, and the law of inertia is not used in such a straightforward way. Nevertheless, essentially the same point can be made. Consider the problem of explaining projectile motion. The motion is divided (conceptually) into two components—a uniform straight line and a uniformly accelerated motion. The first component is explained by subsumption under the law of inertia and the second by subsumption under the law of free fall. The two laws of succession are then used to predict the future velocity components of the projectile. But the two laws have very different status, for while it is thought legitimate to ask why the projectile accelerates toward the center of the earth, it is not thought legitimate to ask why it has a *constant velocity* component as well.

Corresponding to the law of inertia, there is an analogous law in the fields of statics and strength of materials—a law that has similar logical status. This is the law (L_1) that *"every solid object remains undistorted unless it becomes subject to stress."* The close analogy between this law and the law of inertia is worth pursuing. In the first place, if the behavior of a solid object could be explained simply by subsumption under this law, the question of why the object remains undistorted would be rejected as inappropriate. Second, even if the object changes its shape, or remains in a state of strain, we should divide (conceptually) the actual shape of the body into two components—a natural shape and a distortion—and we should attempt to explain only the magnitude and (possibly) variation of the latter component. But we should not postulate the existence of any forces to explain the magnitude or invariance of the former component. To this extent, then, there is a close analogy between L_1 and the law of inertia.

The existence of L_1 raises some interesting possibilities, for might it not be possible to base a criterion for the existence of forces on L_1 (a criterion that would be independent of the law of inertia) and hence to establish the law of inertia inductively? The examination of this suggestion will be the main subject matter of the next section.

Before we proceed with this, however, it is necessary to make some general remarks about the ontological status of forces. According to the proposed criteria for the existence of forces, there is clearly some truth in the idea that the forces we say exist in nature have a kind of *conventional existence,* for a force exists only because we *choose to regard* some succession of states as an unnatural succession. But there appears to be no objective criterion for distinguishing between natural and unnatural successions of states. The fact that a given succession is lawlike (i.e., can be explained by subsumption under a law of succession) clearly does not mean that we should regard this as a natural succession. We have, in fact, drawn certain lines and come to regard certain successions of states as natural and others as unnatural. But in the absence of any clear and objective reasons for drawing these lines as we do, we may wonder whether other lines could be drawn, or whether, indeed, it is necessary to draw any lines at all. Why shouldn't all successions of states be regarded as natural? Or, if that is too radical, why shouldn't all *lawlike* successions be regarded as natural?

Forces are peculiar scientific entities for other reasons, for while the action of a force is supposed to explain (causally) certain patterns of behavior, the occurrence of these patterns is considered a *sufficient* condition for the existence of the precise force required to produce them. If a body accelerates, then it must be acted upon by a force sufficient to produce this acceleration. If a body is distorted, then it must be subject to a stress sufficient to produce this distortion. The nature of this entailment (whether it is physical or logical) will concern us later. But that there *is* such an entailment already marks forces off from other scientific entities, for it is agreed by all that the existence of molecules, genes, and electrons is not entailed by the existence of the effects they are designed to explain.

The Conventionality of the Law of Inertia. It was suggested in the previous section that it may be possible to provide inductive support for the law of inertia, using the law that every body remains undistorted unless it becomes subject to stress to provide us with independent criteria for the existence of forces. Let us now take up this suggestion.

Consider first the case of a solid object sliding on a rough horizontal surface. Such an object will possess a certain shear strain ϕ and a certain deceleration d, and by plotting ϕ against d, for different surfaces it should in principle be possible to determine d for $\phi = 0$. Let us then assume that $d = 0$. But even so, the object is still subject to strain, since its shape will still be a function of its orientation. The object will be compressed in the direction perpendicular to the earth's surface.

Let us now put this object in orbit in such a way that it is not spinning relative to the fixed stars. Under these conditions we should find that its shape is very nearly independent of its orientation. We may be tempted to conclude from this that some kind of accelerated motion must be the natural motion for any physical object. Even under these conditions, however, the object should exhibit tidal distortions. And again, it should be possible (in principle) to plot the distortion ϕ against the acceleration a for similar objects placed in different orbits. Conceivably, we might discover that when $\phi = 0$, $a = 0$, i.e., we might find that every body that is not subject to stress continues in its state of rest or uniform motion in a straight line relative to the fixed stars. Then if finally we are prepared to accept the principle that a body is subject to stress if and only if it is subject to the action of a force, then it follows that if a body is *not* subject to the action of a force it will continue in its state of rest or uniform motion in a straight line relative to the fixed stars.

The argument is certainly tempting—although it is open to a number of criticisms. In the first place, if our present theories are correct, then the observed tidal distortions would not only be a function of the acceleration. They would also be a function of the *convergence* of the gravitational field in which the object is placed, so that if in any small region of space there existed a *uniform* gravitational field, natural motion in this field would turn out to be a uniformly accelerated motion. But it is very doubtful whether we should reject the law of inertia on the evidence of such a discovery—especially if the phenomena described could be explained on the basis of Newtonian theories.

Second, it is not evident that our concept of *uniform straight line* motion is sufficiently precise for these considerations to have any weight at all —even supposing that the measurements of tidal distortion could be made. Moreover, what of our concept of *remaining the same in shape?* Is this a sufficiently precise concept for the purposes of such an investigation?

All of these difficulties need to be discussed. But there is one objection that is fatal to the whole program, for even supposing that these questions can be answered satisfactorily, the demonstration could not completely remove the conventional element from the law of inertia. At best it could only show that our concept of natural motion is cognate with our concept of natural shape. They would be seen to depend upon each other, so that if a new concept of natural motion were introduced, a new concept of natural shape would also be required and vice versa. And there seems to be no reason, in principle, why such new concepts should not be introduced.

To establish that this element of conventionality does indeed exist, let us see how we might proceed to construct a system of dynamics, at least as powerful as Newton's, but using a different principle of natural motion. We begin with the observation that if Newton's law of gravitation is correct, and A and B are any two isolated bodies of mass M_A and M_B, respectively, then:

$$f = G \, \frac{M_A \cdot M_B}{r_{AB}^2}$$

where f is the force of attraction, and r_{AB} is the distance between A and B. From this it follows (by Newton's second law) that the absolute acceleration of A and B must be given by:

$$a_A = G \frac{M_B}{r_{AB}^2} \qquad \text{and} \qquad a_B = G \frac{M_A}{r_{AB}^2}$$

In other words, it is a direct consequence of Newtonian mechanics that every body accelerates toward every other body in the universe with a *relative* acceleration directly proportional to the *sum* of their masses and inversely proportional to the square of the distance between them.

It is important to understand exactly what this means. It does not mean that if measurements of relative accelerations, masses, and distances could be made we should always obtain results exactly in accordance with this formula, for relative accelerations may be compounded. It means only that in working out relative accelerations we should assume that there are relative acceleration *components* that accord with this law.[23] The relative acceleration of the earth and the moon, for example, may be very nearly in accordance with this formula, and so too may be the relative acceleration between the earth and the sun. But the relative acceleration between the moon and the sun would not be even roughly in accordance with this formula.

Now the law derived from Newtonian mechanics is a law of succession. And there appears to be no reason why we should not consider any changes that accord with this law as natural changes. The first law of motion in our new system of mechanics will therefore be that "every body has a component of relative acceleration toward every other body in the universe directly proportional to the sum of their masses and inversely proportional to the square of the distance between them—*unless it is acted upon by a force.*"

For simplicity we shall take the other laws of motion to be identical with Newton's.[24]

To see that this achieves the desired result, let us follow out some of the consequences of accepting this new principle of natural motion. In the first place, it follows that if we have a number of objects of finite mass, randomly distributed and sufficiently removed from each other, then their relative accelerations tend to zero. And assuming our galaxy to be such a system, it follows that any body sufficiently far removed from other bodies in the galaxy will continue in its state of rest or uniform motion in a straight line relative to the various stars in the galaxy unless it is acted upon by a force. The fixed stars may therefore serve as an *absolute* frame of reference in Newtonian mechanics. Adopting this principle of natural motion thus leads to the consequence that it is unnecessary to make any prior distinction between absolute and relative motion in formulating our dynamical principles.

Next, assuming that the system of fixed stars constitutes an absolute frame of reference, and that the sum is sufficiently far removed from other stars in this system, it follows that the center of mass of the solar system must continue in its state of rest or uniform motion in a straight line relative to the fixed stars. Taken individually, the planets must accelerate toward each other, toward the sun, and toward the frame of reference of the fixed stars. But this acceleration would be regarded as a natural acceleration. No force would be required to explain it. Similarly, the parabolic motion of a projectile (as modified by Coriolis's deflection) must be regarded as natural motion, for a body that is accelerating toward the center of the earth with an acceleration g would not be regarded as subject to the action of a force. A force would be required only to arrest this acceleration, e.g., to impart to it an acceleration equal and opposite to its natural acceleration. The measure of this force would be simply the weight of the body.

It follows that in this new system of dynamics, kinematic equilibrium cannot be taken to imply dynamic equilibrium. The fact that a body remains at rest does imply that it is not subject to the action of any *unbalanced* forces. It also follows that in this new system of dynamics no distinction can be drawn between *gravitational* and *inertial* mass, for weighing a body is simply a special way of determining the inertial mass. The weight of a body is only a measure of the force required to impart an acceleration to it equal and opposite to its natural acceleration. The puzzle of the identity of gravitational and inertial mass is therefore simply a puzzle generated by our choosing to regard a certain kind of motion as natural, and it may be resolved without making any assumptions inconsistent with the predictions of Newtonian physics.

On the negative side it must be said that the adoption of this principle of natural motion might force us to revise our concept of natural shape, for tidal effects must either be regarded as *natural effects* produced by the differential natural accelerations of the different parts of orbited bodies, or else they must be regarded as *distortions* set up in these bodies by the action of some stress. If the former alternative were taken (as undoubtedly it would be) then our concept of natural shape would need to be modified. If the latter alternative were adopted, we could retain our concept of natural shape, but we should need to postulate the existence of a kind of inverse square "stress field" surrounding massive bodies.

The adoption of this principle of natural motion would have other ramifications. It would mean, for example, that the law of conservation of momentum would have to be rejected and that a new law relating the *variation* of momentum with the disposition of other massive objects in the universe would come to replace it. But conservation of energy would not be affected, for work would still need to be done to raise a heavy weight.

There is no need, however, to trace through all of the ramifications of adopting the new principle, for since the law of succession upon which this principle is founded is a simple consequence of Newtonian dynamics, the new system is internally consistent and applicable to the world if and only if Newtonian dynamics is. The only difference between the two systems of dynamics is a conceptual one. But it is important that such an alternative system of dynamics can be constructed, for it is proof positive that the law of inertia is not an empirical proposition. It is the sort of proposition that an international gathering of scientists could declare to be false even though they produced not a single fact contradicting it. The possibility of such an alternative dynamic is therefore proof positive of the conventionality of the law of inertia.

The Arbitrariness of the Law of Inertia. The conventionality of the law of inertia has been demonstrated, but its arbitrariness is another matter. We have shown that there is an area of choice, but we have not shown that it is a matter of indifference which choice we make. We shall proceed by making some general remarks concerning the problem of choosing between kinematically equivalent descriptions.

Suppose that the meter rod in Paris were suddenly to burst through the ends (both ends) of its glass case. There are at least two kinematically equivalent ways in which we may describe what has happened. We may say that the universe and everything in it (except the meter rod) has shrunk in size; or we may say that the rod has expanded and the rest of the universe remained the same. Now according to certain positivists, these two descriptions are identical, and it is a matter of complete indifference which we use. But given that we hold the dynamical views which we do— viz., that a body remains the same in size unless it is acted upon by a (compressive or expansive) force—the two descriptions are not dynamically equivalent, and it makes a great deal of difference how we choose to describe what has happened. If the first description were really accepted, then we should want to know what had insulated the rod against this otherwise universal compressive force. Accordingly, we should examine the rod's *present surroundings,* hoping to find some relevant peculiarity. If, on the other hand, the second description were accepted, then we should examine the rod's *past history,* hoping to find some special conditions that might have produced the expansion—although such an inquiry would obviously be *irrelevant* if we believed that the rod had not changed.

Far from its being a matter of indifference which of two kinematically equivalent descriptions we choose, it is a matter of the utmost importance, for the different descriptions make different *theoretical commitments.*

Now this same problem might have arisen in another way. Let us suppose that two scientists, *A* and *B*, view the phenomenon of the meter rod, that *A* holds the principle that every body expands unless it is acted upon by a force, while *B* holds that every body remains the same in size unless it is acted upon by a force. Let us further suppose that *A* and *B* agree on the kinematic description of the phenomenon in question. Let us say that both agree that the rod has expanded, and that the rest of the universe has remained the same. Nevertheless, the two scientists will make different investigations. *A* will examine the rod's present surroundings, hoping to find what has insulated the rod against the otherwise universal compressive force, while *B* will examine the rod's past history, hoping to find the cause of its expansion.

Our choices between different dynamical principles should therefore be governed by the same sorts of considerations as those that govern our choices between kinematically equivalent descriptions, and these choices will depend on what theoretical commitments we wish to make.

The whole course of scientific inquiry is guided by just such choices. Consider, for example, the changeover from the homocentric to the heliocentric universe. *Kinematically,* the heliocentric system of Copernicus is virtually equivalent to the homocentric system of Tycho de Brahé. With slight modifications, the homocentric system of Tycho can be obtained from the heliocentric one of Copernicus by a simple coordinate transformation. But *dynamically,* these two systems are far from equivalent, for they raise very different problems of explanation. So long as a homocentric cosmology is accepted, a homocentric dynamic such as Aristotle's is tenable. But once a heliocentric cosmology is accepted, we must, if we adhere to Aristotelian dynamics, be prepared to say that although the planets move around the sun, they *would* move around the earth if it weren't for the action of certain forces. Hence we must be prepared to explain uniform circular motion about the sun as a *deviation* from uniform circular motion about the earth.

It cannot be maintained therefore that the choice of dynamic principles is arbitrary. To borrow a phrase from Poincaré, they are "conventional, yes; but arbitrary, no."

III. Newton's Second Law of Motion

The essential conventionality of Newton's first law of motion has been established. By similar arguments it should be possible to establish the conventionality of any law whose role in science is to provide us with criteria for the absence of forces, for forces exist only because we choose to regard certain changes or states as natural and others as unnatural. While criteria of simplicity (regarding the format of our laws of distribution and succession of forces) and coherence (between, for example, our concepts of natural motion and natural shape) may guide us in making those choices, it seems that in general there will always remain some area of choice. Consequently, the arguments of the previous section should have general significance and should apply to any law, whether dynamical or otherwise, that attempts to state the conditions under which a system is free from the action of forces.

But what of Newton's second law of motion? What is the logical status of this law? Is it a definition of force? Of mass? Or is it an empirical proposition relating force, mass, and acceleration? In the tradition that has succeeded Mach, Newton's second law of motion has been widely regarded as a definition of force—mass being defined independently via Newton's third law of motion. But is this account correct? By what criteria should we judge it to be correct?

It will be seen that the answers to these questions must depend on the purpose of our discussion. If we are attempting a rational reconstruction of mechanics, then the received account may be satisfactory. But if we are attempting to describe the actual role of Newton's second law in physical science, then it will be seen that this account is highly misleading. And our purpose here is the latter, for a description of the logical status of a law is here understood to be simply a generalized description of its role in physical science. It will therefore be argued that the received account of the logical status of Newton's second law of motion is unsatisfactory. The attempt will then be made to replace this account by a more satisfactory one.

Is Newton's Second Law a Definition of Force? In its original formulation, Newton's second law of motion simply asserted that the motive force acting on a body in (a given time) is proportional to the change of momentum that it undergoes (in that time) and is similarly directed. Thus if a football is kicked, then the magnitude of the kick is proportional to the change of momentum that the football undergoes from the initial to the final moments of contact with the boot. In this form, then, the law was obviously definitional. It was simply a definition erected according to the precept that a cause is proportional to its effect, for certainly Newton had no independent criteria for determining the magnitude and direction of a kick.

But nowadays we have a different conceptual scheme. Instantaneous force is our fundamental dynamical force concept, and impulse is a defined concept. This historical argument therefore says little about the present status of Newton's second law. Nevertheless, the connection between the historical and the modern law is suggestive. Let $I(t)$ be the Newtonian motive force impressed upon a body in time t, and let $M(t)$ be the change of momentum which the body undergoes in that time. Then we have:

$$I(t) = M(t) \qquad \text{(by definition)}$$

Now, provided that $M(t)$ is differentiable with respect to time, we may define the instantaneous force $f(t)$ as:

$$f(t) = \frac{dI(t)}{dt} = \frac{dM(t)}{dt} \tag{1}$$

Hence the modern law is immediately derivable from the original, and it is tempting to argue that it can hardly be any less conventional.

However, the situation is not quite so simple. In the first place, we do not always *in fact* decide what magnitude of resultant force is acting on a given system by measuring its mass and acceleration. And it is seldom if ever true that we *must* use this procedure to determine the magnitude of such a force. Consequently, if the definitional status of (1) is to be maintained, then it must be shown that in the case of conflicting results concerning the magnitude of resultant forces, those results obtained using Newton's second law would always be preferred.

Second, a force is usually thought to be a *cause* of change of momentum, for if a change of mometum $\triangle M$ occurs in a time $\triangle t$, then we should say that this change must be produced by a force whose average value is $\triangle M / \triangle t$. But since causes and effects are always conceived to be independent existences, we should be reluctant to say that the existence of the effect ($\triangle M$) *logically entails* the existence of a resultant force whose average measure is $\triangle M / \triangle t$. If a force of this magnitude does indeed exist, then this ought to be something that, according to our ordinary conception of force, is independently discoverable.

Of course it may be that our ordinary conception of resultant force is at fault, and that what we say about such forces is at variance with what we do. But if so—if (1) is true by definition—then our concept of resultant force must be entirely lacking in explanatory power, for we obviously cannot explain the existence of an effect by postulating the existence of something whose sole raison d'être is that it produces the given effect. In fact, there is a variety of ways of determining the magnitude of the resultant force acting on a given system—ways based on a variety of different force laws. Never mind, for the moment, how these laws themselves are established. The fact is that we can determine the magnitude of the resultant force acting on a given system without *explicitly* relying upon Newton's second law of motion. Thus *prima facie* at least, our concept of resultant force is not lacking in explanatory power, and our ordinary conception of resultant force is not at fault. And if this is the case, then Newton's second law of motion can be held to be true by definition only if we are prepared, if necessary, to divest our concept of resultant force of all explanatory power.

Third, there appears to be a simple category mistake in saying that force is the product of mass and acceleration, for although we may distinguish different kinds of forces, it does not seem that we can also distinguish different kinds of products of mass and acceleration. We might say that the product of the mass and acceleration of a given object defines a scale for the measurement of the resultant force acting on that object. But then the question arises: Why should this scale be taken to be a scale of *force*? Is this an analytic connection? Or is this something that is empirically discoverable? If it is agreed that forces are not to be identified with rates of change of momentum, then how can the measure of the resultant force acting on a given system be identified with the measure of the rate of change of mometum? What kind of connection relates these two quantities? To answer these questions, it is necessary to clarify our ideas concerning scales and quantities.

Scales and Quantities. The existence of a quantity depends upon the existence of an objective linear order, for if the objects (systems, events, states, etc.) $A_1, A_2, A_3 \ldots A_n$ possess a given quantity q, it must always be possible to arrange those things in order of q by some objective ordering procedure—an objective ordering procedure being any that, if perfectly executed, would always lead to the same ordering among the same particulars under the same conditions, independent of who does the ordering. Moreover, if two or more logically independent and objective ordering procedures would always in fact generate the same order among the same particulars under the same conditions, then we should say that they are procedures for ordering those particulars in respect of a quantity q. Neither of these conditions is *both* necessary and sufficient for the existence of a quantity, but the first is necessary, and the second is sufficient.

Of course these conditions are somewhat idealized. Independent procedures for ordering things in respect of the same quantity (e.g., temperature) may differ in both range and definement. But *significant inversions* of order could not be tolerated. If two or more logically independent and objective ordering procedures led to significantly different orders among the same things under the same conditions, then we should say that they were procedures for ordering things in respect of the *different* quantities.

Now, in general, there may be many logically independent procedures for ordering things in respect of the same quantity. In the case of temperature, for example, there are literally dozens of such procedures. Consequently, the criteria for the identity of quantities cannot be tied to the ordering procedures. We cannot, without destroying the whole structure of our science, say that every independent ordering procedure defines a different quantity. Rather we must say that it is the *order,* and not the ordering relationships, which provides us with criteria for the identity of quantities. The order may in fact be *identified* by any of a number of logically independent ordering procedures (just as a man may be identified by any of a number of independent descriptions). But it does not follow from this that any *particular* ordering procedure is essential to the concept of any given quantity (any more than it follows that any particular identifying descriptions belong essentially to the man they identify). Consequently, *it can be an empirical question* whether any given ordering procedure is a procedure for ordering things in respect of a given quantity. For our quantity concepts are, in Gasking's terminology, generally *cluster* concepts.[25]

An ordering procedure does not, of course, define a *scale.* To have a scale we must have some objective procedure for assigning numbers to things. And to have a *scale for the measurement of a given quantity* q, we must have an objective procedure for assigning numbers to things such that if those things are arranged in the order of numbers assigned, they will in fact be arranged in the order of q. Consequently, it is also usually an *empirical* question whether a given scale is a scale for the measurement of some given quantity. To discover, for example, whether a given scale is one for measuring temperature, we should usually have to discover whether or not the above criterion is in fact satisfied, i.e., whether the numerical order corresponds to the temperature order.

Consequently, if force is a quantity, it cannot be absurd in principle to ask whether any particular procedure for assigning numbers to things is a way of measuring the force acting upon them. True enough, to assign the number $dM(t)/dt$ to a given object at a given time is to make a measurement. But is this necessarily a way of measuring the resultant force acting on that body at that time?

To answer this question, we must consider how in fact we are able to order things in respect of the resultant forces acting upon them. One procedure, of course, is to place them in order of the rates at which their momenta are changing. But if this were the *only* procedure, then although the answer to our question would be obvious, it would mean that our concept of resultant force is utterly empty of explanatory power. There are, however, other procedures, for we can often *calculate* the measure of the resultant force acting upon a given system from a knowledge of the measures of the independent forces to which it is subject (assuming the principle of composition of forces). Hence it may yet be a legitimate question to ask whether $f(t) = dM(t)/dt$ necessarily defines a scale for the measurement of resultant forces.

Now there are, in fact, many and various procedures by which the magnitudes of the individual forces acting on a given system may be determined—electrostatic forces by charge and distance measurements, elastic forces by measurement of strain, magnetic forces by current and distance determinations, gravitational forces by mass and distance measurements, and so on. And it is an empirical fact that when all such force measurements are made and the magnitude of the resultant force determined, then the rate of change of momentum of the system under consideration is found to be proportional to the magnitude of this resultant force. It is this fact that justifies us in taking $f(t) = dM(t)/dt$ to define a scale for the measurement of *resultant force*.

Of course if the *calculated* measure of the resultant force did not agree even approximately with the *direct* measure of resultant force, we should not automatically cease to regard $dM(t)/dt$ as defining a scale for the measurement of resultant force. There are many other possible alternatives. We might, for example, question the accuracy of the measurements upon which our calculations were based. We might doubt the validity of applying some law (e.g., Hooke's law) to the particular case in question. We might even doubt the general validity of the principle of composition of forces. Or finally, we might doubt whether we had taken all of the active component forces into consideration. But all such doubts can be reduced. Doubts about the accuracy of our measurements may be reduced by careful repetition. Doubts about the validity of applying, say, Hooke's law could be diminished if the material in question could be shown to be elastic. The principle of composition of forces could be checked statically. And finally, if no side effects were discoverable that were uniquely correlated with the additional component forces required to yield the identity of the calculated and directly measured resultant forces, then we should have no *independent* reason for believing in the existence of any such forces.

However, although we may in this way become more and more doubtful about the legitimacy of taking $dM(t)/dt$ to define a scale for the measurement of resultant force, the rejection of this definition would seem to create an enormous conceptual problem. Many of the primary measurements that must be made to determine the resultant force acting on a given system and many of the principles that must be used in carrying out the necessary calculations will themselves be seen to be justified only on the assumption of the general validity of taking $dM(t)/dt$ to define a scale of resultant force. This does not mean, however, that there is any vicious circularity in the procedure by which we justify our considering $dM(t)/dt$ to be a measure of the resultant force acting upon a system, for it is by no means necessary that the whole complex of inference patterns which we have set up should cohere. On the contrary, it is a truly remarkable fact that they cohere as well as they do. But it is just this fact that would seem to create the enormous conceptual problem of ever rejecting the second law of motion and adopting an entirely different law.

There are some minor adjustments that might be made to Newton's second law without doing great violence to our conceptual framework. Thus we would adopt a scale of force that is nonlinear with respect to our ordinary scales of force without any but the most trivial changes. We could, for example, consider $dM(t)/dt$ to be proportional to the *square* of the resultant force. And then, to retain the coherence of our system of physics, we should merely have to substitute f^2 for f in all physical equations. But such a change would be a mere mathematical manipulation; its possibility tells us very little about the logical status of Newton's second law, for the form of the mathematical expression of *any* law is a function of the kinds of scales on which the related quantities are measured. The really fundamental change, which would demand some kind of conceptual revolution, would result if $dM(t)/dt$ were not considered to define *any* kind of scale for the measurement of resultant force.

If significant differences between the calculated and directly measured resultant forces did in fact exist in a sufficient number and variety of cases, and if, furthermore, there appeared to be no way of explaining these differences without resort to *ad hoc* devices, then no doubt such a fundamental change in our conceptual scheme would be forced upon us. Thus we may say that it is an empirical fact that we are able to maintain Newton's second law in a way that is *methodologically satisfactory*. And we may *express* this point by saying that Newton's second law is empirical. Nevertheless, we may question whether, even in the face of this fact, it would be *irrational* or *methodologically unsound* to accept any other law in its place. If not, then there is a clear sense in saying that it is only conventional that we accept $dM(t)/dt$ as defining a scale of resultant force. And we may express this fact by saying that Newton's second law of motion is only *conventionally* true.

Now in fact, *without* assuming that the results of our calculations and direct measurements of resultant force ever differ significantly or inexplicably, we could cease to consider that the rate of change of momentum of a body provides us with a measure of the resultant force that acts upon it, for if the principle of natural motion described in Part II were adopted, then we should consider that a body may change in momentum even though no forces were acting upon it. The reason is simply that any concept of naturally accelerated motion involves a concept of natural changes of momentum. In place of Newton's second law, therefore, we may postulate that the magnitude of the resultant force acting upon a body is proportional to the *difference* between its natural and its actual rate of change of momentum. This conceptual scheme will then be no less applicable to the physical world than is Newton's.

Consequently, although there is good and sound sense in describing Newton's second law as empirical, there is also good and sound sense in describing it as conventional.

Force and Mass. The problem most usually discussed in connection with Newton's second law is that of the relationship between force and mass, for it is obvious that unless a scale of mass can be set up independently of Newton's second law, we cannot use this law to define a scale of force. Consequently, if Newton's second law is to have any claim to any kind of empirical status, it must be possible to define a scale of mass independently.

But before we proceed to see whether this is possible, let us be more specific about what is required. According to our analysis in the previous section, we have a scale of mass if and only if we have a procedure for assigning numbers to things such that if these things are arranged in order of the numerical assignments, they are thereby arranged in order of mass. What, then, is the order of mass? There are a number of different ordering relationships that serve to identify this order. For ordinary terrestrial objects the order may be identified in any of a number of different ways, e.g., by substitution experiments on a beam balance, by noting velocity changes in collision experiments, or by noting mutually induced accelerations. For microscopic and submicroscopic objects, the first and last of these methods are generally inapplicable and the second is the most useful. But other methods, of varying degrees of directness ranging from mass spectograph readings to quite complex calculations based upon electronic charge measurements, equivalent weights, combining weights, and combining volumes, are also available. And it is hard to say that any one procedure for ordering such things in respect of mass is any more fundamental than any other. They are all regarded as mass ordering (or measuring) procedures because they yield similar results wherever their ranges of applicability overlap. For macroscopic objects, such as the earth or the sun, the primary criterion is undoubtedly dependent upon the noting of mutually induced accelerations, although even here there are other relevant considerations.

Now it is evident that there is no single, universally applicable procedure for determining the mass order. And it is surely questionable whether we should expect to find such a procedure. At least it seems that we cannot assume a priori that there must be such a procedure, unless we are also prepared to accept a Lockean doctrine of real essences. But most philosophers nowadays would reject this doctrine for the very good reason that they have now become clear as to the "cluster" nature of many of our concepts. Consider once again the concept of temperature. As in the case of mass, there are a number of different temperature-ordering procedures. But no one of these is in fact applicable over the whole temperature range to all types and varieties of substances. Nevertheless, we do not feel that there must be a universally applicable temperature-ordering procedure that we could use once and for all to define the temperature order. We are content to allow the unity of the concept of temperature to rest upon the fact that the various temperature-ordering procedures yield similar results wherever their ranges of applicability overlap.

Why, then, should we make other demands of our concept of mass? Yet this demand is constantly made. Time and again, throughout the literature on Newton's second law of motion, the question is asked: what is the definition of mass? And this is almost invariably a request for a unique, universally applicable procedure for assigning numbers to things that will serve the dual purpose of defining the mass order and defining a scale for the measurement of mass. Of course there would be a certain satisfaction in discovering such a definition, for it might well pave the way to a *theoretical* interpretation of mass (just as the discovery of the thermodynamic scale of temperature paved the way for the kinetic theory of temperature). But this has nothing whatever to do with the logical status of Newton's second law of motion. It would only be relevant if the mass order could not be identified in any part of its range except via Newton's second law of motion. And in that case, our concept of mass would be akin to our concept of refractive index—a mere constant of proportionality. But since it is patently false that the various parts of the mass order can be identified only in this way, the quest for a unique, universally applicable procedure for measuring mass, however revealing it may otherwise be, is simply irrelevant to the logical status of Newton's second law.

However, there is some point, relevant to our purposes, in making another kind of investigation, viz., in trying to find a way of defining a scale of mass on which the masses of *macroscopic* objects (such as the sun and the planets) might be determined independently of Newton's second law. It may be held that the cosmologist is justified in regarding the concept of mass that he uses as having something like the status of a constant of proportionality, while the engineer and the chemist are not justified in regarding their concepts as having a similar logical status. It is not clear on what grounds such a view may be held, but presumably they would have to do with the different relevance of empirical considerations to the truth of statements concerning mass relationships in the different fields.

Whether such a view is tenable or not, there is a way of defining a scale of mass on which the masses of macroscopic objects may be determined, and which is independent of Newton's second law of motion. It is easily proved that if $A_1, A_2, A_3, \ldots A_n$ is any set of bodies moving only under the influence of mutually induced gravitational forces, then, assuming that the absolute accelerations $\bar{a}_1, \bar{a}_2, \bar{a}_3, \ldots \bar{a}_n$ of $A_1, A_2, A_3, \ldots A_n$ are determinable, and that the mutual distances and angular displacements are known, the relative masses of these various bodies can be determined on the basis of this information alone.

Let the absolute accelerations $\bar{a}_1, \bar{a}_2, \bar{a}_3, \ldots \bar{a}_n$ of $A_1, A_2, A_3, \ldots A_n$, respectively, be resolved into components such that:

$$\bar{a}_i = C_{i1} I_{i1} + C_{i2} I_{i2} + \ldots + C_{i, i-i} I_{i, i-1} \ldots + C_{i, i+1} I_{i, i+1} \ldots$$
$$+ C_{in} I_{in} \ (i = 1, 2, \ldots n) \tag{1}$$

where I_{ij} is a unit acceleration vector directed from A_i to A_j, and where C_{ij} is the magnitude of the component acceleration directed from A_i to A_j. Now in general it will be possible to achieve such a resolution in a variety of different ways, since the n equations (1) contain $2C_2^n$ unknown magnitudes (the C_{ij}'s). However, if we make the additional restrictions that:

$$M_i C_{ij} = M_j C_{ji} \quad (i, j = 1, 2, \ldots n, i \neq j) \tag{2}$$

M_i and M_j being the masses of A_i and A_j, respectively (this restriction being made in accordance with the kinematic principle of action and reaction—to be discussed in Part IV) and that:

$$C_{ij} + C_{ji} = G \frac{M_i + M_j}{r_{ij}^2} \qquad (i, j = 1, 2, \ldots n, i \neq j) \tag{3}$$

r_{ij} being the distance separating A_i and A_j (this restriction being in accordance with the law of distribution and succession which we have seen is derivable from Newton's law of gravitation), then, if the value of G is arbitrarily fixed, we have, in the sets of equations (1), (2), and (3), $2C_2^n + n$ independent equations involving $2C_2^n + n$ unknowns. There are $2C_2^n$ unknown component acceleration magnitudes (the C_{ij}'s) and n unknown masses, and n equations (1), C_2^n equations (2), C_2^n equations (3). Hence the masses of the various objects moving only under the influence of mutually induced gravitational forces may all be determined.[26]

It may never be necessary, in any field, to rely on Newton's second law to define a scale of mass, for the argument shows that, even in the field of cosmology, it may be possible to set up a mass scale, suitable for determining the masses of macroscopic objects (such as those that comprise the solar system), on the basis of the assumed law that every body accelerates toward every other body in the universe with a component of relative acceleration that is directly proportional to the sum of their masses and inversely proportional to the square of the distance between them, unless they are acted upon by some (nongravitational) forces.

The Logical Status of Newton's Second Law. How, then, should we answer our question concerning the logical status of Newton's second law? Once the cluster nature of our concepts of force and mass becomes evident, it also becomes clear that no short answer is possible. Consider how Newton's second law is actually used. In some fields it is unquestionably true that Newton's second law is used to define a scale of force. How else, for example, can we measure interplanetary gravitational forces? But it is also unquestionably true that Newton's second law is sometimes used to define a scale of mass. Consider, for example, the use of the mass spectrograph. And in yet other fields, where force, mass, and acceleration are all easily and independently measurable, Newton's second law of motion functions as an empirical correlation between these three quantities. Consider, for example, the application of Newton's second law in ballistics and rocketry.

Newton's second law of motion thus has a variety of different roles. Sometimes it is used in one role, sometimes in another. To suppose that Newton's second law of motion, or *any* law for that matter, must have a unique role that we can describe generally and call the logical status is an unfounded and unjustifiable supposition. Many laws simply do not have such a unique function. Even a cursory inspection of the way in which we actually use many of our laws reveals this. Whether in fact we can *ascribe* unique roles to each of our various physical laws and still have a useful conceptual scheme is another and important question. But it is not the question of logical status. It is the question of whether it is possible to achieve a rational reconstruction of physics in which each law has a simple and easily characterizable logical status.

But how can a law be *both* an empirical proposition relating force, mass, and acceleration and, say, a definition of a scale of force? Isn't it self-contradictory to say such a thing? In one way, yes. The one law cannot play two such different roles in a single occurrence within a piece of scientific discourse. But in another way, no. The one law can play two quite different roles in two different pieces of scientific discourse. A contradiction can therefore be derived only if we assume that the one law can have only one role and that this role must be independent of the context of its use. But this is precisely what is being denied. It seems, then, quite evident that we use our laws to do a large variety of different jobs.

Consider an analogy. What is the logical status of the proposition that the angle between a tangent and a chord is equal to the angle in the alternate segment? Well, consider how the sentence can be used. It can be used to express the conclusion that might be drawn from a series of well-executed measurements made on certain kinds of pencil-drawn figures. It can also be used to express the conclusion of a Euclidean geometrical proof. Is the sentence ambiguous? That is certainly one way of thinking about it. But it is not the only way, for we could also say that different sorts of considerations are relevant to the truth of one and the same proposition—the formal considerations of Euclidean geometry as well as the empirical considerations of empirical geometry. And this way of thinking about it is no less natural or more forced than any other.

Now this is roughly the situation with regard to Newton's second law of motion. We have one formula that we use in a variety of different ways. But we do not say that the formula is ambiguous, or expresses a different proposition in each different context of use. Instead, we allow that it expresses one and the same proposition (i.e., Newton's second law) and also plays a variety of different roles. Sometimes it is used to define a scale of force, at other times to define a scale of mass, and at yet other times it expresses an empirical correlation between the results of force, mass, and acceleration measurements.

IV. The Principle of Action and Reaction

If any one of Newton's laws of motion can lay claim to being an empirical law, it is the principle of action and reaction. Let A and B be any two bodies. Then if we define the *free* motion of A without B as the motion that A would have in the absence of the body B, and the *free* motion of the body B without A as the motion that B would have in the absence of A, then it is an empirical fact the change of momentum (gauged relative to the free motion of A) that the body A undergoes in any given time is equal and opposite to the change of momentum (gauged relative to the free motion of B) which the body B undergoes in that same time. Thus if two bodies A and B collide in free fall, then the change of momentum of A (gauged relative to the free motion of A) is equal and opposite to the change of momentum of B (gauged relative to the free motion of B).

Let us call this law the *kinematic principle of action and reaction*. Now this appears to be an empirical law. It appears to be something that could be discovered, using the sorts of techniques that Huyghens used in connection with the law of conservation of momentum. It does not depend upon the adoption of any concept of *natural* motion. Hence it does not derive any element of conventionality from the same source as the first two laws. It is true that we need a concept of free motion for the body A

without *B,* and a concept of the free motion of *B* without *A.* But then these concepts are operationally definable, for we may define the free motion of the body *A* without *B* as the motion that *A* would have when the body *B* was removed to a place infinitely distant from *A.* In fact, infinite removal is not a possibility. But in many cases (sufficiently many, it seems, to justify the universal generalization), infinite removal is not necessary, for as the body *B* is removed, it is seen that the motion of the body *A* (in the given circumstances) is independent of the position of the body *B.*

Consequently (provided only that changes of momentum can be gauged independently), there is a law that is at least a very close relative of the principle of action and reaction and that may well be described as empirical. The law cannot, of course, be tested in all cases. We cannot remove a planet from the solar system in order to gauge the free motions of the remaining planets (in the absence of the given planet). Nevertheless, this hardly counts against its being an empirical law. No law is everywhere testable. The important point is that it *can* be tested for a wide variety of terrestrial objects, composed of a wide variety of different substances, and moving in any of a wide variety of circumstances. And this point seems sufficient to establish the empirical character of the law.

What remains, then, is to see what relationship there is between this law and the principle of action and reaction, and to examine whether changes of momentum can be gauged independently of this law.

Mass, Action, and Reaction. The concepts of free and natural motion are clearly related concepts. Nevertheless, they are not identical, for we may have precisely the same concept of free motion whatever our concept of natural motion. The former is a purely kinematic concept. It is entirely noncommittal about forces. The free motion of *A* without *B* in a given situation can be found simply by removing *B* from that situation to such a position that further removal makes no difference. Whether or not we should say that *A* is then subject to the action of forces is irrelevant. The concept of natural motion, on the other hand, is a dynamic concept, for a body is moving naturally if and only if it is not subject to the action of forces. It would be absurd, therefore, to equate these two concepts, or to maintain that our concept of free motion is in any way dependent on our concept of natural motion.

Yet the principle of action and reaction, as it is usually understood, is a dynamic principle. Hence, since the law we have stated is purely kinematic and depends only on kinematic concepts, it cannot be maintained that the principle of action and reaction is identical with this law. Nevertheless, the connection is very close, for if it is assumed that the Newtonian "motive force" exerted by the body *B* on the body *A* is proportional to the change of momentum that *A* undergoes (gauged relative to the free motion of *A* without *B*) and vice versa, then we have at once that the motive forces exerted by *A* on *B* and *B* on *A* are equal and opposite. This is precisely Newton's third law of motion. Hence Newton's third law is derivable from this kinematic law (provided only that his second law is taken as an additional premise), and consequently, if the kinematic law may be said to be empirical, the principle of action and reaction may also be said to be an empirical law.

The only thing, then, that appears to stand in the way of an unqualified claim that the principle of action and reaction is an empirical law is the question of whether changes of momentum (relative to the free motions of the bodies concerned) can be gauged independently of this law. And since relative velocities are clearly measurable independently of the law, this comes down to the question of whether a scale of mass can be defined independently of the third law of motion.

For ordinary terrestrial objects, like cricket balls and tennis racquets, the answer to this question is obvious. A scale suitable for determining the masses of such things can be set up by the straightforward procedures of fundamental measurement (e.g., by substitution on a beam balance). Hence in the only range in which the law in question might be directly tested empirically, an independent scale of mass can undoubtedly be defined. The empirical character of the principle of action and reaction therefore appears to be established.

Of course it is much less clear that scales of mass suitable for determining the masses of macroscopic objects (such as planets) or microscopic objects (such as electrons and protons) can be set up independently of this principle. Even the scale proposed on p. 58 presupposes at least the kinematic principle of action and reaction. But then since no spatially and temporally unrestricted generalization is empirically testable over the whole of its range, this point does not tell against the empirical claim.

Yet even now the status of Newton's third law of motion is not clear, for how do we know that the mass of an object is independent of its state of motion or its situation? We could never tell this by making experiments with beam balances—unless the masses of different substances were affected differently by changes of situation or motion. Of course if no differential effects of this kind are known to exist, we may as a matter of convention adopt the principle of invariance of mass and then use this principle to test empirically the principle of action and reaction. But then it is doubtful whether we should in fact prefer the principle of invariance of mass to the principle of action and reaction. In other words, if our measurements of mass and velocity change were to conflict with the kinematic principle of action and reaction, we might prefer to reject the principle of invariance of mass and retain the principle of action and reaction. That is, we might prefer to define a scale for the mass of a moving object via the principle of action and reaction, and to reject, if necessary, the principle of invariance of mass.

Thus we may argue that although Newton's third law of motion was initially an empirical discovery, its status today is not that of an empirical proposition, but rather that of a definition, since the mass ratio MA/M_B may be held to be equal to $-(\triangle V_B/\triangle V_A)$ by definition [where $\triangle V_A$ and $\triangle V_B$ are the changes of velocity that A and B undergo (determined with respect to their free velocities) when A and B interact]. Of course it is an empirical fact that a scale of *mass* can be defined in this fashion, for it is an empirical fact that $\triangle V_A$ and $\triangle V_B$ are always oppositely directed and that MA/M_B thus defined accords well with other independent determinations of mass ratios.

However, it seems that neither characterization of Newton's third law is entirely satisfactory. In many cases, in ballistics and rocketry, for example, its role is that of an empirical principle which enables us to predict velocity changes from a knowledge of certain masses and certain other velocity changes. In other cases, however, especially in microphysics, its role is more like that of a definition, for it provides us with one of our principal criteria for comparing the masses of microparticles.

V. Conclusion

It has been argued that Newton's laws of motion were primarily conceptual in origin and that all three of these laws were probably derived directly from Cartesian physics. It cannot, therefore, be claimed that these laws were originally supported by or derived inductively from detailed experimental work. It has also been shown that this lack of empirical support, at least in the case of the law of inertia, was not just an accidental feature, for no amount of experimental work could ever establish a principle of natural motion, or even provide inductive support for such a principle. The reason for this, it was argued, derives from our concept of a force. Forces exist only to explain effects. Since we are at liberty to choose to regard whatever changes or states we please as natural (i.e., as *not* being effects), it follows that the forces which we should say are operative in nature depend upon the choices we actually make.

For this reason, then, a principle of natural motion, such as the law of inertia, cannot be an empirical principle. Of course it does not follow that our choice of principle of natural motion is arbitrary. On the contrary, it has been shown that it is just such choices which govern the whole course of physical inquiry. The choices we actually make are always theory committed. Nevertheless, it has been seen that there still remains some area of choice. And, in fact, it is possible to construct a system of dynamics that appears to be the equal of Newton's, but that employs an entirely different principle of natural motion.

Newton's second law of motion was seen to present different problems. Is it an empirical principle relating force, mass, and acceleration?

Does it define a scale for the measurement of resultant force? Or does it, perhaps, define a scale for the measurement of mass?

It has been argued that no clear affirmative or negative answers can be given to any of these questions. Any answer we give will misrepresent the role that the second law of motion actually has within the body of science. Does the engineer ever predict the acceleration of a given body from a knowledge of its mass and of the forces acting upon it? Of course. Does the chemist ever measure the mass of an atom by measuring its acceleration in a given field of force? Yes. Does the physicist ever determine the strength of a field by measuring the acceleration of a known mass in that field? Certainly. Why then, should any one of these roles be singled out as the role of Newton's second law of motion? The fact is that it has a variety of different roles. In some fields, where mass is particularly difficult to measure, Newton's second law may provide us with a scale for the measurement of mass. In other fields, where force measurement is otherwise difficult or impossible, Newton's second law may provide us with a suitable scale for the measurement of force. And in yet other fields, where force, mass, and acceleration are all easily and independently measurable, Newton's second law appears as an empirical principle relating these three quantities. To suppose that there must be a *central* role is an unwarranted metaphysical postulate. The important question is: "How is the law in question actually used in science?"

Similar considerations were seen to apply to Newton's third law of motion. Is it an empirical proposition relating momentum changes? Or is it a proposition whose role in science is to define a scale of mass? The most appropriate answer is "both." Consider its role in science. Clearly, it sometimes has the role of an empirical proposition, for momentum changes are often determinable independently of this law. But also clearly, its role is sometimes to define a scale of mass. To maintain that its role *must* be one or the other is an unwarranted and unjustifiable assumption.

In discussing the three laws of motion, the framework of Newtonian physics has been assumed throughout. No questions concerning *absolute* motion, *uniform* motion, or *straight line* motion have been asked or considered. This is indeed a serious omission. But in that most discussions of Newtonian dynamics have concentrated on these questions to the neglect of those concerning the concepts of force and mass, it is hoped that this essay may serve as a useful complement. To embark now on a discussion of these other concepts would require an essay at least as long as the present one.

Notes

1. H. Butterfield, *The Origins of Modern Science, 1300-1800* (London: G. Bell & Sons, Ltd., 1949).

2. Galileo Galilei, *Dialogues Concerning Two New Sciences,* H. Crew and A. de Salvio, trans. (New York: Dover Publications, Inc., 1914).

3. *Ibid.,* p. 166.

4. *Ibid.,* p. 244.

5. *Ibid.,* p. 251.

6. *Ibid.,* pp. 165-166.

7. Galileo Galilei, *Dialogue Concerning the Two Chief World Systems— Ptolemaic and Copernican,* Stillman Drake, trans. (Berkeley: University of California Press, 1953).

8. *Ibid.,* pp. 19, 31.

9. René Descartes, *A Discourse on Method, etc.,* John Veitch, trans. (London: J. M. Dent & Sons, Ltd., 1912), p. 16.

10. René Descartes, *Philosophical Writings,* E. Anscombe and P. Geach, trans. and eds. (London: Thomas Nelson & Sons, 1954), p. 215.

11. *Oeuvres de Descartes,* Charles Adam and Paul Tannery, eds., Vol. XI (Paris: Cerf, 1909).

12. For an English translation of most of the relevant sections, see *Philosophical Writings, op. cit.,* pp. 199-228.

13. *Ibid.,* p. 217.

14. Sir Isaac Newton, *Philosophiae Naturalis Principia Mathematica,* 2nd ed. (1713), p. 12. Emphasis added.

15. Sir Isaac Newton, *Mathematical Principles of Natural Philosophy,* Andrew Motte, trans., rev. by F. Cajori (London: Cambridge University Press, 1934). Emphasis added.

16. M. Jammer, *Concepts of Force* (Cambridge: Harvard University Press, 1957), p. 124.

17. *Principia,* p. 13. Emphasis added.

18. *Ibid.,* p. 14. Emphasis added.

19. *Ibid.,* p. 21. Emphasis added.

20. *Ibid.,* p. 2.

21. It was a universally received doctrine that forces are conserved, and hence that they are in some sense equal to the effects they produce. The only point of disagreement concerned the true measure of force. The Cartesians measured force by the product of mass and velocity, the Leibnizians by the product of mass and velocity squared (*vis viva*). But all were agreed that forces are conserved. For an account of this dispute, see W. F. Magie, *A Sourcebook in Physics* (McGraw-Hill Book Company, 1935).

In any case, the doctrine, *causa aequat effectam,* must be supposed to lie behind Newton's second law of motion however it is interpreted. *Cf.* d'Alembert's comment on Newton's second law of motion that it is "a principle based on that single, vague and obscure axiom that the effect is proportional to its cause."

22. *Oeuvres de Descartes, op. cit.,* Vol. XI, p. 43.

23. The precise meaning of this law is explained more fully in Section III, where it is applied to show that an analytic solution of the *n*-body problem is in principle possible—a conclusion that incidentally demonstrates the possibility of setting up a purely *kinematic* definition of mass.

24. Identical in formulation—although not identical in application, for "acceleration" must here be taken to mean "acceleration relative to the natural ac-

celeration" rather than "acceleration relative to the natural *unaccelerated* motion."

25. D. A. T. Gasking, "Clusters," *Australasian Journal of Philosophy,* Vol. XXXVIII, No. 1 (May, 1960).

26. Some comments on this demonstration are in order. First, the resolution of the absolute accelerations into component accelerations of *finite* magnitude may not always be possible. If, for example, the n bodies A_1, A_2, A_3 . . . A_n are all coplanar, but the absolute acceleration a_1, a_2, . . . a_n are not similarly coplanar, then the resolution will not be possible. Hence a necessary condition for the general applicability of this procedure to determine a set of M's is that there should exist in the set A_1, A_2, A_3 . . . A_n at least four bodies that are not coplanar. Second, the M's determined by this procedure may not all be positive. They must all be real, since all of the constants must be real, and all of the equations are linear in each variable. But the equations themselves give no guarantee that the M's will be > 0, or that they will remain invariant, in time. Nevertheless, if the equations of Newtonian mechanics accurately describe the motions of a set of objects that move only under the influence of mutually induced gravitational forces, then such results should not *in fact* be obtained. And if, in isolated cases, such results were obtained, then we may suspect either that the chosen frame of reference (for the determination of absolute accelerations) was not an inertial frame, or that some forces other than gravitational forces were operative.

NORWOOD RUSSELL HANSON
Yale University

A Response
to Ellis's Conception
of Newton's First Law

In an ingenious exposition, Brian Ellis sets out the conceptual content of the first law of motion—first by analogy with a "strength of materials" generalization, and then via a kinematical restatement of the law of universal gravitation. En route he makes ancillary points of independent interest. I will react first to the structure of his argument and then to its details.

Section I

There can be no doubt about it—the philosophical thrust of Ellis's essay is just as profound as it is ingenious. His central theme is that forces were invented in the history of science just to *explain* effects. Over this Ellis would encounter little opposition from historians and philosophers. His subsidiary theme, however, is to the effect that these "force-ful explanations" are illusory—or, rather, man's need for an understanding of the "insides" of mechanical phenomena is only apparently provided by "force talk." Such talk sidesteps altogether the necessity of providing explanations. Thus his analogical restatement of the law of inertia in terms of *undistorted* bodies and distorting forces: every body free of distorting forces will retain its natural (undistorted) shape. When one perceives the completeness of the analogy between this and classical statements of the first law, it transpires that what one is really discussing, in *both* cases, is the "natural kinematics" as against the "unnatural kinematics" of bodies in motion. Thus Newton's laws, and the law of inertia in particular, do not really provide force-orientated explanations of mechanical motion; rather they enable us to set out analyses of the *un*natural kinematics of a body's translation (or configuration) in terms of its natural kinematics and *un*distorted configurations.

Once this conceptual bridgehead is secured—once one sees that mechanical explanations in terms of forces are (operationally) no more than substitutions of one kind of kinematical description (the "natural" one) for another (the "unnatural" one)—one is then free to attack more extended formulations of the laws of physics. This attack would seek to level all references to hidden "explanatory" parameters (like forces), the very mention of which seems somehow to make natural happenings more intelligible. In a purely kinematical presentation of classical mechanics, the first law of motion might have been made to read: *Every body has a component of relative acceleration toward every other body in the universe directly proportional to the sum of their masses and inversely proportional to the square of the distance between them—unless it is acted upon by a force* (i.e., unless it is unnaturally disturbed from moving in terms of its natural kinematics). Notice what Ellis's reformulation is capable of. All of the "classical" problems about universal gravitational forces, about *actio in distans,* and the troubling aspects of classical perturbation theory—these deliquesce as one describes the motions that pairs of bodies are observed naturally to describe with respect to each other. This, then, becomes *the* natural-kinematical statement of moving bodies' properties. This is what they *do*. There is nothing further to be "explained," and certainly not in terms of more fundamental forces. Indeed, such a purified, ghost-cleansed account would render it unnecessary to "explain" freely falling bodies or the conic paths of planets—although now the rectilinear Huyghensian trajectory described by projectiles released from slings *may* require elaboration. As usually understood, the law of inertia treats this last as "natural," requiring no further elaboration, as against the more ancient view that planetary circular motion was natural. Now, of course, the planetary orbits are construed as delicately balanced conspiracies between the law of inertia and the law of universal gravitation, perturbations in terms that make for some remarkably intricate computational quirks.

So once the kinematical game's afoot, it becomes a matter of convention, of what one wishes to achieve with a theory, that one physicist should take *these* as the natural kinematics of moving bodies, while another physicist takes *those*. Once classical gravitation is laid out kinematically as a "first law of motion," then further adjustments are theoretically inevitable. *Now* one can examine further the semantical content of "toward every other body," which figured in Ellis's kinematical version of the law of gravitation. Will bodies in the universe naturally move toward each other (directly as their masses and inversely as their distance squares) along Euclidean straight lines, Riemannian straight lines, Lobachevskian straight lines, or what? There may be theoretical advantages in leaving this matter wide open, in making the described path simply the *shortest*, given the physico-spatial context within which such body pairs are found. *Paths of body pairs, then, will be geodesic toward each other in some space.* Thus planets will move along geodesics quite naturally when their associated spatial frameworks have been "determined" by very large masses, such as our sun. The door to general relativity is now well ajar. The paths of planets are no longer to be explained in terms of fundamental forces and dynamical laws—ghosts in the celestial machinery. Rather it is simply a kinematical fact that small bodies will move along geodesic paths through spaces determined by large bodies. When this fact is mapped back into our classical Euclidean conceptions, it appears that our planets traverse ("almost") closed, plane, conic sections. But this last restatement is not really any simplification at all. Even in the nineteenth century, the apparent simplification of the Keplerian assignment of orbits to the planets was becoming unmanageably complex. The apsidal lines connecting aphelia and perihelia are always in motion, so that even in a plane the planetary curves are not really closed. Even Euler's concept of the *variational orbit* cannot make conic sections out of curling hairsprings. And when this is further complicated by the recognition that our sun itself has a "proper motion" through space, the planetary paths had to be seen as complex helices elliptical in section, twisting from west to east, pulled by the sun through all eternity. Moreover, the planetary proper motions are describable only in n-body perturbation terms requiring partial, nonlinear differential equations (and statistical approximations) for anything even approaching a satisfactory analysis. In short, there is no simplificatory virtue in mapping a planet's geodesic path, its natural motion through a space "shaped" by other masses, back into the Euclidean envelope of yore—within which closed plane curves and rectilinearities are *de rigueur*, and from which all deviations are to be explained in terms of dynamical forces and other explanatory apparitions. Better to leave things kinematic and geodesic. After all, how can *any* translation of an object in nature constitute "unnatural motion"? Why not simply leave it that all bodies move naturally in accordance with the spatial envelopes through which their translations occur, but that these envelopes have their own geometrical properties as determined in a highly context dependent way (i.e., a mass distributive way)?

Ellis has encapsulated most of this speculative history within his reflections concerning the kinematical versus the dynamical components of Newton's laws of motions. In his recognition that fundamental physical explicantia are kinematical, he suggests that major advances within theoretical physics often consist in the choice of more comprehensive, and algorithmically more powerful, clusters of "natural" kinematical commitments. This is an insight at once profound and ingenious—and beautifully expressed in Ellis's paper.

Section II

Some details of Ellis's exposition, however, are more vulnerable. In his very first paragraph he contrasts kinematics—when we "speak of the way in which bodies actually move," with dynamics—when "we speak of the way in which they *would* move, given that they are subject to the action of certain forces." This must be a merely linguistic point. There is nothing to prevent me from speaking of how some Cape Kennedy missile is moving *now* in wholly *dynamical* terms. Nor would one be prevented from referring *kinematically* to how bodies *would* move, given certain suitably described contexts for that motion.

Again, one could imagine opposition to Ellis's claim that "laws of motion do not distinguish between different kinds of forces." It might be argued that their very structural differences are quite sufficient to distinguish laws of electricity from those of celestial mechanics. But Ellis might well retort that all these structures and forms can themselves be "cashed operationally" in kinematical terms. *Beyond that*, there is no mystical subterranean force of a peculiarly electrical nature whose properties are to be contrasted with some similarly obscure celestial mechanical force. This retort is consistent with Ellis's general theme—but some readers might have wished he had stressed this more clearly.

Ellis's central historical contribution consists in enlarging Descartes' importance in the conceptual evolution leading ultimately to the law of inertia. Doubtless René Descartes was the first to state this law explicitly. Doubtless his physical inquiries were genetically important in the later work of Newton. But in his enthusiasm to stress this point, Ellis makes Galileo sound like a mere empiricist. Galileo's episodic contributions to the total story of inertia seem to Ellis to have been set largely in trials, experiments, and observations—none of which (Ellis suggests) could be relevant to any full comprehension of the semantical content of the law of inertia. This is surely questionable. As my own earlier paper was meant to show, Galileo's work was almost exclusively within the area of the *Gedankenexperiment*. The methodologically tricky and operationally novel features of this first law seem to have been quite apparent to Galileo in his own writings. He uses expressions like *"ad infinitum"* so as to indicate he could not have felt any single set of experiments would either confirm or disconfirm his general conclusions. In short, Galileo says much that goes well beyond mere kinematics (as classically understood).

Moreover, I wonder how much should be made of the fact that the first statement of the law of inertia is due to Descartes. Remember the first statement of Occam's razor is demonstrably not due to William of Occam. On the other hand, however, Occam is quite properly thought to have been responsible for the razor—the principle of parsimony. Indeed, in the works of Buridan and Orêsme [well sampled in Clagett's *Science of Mechanics in the Middle Ages* (Madison: University of Wisconsin Press, 1959)] one finds many references to motion that go well beyond any simple mediaeval impetus theory—the only sort Ellis remarks. Perhaps one should not contrast Buridan and Orêsme with the final Cartesian form of the law of inertia; rather the elements of that authoritative statement are to be found substantially in the writings of Buridan and Orêsme, even though the *final* statement is a Cartesian triumph.

The interesting calculations on pages 60-65 above are subject to considerable dubiety and re-examination. It would appear, at first glance, that Ellis is attempting an analytical solution for the n-body problem. It is perhaps demonstrable that no such purely analytical, algorithmically complete proof can be generated. The reason for this seems simple: whatever the mechanical parameters—be they accelerations, masses, trajectories— after the perturbational influences of any two bodies are calculated with respect to each other, that calculation must be completely overhauled when the perturbational influence of a third body *on both* is introduced. But what *is* that third perturbational influence? This can be completely determined only by plugging in accurate values for the perturbational influence of the first two bodies on each other. But this cannot be settled without reckoning on disturbances from without. *Ad indefinitum.* But then all this is also the case when one seeks to "zero in" on that third perturbational description in the presence of a fourth body, and a fifth, and a sixth. And the perturbational effect on an nth body can never be more than "crowded" by successive approximations, because however one assigns a value for the perturbational variables in that nth body's description, it will necessitate a continual redescription of all the interactions that took place between the first and second bodies, the first and second as against the third, and so on. A formal "proof" in this context can never be more than successive and approximate—which is certainly not what a host of classical mathematicians would ever have identified as proof. Ellis sidesteps this exciting (and historically quite depressing) issue by references to "absolute accelerations" and "additional restrictions." The "operational cash value" of these will, on analysis [see my "Laplace and Hilbert," *Philosophy of Science* (1964)], turn out to beg the very question of whether or not one could analytically generate a possible solution for the n-body problem.

But these small darts are as nothing when compared with the heavy intellectual artillery Ellis has put at our disposal. A mark of my high estimation of Ellis's foregoing contribution is that I wish I could have written it myself as the second half of my own paper. As for the latter, I shall henceforth always amplify the argument therein with sundry Ellisian references.

HILARY PUTNAM
Massachusetts Institute of Technology

A Philosopher Looks
at Quantum Mechanics*

Those defile the purity of mathematical and philosophical truth, who confound real quantities with their relations and sensible measures.

—ISAAC NEWTON, *Principia*

Before we say anything about quantum mechanics, let us take a quick look at the Newtonian (or "classical") view of the physical universe. According to that view, nature consists of an enormous number of particles. When Newtonian physics is combined with the theory of the electromagnetic field, it becomes convenient to think of these particles as dimensionless (even if there is a kind of conceptual strain involved in trying to think of something as having a *mass* but not any *size*), and as possessing electrical properties—negative charge, or positive charge, or neutrality. This leads to the well-known "solar system" view of the atom—with the electrons whirling around the nucleons (neutrons and protons) just as the planets whirl around the sun. Out of atoms are built molecules; out of molecules, macroscopic objects, scaling from dust motes to whole planets and stars. These latter also fall into larger groupings—solar systems and galaxies—but these larger structures differ from the ones previously mentioned in being held together exclusively by gravitational forces. At every level, however, one has trajectories (ultimately that means the possibility of continuously tracing the movements of the elementary particles) and one has causality (ultimately that means the possibility of extrapolating from the history of the universe up to a given time to its whole sequence of future states).

75

When we describe the world using the techniques of Newtonian physics, it goes without saying that we employ *laws*—and these laws are stated in terms of certain *magnitudes,* e.g., distance, charge, mass. According to one philosophy of physics—the so-called *operationalist* view so popular in the 1930's—statements about these magnitudes are mere shorthand for statements about the results of measuring operations. Statements about distance, for example, are mere shorthand for statements about the results of manipulating foot rulers. I shall here assume that this philosophy of physics is *false.* Since this is not a paper about operationalism, I shall not defend or discuss my "assumption" (although I *do* refer the interested reader to the investigations of Carnap, Braithewaite, Toulmin, and Hanson for a detailed discussion of this issue). I shall simply state what I take the correct view to be.

According to me, the correct view is that when the physicist talks about electrical charge, he is talking quite simply about a certain magnitude that we can distinguish from others partly by its "formal" properties (e.g., it has both positive and negative values, whereas mass has only positive values), partly by the structure of the system of laws this magnitude obeys (as far as we can presently tell), and partly by its *effects.* All attempts to "translate" *literally* statements about, say, electrical charge into statements about so-called observables (meter readings) have been dismal failures, and from Berkeley on, all a priori arguments designed to show that all statements about unobservables must ultimately reduce to statements about observables have contained gaping holes and outrageously false assumptions. It is quite true that we "verify" statements about unobservable things by making suitable *observations,* but I maintain that without imposing a wholly untenable theory of meaning, one cannot even *begin* to go from this fact to the wildly erroneous conclusion that talk about unobservable things and theoretical magnitudes *means the same* as talk about observations and observables.

Now then, it often happens in science that we make inferences from measurements to certain conclusions couched in the language of a physical theory. What is the nature of these inferences? The operationalist answer is that these inferences are *analytic*—that is, since, say, "electrical charge" *means by definition* what we get when we measure electrical charge, the step from the meter readings to the theoretical statement ("the electrical charge is such-and-such") is a purely conventional matter. According to the nonoperationalist view, this is a radical distortion. We know that this object (the meter) measures electrical charge, *not* because we have adopted a "convention," or a "definition of electrical charge in terms of meter readings," but because we have accepted a body of theory that includes a *description of the meter itself in the language of the scientific theory.* And *it follows from the theory,* including this description, that the meter measures electrical charge (approximately, and under suitable circumstances).

The operationalist view disagrees with the actual procedure of science by replacing a probabilistic inference within a theory by a nonprobabilistic inference based on an unexplained linguistic stipulation. (Incidentally, this anti-operationist view has sometimes been termed a "realist" view, but it is espoused by some positivists—in particular by Carnap in his well-known article "The Interpretation of Physical Calculi." It is well, for this reason, to avoid these "-ist" words entirely, and to confine attention to the actual methodological theses at issue in discussing the philosophy of the physical sciences.)

If the nonoperationist view is generally right (that is to say, correct for physical theory in general—not just for Newtonian mechanics), then the term "measurement" plays *no fundamental role in physical theory as such*. Measurements are a subclass of physical interactions—no more or less than that. They are an important subclass, to be sure, and it is important to study them, to prove theorems about them, etc.; but "measurement" can never be an *undefined* term in a satisfactory physical theory, and measurements can never obey any "ultimate" laws other than the laws "ultimately" obeyed by *all* physical interactions. It is at this point that it is well for us to shift our attention to quantum mechanics.

The main fact about quantum mechanics—or, at any rate, the main fact as far as this paper is concerned—is that the state of a physical system can be represented by a set of waves (more precisely, by a "ψ-function"; however, nontechnical language will be employed throughout this paper, so I shall stick to the somewhat inaccurate, but more popularly intelligible, phrase, "set of waves"). Of course any set of waves (superimposed on one another) may also be regarded as a single wave; thus if we have a system of, say, three particles, we may represent the "state" of the entire system by a single wave. A "wave" is simply a magnitude, with an intensity at every point of space, whose intensity has certain periodicity properties (normally expressed by differential equations of a certain kind). For example, the intensity may (moving in space in a certain direction) rise to a peak, then fall, then rise to a peak again, etc. Normally the places at which these peaks are located are more or less evenly spaced—at least in the case of simple harmonics; all regularity may be lost when a great many different waves are superimposed in a jumbled way—and these places change location with time in the case of a moving "wave front." All of this (which is, of course, purely qualitative) should be familiar to the reader from a consideration of sound waves in air or pressure waves in water.

The quantum mechanical story becomes complicated immediately, however. In the first place, the waves treated in quantum mechanics—the waves used to represent the state of a physical system—are not waves in ordinary space, but waves in an abstract mathematical space. If we are dealing with a system of three particles, for example (with position coordinates x_1, y_1, z_1 for the first particle, x_2, y_2, z_2 for the second particle, and x_3, y_3, z_3 for the third particle), then we employ a space with nine "dimensions" to represent the system (one dimension for each of the coordinates $x_1, \ldots z_3$). In the second place, the amplitude of the waves employed in quantum mechanics can have such recherché values as the square root of minus one. (In technical language, the ψ-function is a *complex* valued function of the dimensions of the space, whereas "ordinary" waves are real valued.) Finally, the dimensions of the space need not be thought of as corresponding to the position coordinates of the various particles—one can instead think of the nine dimensions (in the case of a three-particle system) as corresponding to, say, the nine momentum coordinates (or to certain other sets of physical magnitudes, or "observables," in the jargon of quantum mechanics); this will make a difference, in the sense that one will employ a different *wave* to represent the state. (This is called using "momentum representation" instead of "position representation." The two representations are equivalent, in the sense that there exist mathematical rules for going from one to the other.)

At any rate, ignoring these complications for one moment, let us repeat the "simple-minded" statement we made at the outset of this part of the paper: the state of a physical system can be represented by a set of waves. What are we to make of this fact? Why is this technique of representation successful? What is it about physical systems that makes them lend themselves to representation by systems of waves? In short, what is the significance of the "waves"? Answers to this question are usually known as "interpretations" of quantum mechanics; and we shall now proceed to consider the most famous ones.

Note at the outset that there is one answer that we can dismiss, namely, that the success of this technique is just an accident. No physicist, in fact, no person in his right mind, has ever proposed *this* answer—and for good reason. Not only does the quantum mechanical formalism yield correct answers to too many decimal places, but it also yields too many predictions of whole classes of effects that would not have been anticipated on the basis of older theory, and these predictions are correct.

The historically first answer to our question might be called the "De Broglie interpretation of quantum mechanics." It is simply this: physical systems *are* sets of waves. The waves spoken of in quantum mechanics do not merely "represent" the state of the system; they *are* the system.

Unfortunately, this interpretation runs at once into insuperable difficulties (at least, almost all physicists consider them insuperable; a small minority, including De Broglie, continue to defend it). We have, in effect, stated these difficulties in pointing out the differences between quantum mechanical waves and ordinary waves: that the amplitudes are complex and not real; that the space involved is a mathematical abstraction, and not ordinary space; that the waves depend not just upon the system, but also upon the "representation" used—i.e., upon the set of "observables" we are interested in measuring (position, or momentum, or whatever). One further difficulty is important: the so-called "reduction of the wave packet."

The "state" of a physical system changes, of course, through time. This change is represented in two ways (not just one—motion along a trajectory, or collection of trajectories—as in classical physics). When one wishes to represent the state of a system as it is changing during an interval in which the system remains isolated, one allows the waves to "expand" in a continuous fashion. This process is called "motion" (this "expansion" or "spreading out" of the waves is governed by the famous Schrödinger equation). When one wishes to represent the change in the state of a system induced by a *measurement,* one simply "puts in" the new state (as determined by the measurement). In Von Neumann's axiomatization of quantum mechanics, this putting in of a state is allowed for in the simplest (one is tempted to say "crudest") way possible: an axiom is introduced which says, in effect, that a measurement throws a system *discontinuously* into a new state. Axiomatization aside, the fact remains that in the quantum mechanical formalism one sometimes speaks of the waves as undergoing certain kinds of continuous changes, and sometimes as *abruptly* or *discontinuously* jumping into a new configuration.

An example may make this clear: uncertainty concerning the position of a particle is represented by the volume occupied by the wave corresponding to the particle. The particle may be located (if a position measurement comes to be made) anywhere in the region corresponding to the volume occupied by the wave. Thus a particle whose position is known with high accuracy corresponds to a wave that is concentrated in a very small volume—a wave packet—whereas a particle whose position is known with low accuracy corresponds to a wave that is spread out over a large volume. Now suppose a position measurement is made on a particle of the latter kind—one whose position was known with low accuracy before the measurement. After the measurement, the position will be known with high accuracy. So the change of state produced by the measurement will be represented as follows: a very spread out wave suddenly jumped into the form of a wave packet. In other words, the wave suddenly vanished almost everywhere, but at one place the intensity suddenly increased (to compensate for the vanishing elsewhere). Needless to say, this reduction of the wave packet constitutes very strange behavior if this is really to be thought of as a physical wave.

The second "interpretation of quantum mechanics" we shall consider will be called the "Born interpretation" (perhaps I should call it the *original* Born interpretation, to distinguish it from the "Copenhagen interpretation," which also depends in part on Born's ideas, and which is discussed below). This interpretation is as follows: the elementary particles *are* particles in the classical sense—point masses, having at each instant both a definite position and a definite velocity—though not obeying classical laws. The wave corresponding to a system of particles does *not* represent the state of the system (simultaneous position and velocity of each particle), but rather our *knowledge* of the state, which is always incomplete. That our knowledge of the state must always be incomplete, that one cannot, for example, simultaneously measure position and momentum with arbitrarily high accuracy (the famous "uncertainty principle"), is not an independent assumption. It is a mathematical consequence of the basic assumption that any possible state of a system—or, rather, any physically obtainable *knowledge* of the state of a system—can be represented by a set of waves according to the rules of quantum mechanics, for a particle whose position and momentum were both known with virtually perfect accuracy would have to correspond to a wave that had the property of being "packet-like" (occupying a very small volume) in *both* "position representation" and "momentum representation," and as a fact of pure mathematics, there are no such waves.

This interpretation is able to deal very easily with all the difficulties mentioned in connection with the De Broglie interpretation. For example, it is no difficulty that the spaces used in quantum mechanics[1] are multi-dimensional, for they are not *meant* (according to this interpretation) to represent real physical space; the so-called "dimensions" are nothing but sets of "observables" (position coordinates, momentum coordinates, or whatever). The complex amplitudes are handled by a simple device: *squaring* (or, rather, taking the square of the absolute value). In other words, the probability that the particle is inside a given volume is measured not by the intensity of the wave inside that volume, but by the squared absolute value of the intensity, which is always a nonnegative real number not greater than one (and so can be interpreted as a probability, as the square root of minus one obviously cannot be). This way of calculating probabilities—squaring the intensity of the wave—leads to experimentally correct expectation values, and has become fundamental to quantum mechanics (as we shall see, it is taken over in the Copenhagen interpretation). Since the wave is not supposed to be a physical wave, but only a device for representing knowledge of probabilities, it is not serious that we have to perform this operation of squaring to get the probabilities, or that the intensity (amplitude) is a complex number before we square.

Finally, the reduction of the wave packet is no puzzle, for it represents *not* an instantaneous change in the state of a whole, spread out physical wave (which is supposed to take place the moment a measurement is over), but an instantaneous change in our *knowledge* of the state of a physical system, which takes place the instant we learn the result of the measurement. If I know nothing concerning the position of a particle except that it is somewhere in a huge volume of space, and then I make a position measurement that locates the particle *here,* the wave that represents my knowledge of the position of the particle, and that occupies the entire huge volume until the result of the measurement is learned, has to be replaced by a wave packet concentrated in the appropriate submicroscopic place. But a physical "something" does not thereby contract from macroscopic to submicroscopic dimensions; all that "contracts" is human ignorance.

In view of all the attractive features of the (original) Born interpretation, it is sad that it, too, encounters insuperable difficulties. These difficulties have to do with two closely related phenomena (mathematically, they are virtually one phenomenon)—"interference" and the "superposition of states." Interference may be illustrated by the famous "two-slit experiment." Here a particle is allowed to strike a surface—say, a sensitive emulsion—after having first passed through one or the other of two suitably spaced narrow slits (we do not know which one it passes through in such an experiment, but it *has* to pass through one or the other, on the particle interpretation, to reach the emulsion, because the emitter is placed on the other side of the barrier from the emulsion). What the experiment reveals is that the mathematical interference of the quantum mechanical waves shows up *physically* in this case as an interference pattern. The various particles—say, photons—are not able to interact (this can be insured by reducing the intensity of the radiation until only one or two photons are being emitted per second), so this is a case in which the wave that represents the whole system can be obtained by simply superimposing the waves of the individual particles. Also, each particle is represented by a wave in three-dimensional space (since there are three position coordinates), so the whole system of particles (in this special case) can be represented by a wave in three-dimensional space. This facilitates comparing the wave with a physical wave—in fact, if we identify the space used in this representation with ordinary physical space, we are led to predict exactly the correct interference pattern. Thus we have the following difficult situation: the reduction of the wave packet makes no sense unless we say that the waves are not physical waves, but only "probability waves" (which is why they collapse when we obtain more information); but the interference pattern makes no sense unless we say the waves (in this very special three-dimensional case) *are* physical waves.

If we analyze the interference mathematically (which we are able to do, since it is correctly predicted by the quantum mechanical formalism), we find that what the *mathematics* reveals is even stranger than what the experiment reveals. Mathematical analysis rules out even more conclusively than the experiment the intuitive explanation of the phenomenon —that particles going through one slit somehow interfere with *different* particles going through the other slit. Rather, different particles correspond to "incoherent" or unrelated waves, and these produce no detectable interference. Each particle corresponds to a wave that is split into two halves by passing through the two slits; the two halves of the wave belonging to a *single* particle are "coherent," or have intimately related wave properties, and all the detectable interference is produced by the interference between the two coherent halves of the waves corresponding to single particles. In other words, we get an interference pattern because *each* photon *interferes with itself* and *not* because different photons somehow interfere.

Superposition of states may be described as follows: let S be a system that has various possible states $A, B, C, \ldots$ according to classical physics. Then in *addition* to these states, there will also exist in quantum physics certain "linear combinations" of an arbitrary number of these states, and a system in one of these may, in certain cases, behave in a way that satisfies no classical model whatever.

I have stressed from the beginning the *unity* of the quantum mechanical formalism. Mathematically, almost all there is is a systematic way of representing physical systems and situations by means of a *wave* in a suitable *space*. The uncertainty principle, we noted above, is not an independent assumption, but follows directly from the formalism. Similarly, superposition of states is not something independent, but corresponds to the fact that any two permissible waves (representing possible states of a physical system) can literally be superimposed to obtain a new wave, which will also (according to the theory) represent a possible state of the system. More precisely, if ψ_1 and ψ_2 are permissible wave functions, then so is $c_1\psi_1 + c_2\psi_2$, where c_1 and c_2 are arbitrary complex constants.

To illustrate the rather astonishing physical effects that can be obtained from the superposition of states, let us construct an idealized situation. Let S be a system consisting of a large number of atoms. Let R and T be properties of these atoms which are incompatible. Let A and B be states in which the following statements are true according to both classical and quantum mechanics:

1. When S is in state A, 100 per cent of the atoms have property R.
2. When S is in state B, 100 per cent of the atoms have property T—and we shall suppose that suitable experiments have been performed, and (1) and (2) found to be correct experimentally. Let us suppose there is a state C that is a "linear combination" of A and B, and that can somehow be prepared. Then classical physics will not predict anything about C (since C will, in general, not correspond to any state that is recognized by classical physics), but quantum mechanics can be used to tell us what to expect of this system. And what quantum mechanics will tell us may be very strange. For instance, we might get:
3. When S is in state C, 60 per cent of the atoms have property R, and *also* get:
4. When S is in state C, 60 per cent of the atoms have property T—and these predictions might be borne out by experiment. But how can this be?

The answer is that, just as it turns out to be impossible to measure *both* the position and the momentum of the same particle at the same time, so it turns out to be impossible to test *both* statement (3) *and* statement (4) experimentally in the case of the same system *S*. Given a system *S* that has been prepared in the state *C,* we can perform an experiment that checks (3). But then it is physically impossible to check (4). And similarly, we can check statement (4), but then we must disturb the system in such a way that there is then no way to check statement (3).

We can now see just where the Born interpretation fails. It is based (tacitly) on the acceptance of the following principle:

THE PRINCIPLE OF NO DISTURBANCE (*ND*)

The measurement does not disturb the observable measured—i.e., the observable has almost the same value an instant before the measurement as it does at the moment the measurement is taken.

But this assumption is incompatible with quantum mechanics. Applied to statements (3) and (4) above, the incompatibility is obvious, and Heisenberg and Bohr have given quite general arguments (Heisenberg's of a more precise and mathematical nature, Bohr's of a more general and philosophical nature) to show this incompatibility. Even in the case of the two-slit experiment, the falsity of *ND* is indirectly involved. This enters in the following way: if the Born interpretation were correct and *ND* true, it would make no difference if we modified the experiment by placing in each slit a gadget that determined through *which* slit the particle passed —for the interference pattern would be the same whether we used two slits or two emitters (assuming the history of the particle *before* reaching whichever slit it went through to be irrelevant), and the gadget could be so designed that the distribution of "hits" on the emulsion would be the same in the case of a *single* slit whether the gadget was used or not (which shows that the history of the particle *after* leaving the slit is not being affected statistically). But, in fact, such a gadget always destroys the interference pattern. This argument is not rigorous, because even if the measurement determines the slit through which the particle was going to pass "in any case" ("even if the measurement had not been made"), the gadget might disturb its *subsequent* behavior in a way too subtle to show up in the case of an experiment with a *single* slit. However, the impossibility of (3) and (4)'s *both* being true *before* any *R*-measurement or *T*-measurement is made is a matter of simple arithmetic (plus the incompatibility of the properties *R* and *T*), and if the Born interpretation were correct, the very meaning of the wave (considered in two different "representations," corresponding to *R*-measurement and *T*-measurement) would be that, in the case of a system *S* in the state *C,* (3) and (4) are *both* true. Thus the principle *ND* and the (original) Born interpretation must both be abandoned.

In the literature of quantum mechanics, interpretations according to which the elementary particles have *both* position and momentum at every instant (although one can know only the position or the momentum, but never both at the same instant) are called "hidden variable" interpretations. The falsity of *ND* has serious consequences for these hidden variable theories. They are required to postulate strange laws whereby each measurement somehow disturbs the very thing it is measuring—e.g., letting a particle collide with a plate produces a speck on the plate, but at a place where the particle *would not have been* but for the presence of the plate. Actually, such a disturbance of the thing observed by the measurement need not be postulated in the case of *every* measurement, but it does have to be introduced in a great many cases. For example, in the two-slit experiment, *one* of the two measurements—the measurement that takes place when a speck appears on the emulsion, showing that a particle has hit, or the measurement that takes place when a gadget is introduced at the slits to determine which particles go through which slit—must disturb the particle. Similarly, in the case of (3) and (4), at least one of the two measurements—*R*-measurement or *T*-measurement—must disturb the system. In the best-known hidden variable theory—that due to David Bohm—an unknown physical force (the "quantum potential") obeying strange laws is introduced to account for the disturbance by the measurement.

The most famous interpretation of quantum mechanics, and the one that "works" (in the opinion of most contemporary physicists), is usually referred to as the "Copenhagen interpretation" (hereafter abbreviated "*CI*"), after Bohr and Heisenberg (who worked in Copenhagen for many years), although it is in many ways a modification of the Born interpretation, and Born's principle (that the squared amplitude of the wave is to be interpreted as a probability) is fundamental to it. What the *CI* says, in a nutshell, is that "observables," such as position, *exist* only when a suitable measurement is actually being made. Classically, a particle is thought of as having a position even when no position measurement is taking place. In quantum mechanics (with the *CI*) a particle is something that has, at most times, no such property as a definite position (that could be represented by a set of three position coordinates), but only a *propensity* to have a position if a suitable experimental arrangement is introduced.

The first effect of the *CI* is obvious: principle *ND* has to be abandoned. Principle *ND* says that an observable has the same value (approximately) just *before* the measurement as is obtained by the measurement; the *CI* denies that an observable has *any* value before the measurement. Born's principle can be retained, but with a modification: the squared amplitude of the wave measures not the probability that the particle *is* in a certain place (whether we look or not), but the probability that it *will be* found in that place if a position measurement is made at the appropriate time. (If no position measurement is made at the relevant time, then it *does not* make sense to ascribe a position to the particle.) It has often been said that this view is intermediate between views according to which the particle is a (classical) particle and views according to which the particle is really a physical wave, since sometimes the particle (say, an electron) has a definite, sharply localized position (when a precise position measurement is made), whereas at other times it is spread out (can only be assigned a large region of space as its position).

The effect of the *CI* on statements (3) and (4) is also straightforward. Under the *CI*, these statements get replaced by:

3'. When *S* is in state *C and an* R-*measurement is made,* 60 per cent of the atoms have property *R,* and

4'. When *S* is in state *C and a* T-*measurement is made,* 60 per cent of the atoms have property *T.*

Thus the incompatibility vanishes. Of course incompatibility is avoided only because the following statement is also true:

5. An *R*-measurement and a *T*-measurement cannot be performed at the same time (i.e., the experimental arrangements required are mutually incompatible).

This replacement of classically incompatible statements such as (3) and (4) by corresponding compatible statements such as (3') and (4'), as a result of the *CI,* is often referred to by the name *complementarity:*[2] the most famous case is one in which the one of the two incompatible properties is a "wave" property—such as momentum[3]—while the other is a "particle" property—especially (sharply localized) *position*. In all cases of complementarity, the idea is the same: the principle *ND* would require us to assign incompatible properties to one and the same system at the same time, whereas the incompatibility disappears if we say that a system has one or the other property only when a measurement of that property is actually taking place.

Even if the *CI* does meet all the difficulties so far discussed, it seems extremely repugnant to common sense to say that such observables as position and momentum exist only when we are measuring them. A hidden variable theory would undoubtedly fit in better with the preconceptions of the man in the street. It seems worthwhile for this reason to describe in a little more detail the difficulties that have led almost all physicists to abandon the search for a successful hidden variable theory and to embrace the extremely counterintuitive *CI.*

First of all, consider the phenomena for which a hidden variable theory must account. On the one hand, there are diffraction experiments, interference experiments, etc., which suggest the presence of physical waves (although it should be emphasized that such waves are never directly detected—the interference patterns, for example, in the case of a two-slit experiment, can be shown to consist of a myriad of tiny specks built up by individual particle collisions). On the other hand, when we emit some particles, and then interpose a "piece of flypaper"—i.e., a plate with a suitably sensitized surface—we do, indeed, get sharply localized collisions with the flypaper, which seems to confirm the view that what we emitted *were* "particles," with sharply localized positions at each instant and with continuous trajectories. (If the position were a discontinuous function of the time, the particle would not have to hit the flypaper at all; it could first exist on one side of it and then on the other side without leaving a mark.) In view of these phenomena, it is not surprising that existing hidden variable theories all make the same assumption: that there are *both* waves *and* particles. Some success has been encountered in elaborating this idea, particularly by De Broglie and his students at the Institut Henri Poincaré. Before explaining what goes wrong (or seems to go wrong), let me first describe a case in which the "pilot wave" idea—the idea that there are both particles (which are what we ultimately detect) and (indetectable) waves "guiding" the particles—has been successfully worked out. This case is the two-slit experiment that we have already described.

Briefly, the idea is that the particle is nothing but a singularity in the wave—that is, a point in the wave at which a certain kind of mathematical discontinuity exists. The energy of the wave particle system is almost wholly concentrated in the particle, and the relation between the wave and the particle is such that the *probability* that the particle is at any given place is proportional to the squared amplitude of the wave at that place. (Thus on this theory, the original Born interpretation is correct for "position representation." One feature of quantum mechanics that is sacrificed by this theory—and, indeed, by any hidden variable theory—is the nice symmetry between position representation and any other representation; only the wave used in position representation is a physical wave according to this interpretation, and then only in the three-dimensional case.) Singularities having the right properties—the ratio between the energy of the singularity and the energy of the wave is of the order c^2, where c is the speed of light—have been constructed by De Broglie.

Let us now consider how this interpretation affects the two-slit experiment. The fact that we can detect the wave only indirectly—through its effect on the particle that it guides—is explained on this theory by the fantastically low energy of the wave. The interference we detect is *real* interference—the wave corresponding to a single particle is split on passing through the slits, and the two halves of the wave (which still cohere, as explained above) then interfere. But what about the reduction of the wave packet?

De Broglie's answer is that the wave packet is *not* reduced. A wave that has lost the particle it guides does not collapse—it just goes on. But although it goes on, we have no way of detecting it, which is why in the usual theory we treat it as no longer existing. We cannot detect it directly because of its low energy. And we cannot detect it indirectly (through interference effects) because it does not cohere with the wave of any other particle, and there is no experimental procedure to detect the random interference caused by the interference of two mutually incoherent waves.

My opinion is that De Broglie, Bohm, and others who are working along these lines *have* indeed constructed a classical model for the two-slit experiment. But the model runs into difficulties—insuperable difficulties, I believe—in connection with *other* experiments.

In the first place, observe that, according to the theory just outlined, the principle *ND* is correct for the special case of position measurement. The principle *ND* cannot be true for an arbitrary measurement, as we showed before. But if we are really dealing with a particle whose position varies continuously with time, then, of course, we can measure the position of that particle by allowing it to smack a suitable kind of flypaper and then looking to see where the mark is on the flypaper. Wherever the mark may appear, we can say, "an instant before, the particle cannot have been very far away from here because of the assumed continuity of the particle's motion." However, I believe that the principle *ND* cannot be correct even for the case of position measurement. In order to explain why I believe this, I have to describe another phenomenon—"passage through a potential barrier."

Imagine a population P that is simply a huge collection of hydrogen atoms, all at the same energy level e. Let D be the relative distance between the proton and the electron, and let E be the observable, "energy." Then we are assuming that E has the same value, namely e, in the case of every atom belonging to P, whereas D may have different values d_1, d_2, . . . in the case of different atoms A_1, A_2. . . .

The atom is, of course, a system consisting of two parts—the electron and the proton—and the proton exerts a central force on the electron. As an analogy, one may think of the proton as the earth and of the electron as a satellite in orbit around the earth. The satellite has a potential energy that depends upon its height above the earth, and that can be recovered as usable energy if the satellite is made to fall. It is clear from the analogy that this potential energy P, associated with the electron (the satellite), can become arbitrarily large if the distance D is allowed to become sufficiently great. However, P cannot be greater than E (the total energy). So if E is known, as in the present case, we can compute a number d such that D cannot exceed d, because if it did, P would exceed e (and hence P would be greater than E, which is absurd). Let us imagine a sphere with radius d and whose center is the proton. Then if all that we know about the particular hydrogen atom is that its energy E has the value e, we can still say that wherever the electron may be, it cannot be outside the sphere. The boundary of the sphere is a "potential barrier" that the electron is unable to pass.

All this is correct in classical physics. In quantum physics, if we use the (original) Born interpretation, then in analogy to (3) and (4) we get (the figure 10 per cent has been inserted at random in the example):

6. Every atom in the population P has the energy level e.
7. Ten per cent of the atoms in the population P have values of D which exceed d.

These statements are, of course, in logical contradiction, since, as we have just seen, they imply that the potential energy can be greater than the *total* energy. If we use the *CI*, then, just as (3) and (4) went over into (3′) and (4′), so (6) and (7) go over into:

6′. If an energy measurement is made on any atom in P, then the value e is obtained, and
7′. If a D-measurement is made on any atom in P, then in 10 per cent of the cases a value greater than d will be obtained.

These statements are consistent in view of

8. An E-measurement and a D-measurement cannot be performed at the same time (i.e., the experimental arrangements are mutually incompatible).

Moreover, we do not have to accept (6′) and (7′) simply on the authority of quantum mechanics. These statements can easily be checked, not, indeed, by performing both a D-measurement and an E-measurement on each atom [that is impossible, in view of (8)], but by performing a D-measurement on every atom in one large, fair sample selected from P to check (7′) and an E-measurement on every atom in a different large, fair sample from P to check (6′). So (6′) and (7′) are both known to be true.

In view of (6′), it is natural to say of the atoms in P that they all "have the energy level e." But what (7′) indicates is that, paradoxically, some of the electrons will be found on the wrong side of the potential barrier. They have, so to speak, "passed through" the potential barrier. In fact, quantum mechanics predicts that the distance D will assume arbitrarily large values even for a fixed energy e.

Since the measurement of D is a distance measurement, on the hidden variable theory, the assumption ND holds for it (in general, distance measurement reduces to position measurement, and we have already seen that position measurement satisfies ND on this theory). Thus we can infer from the fact that D is very much greater than d in a given case, that D must also have been greater than d before the measurement—and hence before the atom was disturbed. So the energy could not have been e. In order to square this with (6′), the hidden variable theory assumes that the E we measure is not really the *total* energy. There is an additional energy, the so-called "quantum potential," which is postulated just to account for the fact that the electron gets beyond the potential barrier. Needless to say, there is not the slightest direct evidence for the existence of quantum potential, and the properties this new force must have to account for the phenomena are very strange. It must be able to act over arbitrarily great distances, since D can be arbitrarily large, and it must not weaken over such great distances—in fact it must get stronger.

On the other hand, the CI takes care of this case very nicely. In order to get a population P with the property (6′), it is necessary to make a certain "preparation." This preparation can be regarded as a "preparatory measurement" of the energy E; so it is in accord with the CI, in this case, to replace (6′) by the stronger statement (6). However, a D-measurement disturbs E (according to both classical and quantum physics). So when a D-measurement is made, and a value greater than d is obtained, we can say that D exceeds d, but we can make no statement about the energy E. So there is no potential barrier, and a fortiori no passing of a potential barrier. If we had the assumption ND (for distance measurement), we could conclude from the fact that D exceeded d by some large amount that D must also have exceeded d *before* the disturbance of E by the D-measurement, and then we would be back in trouble. But the CI rejects the idea that there is such a thing as the "value" of D an instant before the D-measurement. Thus no quantum potential is necessary (on the CI). If the energy of the atom exceeds E (when D is found to be greater than d), it is not because some mysterious quantum force is at work, but simply because energy has been *added by the* D-*measurement*.

In view of this rather disappointing experience with the hidden variable theories, I believe that it would be natural to impose the following three *conditions of adequacy* upon proposed interpretations of quantum mechanics:

A. The principle ND should not be assumed even for position measurement.
B. The symmetry of quantum mechanics, represented by the fact that one "representation" has no more and no less physical significance than any other, should not be broken. In particular, we should not treat the waves employed in one representation (position representation in the case of the hidden variable theorists) as descriptions of physically real waves in ordinary space.
C. The phenomenon of superposition of states, described at the beginning of this paper, must be explained in a unitary way.

The hidden variable theories violate all three of these principles. In particular, both the two-slit experiment and the passage of a potential barrier represent superpositions of classical states, as do all the so-called anomalies encountered in quantum mechanics. Yet the theories just described handle the one (the two-slit experiment) by means of a classical model with waves guiding particles, and the other (passage of a potential barrier) by means of quantum potential, in flagrant violation of (C). [The classical states superimposed in the two-slit experiment correspond to the two situations: (a) just the left-hand slit open, and (b) just the right-hand slit open. These are "classical" in the sense that the same predictions can be derived from classical and from quantum physics in these states. The nonclassical state that is represented by superimposing the wave representations of (a) and (b)—i.e., represented in the formalism of quantum mechanics as a linear combination of the states (a) and (b)—is the state (c), which is prepared by having both slits open. In the potential barrier experiment we have an *infinite* superposition of states—for the state corresponding to the energy level *e* is a linear combination of all possible states of the distance *D*, in the formalism.]

What about the *CI?* As already noted, the *CI* rejects the principle *ND* in all cases. Thus (A) is satisfied. Also, the waves are interpreted "statistically," much as in the Born interpretation, and one representation *is* treated just like any other. Thus (B) is satisfied. (C) is the doubtful case; superposition of states is not *explained* at all, but simply assumed as primitive. However, this is certainly a unitary treatment, and the *CI* theorist would say that our demand for some further explanation is just an unsatisfiable hankering after classical models.

We see now, perhaps, why the *CI* has been so widely accepted in spite of its initial counterintuitiveness. In view of the need for rejecting *ND,* it seems pointless to talk of "values" of observables at moments when no measurement of these observables is taking place. For if such "values" exist, they cannot be assumed to be related to the values we find upon measurement even by the very weak assumption of *continuity.* So why assume that these values (hidden variables) exist? Even if such an argument may appear convincing to most physicists, I shall now try to show that the *CI* runs into difficulties of its own which are just as serious as the difficulties that beset previous interpretations.

To begin with, it is necessary to grant at once that the *CI* has made an important and permanent contribution to quantum mechanics. That the principle *ND* must be rejected, and that such statements as (3) and (4) must be reformulated in the way illustrated by (3′) and (4′), is a part of quantum mechanics itself. To put it another way, it is a part of quantum mechanics itself as it stands today that the proper interpretation of the wave is statistical in this sense: the squared amplitude of the wave is the probability that the particle will be found in the appropriate place *if a measurement is made* (and analogously for representations other than position representation). We might call this much the *minimal statistical interpretation* of quantum mechanics, and what I am saying is that the minimal statistical interpretation is a contribution of the great founders of the *CI*—Bohr and Heisenberg, building, in the way we have seen, on the earlier idea of Born—and a part of quantum mechanical theory itself. However, the minimal statistical interpretation is much less daring than the full *CI.* It leaves it completely open whether there are any observables for which the principle *ND* is correct, and whether or not hidden variables exist. The full *CI,* to put it another way, is the minimal statistical interpretation *plus* the statement that hidden variables do not exist and that the wave representation gives a *complete* description of the physical system.

Before actually discussing the *CI*, let us consider for a moment what a formalization of quantum mechanics might look like. The famous von Neumann axiomatization takes the term "measurement" as primitive, and postulates that each measurement results in a "reduction of the wave packet." That the term "measurement" is taken as primitive is no accident. Any formalization of quantum mechanics must either leave the question of interpretation open—in which case no testable predictions whatsoever will be derivable within the formalization, and we will have formalized only the *mathematics* of quantum mechanics, and not the physical theory—or we must include in the formalization at least the minimal statistical interpretation, in which case the term "measurement" automatically enters.

As we remarked at the outset, however, it is not something that we can accept in the long run that the term "measurement" should simply remain primitive in physical theory. Measurements are only a certain subclass of physical interactions; we have to demand that this subclass should be specified with sufficient precision, and without reference to anything subjective. To define a measurement as the apprehension of a fact by a human consciousness, for example, would be to interpret quantum mechanics as asserting a dependence of what exists on what human beings happen to be conscious of, which is absurd.

Bohr has proposed to meet this difficulty by saying, in effect, that an interaction (between a microsystem and a macrosystem) is a measurement if the interaction is a measurement according to *classical physics*. Besides pushing the problem back to exactly the same problem in classical physics (where it was never solved because it never had to be solved—it never became necessary to have a definition of "measurement" in *general*), the suggestion is completely unacceptable in the context of axiomatization. We can hardly refer to one theory (classical physics) in the axiomatization of another if the first theory is supposed to be incorrect and the second is designed to supersede it.

A more acceptable alternative might be as follows: let us define a measurement as an interaction in which a system *A* (the "measured" system), which was previously isolated, interacts with a system *B* (the "measuring" system) in such a way as to cause a change in *B* that affects some "macro-observable"—some macroscopically observable property of *B*. This definition is, of course, incomplete. We also need to be told how to *interpret* measurements—i.e., how to determine what the change in a particular macroscopically observable property of *B* signifies. It is possible to do this, however, and in a way that does not go outside the formalism of quantum mechanics itself.[4] Thus we might arrive at a formalization of quantum mechanics (including the minimal statistical interpretation) that does not contain the term "measurement" as primitive. But the term "macro-observable" *will* now appear as primitive. Perhaps, however, this term will prove easier to define. It seems plausible that the macro-observables can be defined as certain averages of large numbers of micro-observables in very big systems.

Two more points before turning to the *CI* itself:

i. Quantum mechanics must, if correct, apply to systems of arbitrary size, since it applies both to the individual electrons and nuclei and also to the interactions between them. In particular, it must apply to macrosystems.[5]

ii. Since we are considering only elementary, nonrelativistic quantum mechanics, we have no reason to consider the nature of the universe "in the large." We can perfectly well idealize by neglecting all forces originating outside our solar system and by considering that we are dealing with experiments that in the last analysis take place in *one* isolated system—our solar system.

Let us now consider the question: how might the *CI* be incorporated in the formalization of quantum mechanics itself? The simplest method is to modify the formation rules so that an observable does not necessarily exist (have any numerical value) at a time, and then to add a postulate specifying that an observable exists in a system A at a time if and only if a measurement of that observable takes place at that time. However, if A is the solar system, then no observable exists in A since we are assuming that our solar system is all there is. In fact, since not even macro-observables exist, if this is taken literally, then the solar system itself does not exist.

Let us consider the difficulty a little more closely. The difficulty is that measurement, as we proposed defining it, requires interaction with an *outside* system. If we introduce the assumption that all measurements ultimately take place in some "big," isolated system (the "universe," or the solar system, or whatever), then we immediately arrive at a contradiction, for macro-observables in the "big" system are supposed to exist (and thus be measured), but there is no *outside* system to measure them.

We might try to avoid the difficulty by rejecting the idea of a "biggest system," although given the present formalism of quantum mechanics, this also leads to mathematical contradictions.[6] But this is not much happier. What of macro-observables that are isolated for a long time, say, a system consisting of a rocket ship together with its contents out in interstellar space? We cannot seriously suppose that the rocket ship begins to exist only when it becomes once again observable from the earth or some other outside system.

In view of these difficulties, we very quickly see that there is only one reasonable way of formalizing quantum mechanics that does justice to the ideas of the founders of the *CI*. We cannot really suppose that the founders of the *CI* meant that *no* observable has a sharp value unless measured by an outside system. Indeed the repeated insistence of Bohr and Heisenberg on macrophysical realism, combined with their equally strong insistence that realism is untenable as a philosophical point of view with respect to micro-observables, leads us to the following, more careful formulation of the *CI*. Instead of saying henceforth that, according to the *CI*, observables do not exist unless measured, we shall have to say that, according to the *CI*, *micro*-observables do not exist unless measured. We shall take it as an assumption of quantum mechanics that *macro*-observables retain sharp values (by macroscopic standards of sharpness) at all times. The formulation of the *CI* that I am now suggesting, then, comes down to just this: that macro-observables retain sharp values at all times in the sense just explained, while micro-observables have sharp values only when measured, where measurement is to be defined as a certain kind of interaction between a micro-observable and a macro-observable.

This formulation would in all probability be unacceptable to Bohr and Heisenberg themselves. They would accept the statement that micro-observables do not have sharp values unless a measurement takes place. And they would agree that, as we have just seen, this principle has to be restricted to micro-observables. In this way they differ sharply from London and Bauer and, possibly, Von Neumann, who would hold that all observables can have unsharp values unless measured by a "consciousness." Bohr and Heisenberg are macrophysical realists who hold that the *nature of the microcosm* is such that we cannot succeed in thinking about it in realistic terms. We are dealing with what has been called a "nonclassical relation between the system and the observer." But this has to do, in their view, with the special nature of the microcosm, not with any special contribution of human consciousness to the determination of physical reality. What Bohr and Heisenberg do not see, in my opinion, is that their interpretation has to be restricted to *micro*-observables from the very outset, and that the relative sharpness of macro-observables is then an underived assumption of their theory.

I shall say something later about the rather ambiguous position of the Copenhagen theorists (especially Heisenberg) with respect to the reduction of the wave packet. The central question in connection with the *CI* is whether the special character of macro-observables can be *derived* from some plausible definition of macro-observable together with a suitable formulation of the laws of quantum mechanics. In classical physics macro-observables were simply certain averages defined in very big systems. This definition will not do in quantum mechanics because the quantities of which macro-observables are supposed to be averages are not required always to "exist," i.e., have definite numerical values, in quantum mechanics. A macro-observable in quantum mechanics has to be defined as an observable whose corresponding operator is in a certain sense an average of a number of operators. The question we face is whether from such a quantum-mechanical characterization of a macro-observable together with the laws of quantum mechanics it is possible to deduce that macro-observables always retain sharp values whether a measurement interaction involving them is going on or not. If we can do this, then the appearance of paradox and the *ad hoc* character of the *CI* will disappear.

In spite of a number of very ingenious attempts, it does not appear that this can be done. Briefly, what is needed is some demonstration that superpositions of states in which macro-observables have radically different values cannot be physically prepared. Unfortunately, it does not seem that there is anything in quantum mechanics to give this conclusion.

One fallacious attempt to obtain this conclusion has been via an appeal to what are known as the "classical limit theorems" of quantum mechanics. These say that a system obeying the laws of classical physics and not subject to relevant microscopic uncertainties will behave in the same way in quantum mechanics as in classical physics. In other words, if we take a wholly classical system, say, a machine consisting of gears and levers, and translate the description of that system with great pains into the mathematical formalism of quantum mechanics, then the wave representation will yield the information that the various macro-observables defined in that system have sharp values. Moreover, as the state of the system is allowed to change with time, although the wave will spread out in the relevant space, this spreading out will happen in such a way that the macro-observables in question will retain sharp values for long periods of time. In short, if all macro-observables had sharp values to begin with, then they will retain sharp values during the period of time under consideration without our having to assume that any measurement took place.

This result shows that *in some cases* it follows from quantum mechanics that macro-observables will retain sharp values whether interacted with or not. Unfortunately, there are other cases in which a diametrically opposed conclusion follows, and these are of great importance. The most famous of these cases is the so-called "Schrödinger's cat" case. Schrödinger imagined an isolated system consisting of an apparatus that contains a cat together with a device for electrocuting the cat. At a certain preset time, an emitter emits a single photon directed toward a half-silvered mirror. If the photon is deflected, then the cat will live; if the photon passes through the half-silvered mirror, then the switch will be set off and the cat will be electrocuted. The event of the photon passing through the half-silvered mirror is one whose quantum mechanical probability is exactly $\frac{1}{2}$. If this entire system is represented by a wave function, then prior to the time at which the emitter emits the photon—say, 12:00—the wave that represents the state of the system will evolve in accordance with the classical limit theorems, that is to say, all macro-observables, including the ones that describe the behavior of the cat, will retain sharp values "by themselves." When the photon is emitted, however, the effect of the half-silvered mirror will be to split the wave corresponding to the photon into two parts. Half of the wave will go through the half-silvered mirror and half will be reflected. From this point on, the state of the system will be represented by a linear combination of two waves, one wave representing the system as it would be if the photon passed through the half-silvered mirror and the cat were electrocuted, and the other wave representing the state of the system as it would be if the photon were reflected and the cat were to live. Thus if the system is not interfered with prior to 1:00 P.M., then we will predict that at 1:00 P.M. the system will be in a state that is a superposition of "live cat" and "dead cat." We then have to say that the cat is *neither alive nor dead* at 1:00 P.M. unless someone comes and looks, and that it is the act of looking that *throws the cat into a definite state.* This result would, of course, be contrary to the macro-physical realism espoused by the *CI.*

It should be observed that for all its fanciful nature, Schrödinger's cat involves a physical situation that arises very often, in fact one that arises in all quantum mechanical experiments. Almost every experiment in which quantum mechanics is employed is one in which some micro-cosmic uncertainty is amplified so as to affect something detectable by us human beings—hence detectable at the macro-level. Consider, for example, the operation of a Geiger counter. A Geiger counter affects a macro-observable, the audible "clicks" that we hear, in accordance with phenom-ena (individual elementary particle "hits") that are subject to quantum mechanical uncertainties. It is easily seen that if we describe any situation in which a Geiger counter is employed by a wave, then the result is analog-ous to the Schrödinger's cat case. Since the Geiger counter will click if a particle hits it and will not click if a particle does not hit it, then its state at the relevant instant will be represented by a wave that is a superposition of the waves corresponding to "click" and "no-click."

It must be admitted that most physicists are not bothered by the Schrödinger's cat case. They take the standpoint that the cat's being or not being electrocuted should itself be regarded as a measurement. Thus in their view, the reduction of the wave packet takes place at 12:00, when the cat either feels or does not feel the jolt of electric current hitting its body. More precisely, the reduction of the wave packet takes place pre-cisely when if it had not taken place a superposition of different states of some macro-observable would have been predicted. What this shows is that working physicists accept the principle that macro-observables always retain sharp values (by macroscopic standards of sharpness) and deduce when measurement *must* take place *from* this principle. But the intellec-tual relevance of the Schrödinger's cat case is not thereby impaired. What the case shows is that the principle that macro-observables retain sharp values at all times is not *deduced* from the foundations of quantum me-chanics, but is rather dragged in as an additional assumption.

A number of physicists, most notably Ludwig, have reasoned as follows. "All right," they say, "let us suppose that it is possible to prepare a superposition of states in which macro-observables have radically different values. We know, as already remarked in connection with the two-slit experiment, that detectable interference effects require mutually coherent waves. Since the waves corresponding to different states of a macro-observable will be incoherent, no *detectable* interference effects will result, and hence (sic) macro-observables do always retain sharp values."

To make this argument a little more respectable, we may construe Ludwig as unconsciously advocating that certain further assumptions be added to quantum mechanics. The *CI* involves what has been called "the completeness assumption," that is to say, the assumption that the wave representation gives a *complete* description of the state of the physical system. If the completeness assumption is accepted without any restriction, then a state that is a superposition of two states assigning different values to some macro-observable could not and should not be confused with a state in which the system is really in one of the two substates but we do not know in which. *Physical* states should not be confused with states of ignorance, and there is all the difference in the world between a state that is a superposition of two states A and B (the superposition being construed as a third state C in which the macro-observable in question has no sharp value) and a *state of knowledge,* call it D, in which *we* are when we know that the system in question is either in state A or in state B, but we do not know which. Ludwig in effect argues: "If a system is in state C, then since the interference effects are not detectable, we will get the same predictions for the outcomes of all practically possible observations as if we supposed that the system is either in state A or in state B, but we do not know which. So in that case, being pragmatic, let us say the system either is in state A or is in state B but we don't know which." This argument is incoherent since it tacitly assumes the usual statistical interpretation of quantum mechanics, which in turn assumes that looking at the system throws it into some definite state.

If we say, for example, that the cat might have been in a neither-alive-nor-dead state until 1:00 P.M., but that due to the mutual incoherence of the waves corresponding to "dead cat" and "live cat" this leads to exactly the same predictions as the theory that the cat was thrown into a definite state by the event that took place at 12:00, then we are still left with the question with which we started: how does our looking affect the state of the cat? What is it about us and about our looking that makes this interaction so special? If the answer is that it is just that we are macrophysical systems, and that macro-observables always retain sharp values, then it is necessary to object that it is just this assumption which is abandoned when we assume that it may be possible to prepare superpositions of states that assign different values to some macro-observable. In other words, the assumption is rejected when it is assumed that prior to 1:00 P.M. the cat really is in a superposition of states—"live cat" and "dead cat"—but then this very assumption is used when it is assumed that the observer, upon looking, will find the cat either alive or dead. Why should not the result of this new interaction—my looking—rather be to throw me into a superposition of two states: "Hilary Putnam seeing a live cat" and "Hilary Putnam seeing a dead cat"?

The "coherence" or "destruction of interference" theory might perhaps be repaired in the following way. Suppose that we add to quantum mechanics the following assumption: that some wave descriptions are *complete* descriptions of the state of a physical system while others are *not*. In particular, *if a wave is a superposition of mutually incoherent waves* corresponding to two or more possible states of a system, *then that wave description gives incomplete information concerning the state of the system in question.* What we must say is that the system in question is really in one of the more precisely specified states, only we do not know which. Applied to the Schrödinger's cat case, this would mean saying that the interaction that took place at 12:00 sent the system into a superposition of two states, "live cat" and "dead cat." This is a superposition of two states *A* and *B* whose corresponding waves are incoherent. Since the wave function that represents the state of the system after 12:00 is a superposition of this kind, we are required to say, by the new theory, that this wave function is an *incomplete* description of the state of the system, and that the cat is really either alive or dead. I think this is the most respectable interpretation of what the coherence theorists, Ludwig, *et al.,* are really trying to do.

But there are grave difficulties with this suggestion. In particular, to say that a wave gives an incomplete description of the state of a system is to say that systems whose state can be described by that wave function may really be in several quite different physical conditions. Completeness versus incompleteness is not a matter of *degree*. On the other hand, coherence and incoherence of waves *are* a matter of degree. It seems that in principle a postulate of the kind suggested is going to be unworkable. Exactly what degree of incoherence of the superimposed waves is required before we are to say that a wave function gives an incomplete description of the state of a system? Secondly, it would have to be shown that the proposed added principle is consistent. Perhaps every wave function can be written as the superposition of two mutually incoherent waves. What is worse is that we might well required to say that the descriptions were incomplete in *mutually incompatible ways.*

At this point, I would like briefly to discuss Heisenberg's remarks on the subject of the reduction of the wave packet. These remarks are puzzling, to say the least. First of all, Heisenberg emphatically insists that at least when macro-observables are involved the reduction of the wave packet is not a physical change, but simply an acquisition of information. What Heisenberg says is that any system which includes a macroscopic measuring apparatus *is never known to be in a particular pure state but only known to be in a mixture.* He does not elaborate the exact significance of this remark, and therefore it becomes incumbent upon us to ask just what it might mean and what its relevance is supposed to be in the present context.

Usually the statement that a system is in a mixture of states *A, B, C* is interpreted as meaning that it is really in one of the states *A, B,* or *C,* but we do not know which. I shall henceforth refer to this as the "ignorance interpretation" of mixtures. If the ignorance interpretation of mixture is correct, then Heisenberg's remark is simply irrelevant. What Heisenberg is saying is that Schrödinger's cat starts out in some pure state and that we can give a large set of pure states *A, B, C, . . .* so that the state it starts out in must be one of these, but we cannot say which one. This is irrelevant because Schrödinger's argument did not depend on our *knowing* the state in which the cat started out. Suppose, however, the ignorance interpretation of mixtures is *incorrect.* In that case quantum mechanics involves two fundamentally different kinds of states, *pure states* and *mixtures.* We may also say that states may be combined in essentially different ways: on the one hand, two pure states *A* and *B* may be superimposed to produce a new state that will be a linear combination of *A* and *B;* and, on the other hand, the two states *A* and *B* may be "mixed" to produce a state that will depend on *A* and *B.* To say that a system is in some particular mixture of the two states *A* and *B* will not be to say that it is really in one of the cases of the mixture, but to say that it is in a new kind of state not exactly representable by any wave but only by this new mathematical object—a mixture of wave functions. It is easily seen that giving up the ignorance interpretation of mixtures will not help us, for now we have the difficulty that when the observer looks at 1:00 P.M. he does not find the cat in a mixture of the states "live cat" and "dead cat" but in *one state or the other.*

If to say that the cat is in a mixture does not *mean* that the cat is already either alive or dead, then we have again the result that the observer's looking throws the cat from one kind of state—a mixture of superpositions of "live cat" and "dead cat"[7]—into another state, say "live cat." Put in another way, *selection* from a mixture is not usually regarded as analogous to reduction of the wave packet, because it is generally accepted that if a system is in a mixture then it must already be in one of the pure cases of the mixture. On this interpretation, if a system is known to be in a mixture and we look to see which of the cases of the mixture it is in, our looking does not in any way disturb the system. As soon as we give up the ignorance interpretation of mixtures, however, this process of selection becomes a discontinuous change in the state of the system, and all the anomalies connected with the reduction of the wave packet now arise in connection with the idea of selection from a mixture.

Let us now conclude by seeing what has happened. The failure of the original Born interpretation, we recall, led us to the conclusion that the principle *ND* is incorrect. The falsity of the principle *ND* was in turn the crucial difficulty for the hidden variable interpretations. In order to account for the falsity of the principle *ND*, these interpretations had to introduce special forces, e.g., quantum potential, for which no real evidence exists and with quite unbelievable properties. Since we cannot assume that the principle is true and that micro-observables, if they exist, are related to measured values even by the very weak assumption of continuity, we decided not to assume that micro-observables exist—i.e., have sharp numerical values—at all times and to modify such statements as (3) and (4) in the way illustrated in (3′) and (4′)—that is to say, to adopt at least the minimal statistical interpretation of quantum mechanics. At this point we found ourselves in real difficulty with macro-observables. The result we wish is that although micro-observables do not necessarily have definite numerical values at all times, macro-observables do. And we want this result to come out of quantum mechanics in a natural way. We do not want simply to add it to quantum mechanics as an *ad hoc* principle. So far, however, attempts to derive this result have been entirely unsuccessful.

What, then, is to be done? We might try giving up the idea that micro-observables can have such things as unsharp values. However, this would not be of much help, for if we accept the minimal statistical interpretation and the falsity of principle *ND*, then we are not much better off than with the Copenhagen interpretation. It is true that we would then know that micro-observables always exist, and hence that certain macro-observables will always exist, namely, such macro-observables as are simply averages of large numbers of micro-observables; but not all macro-observables are so simple. The surface tension of a liquid, for example, is a macro-observable that ceases to exist if the swarm of particles in question ceases to be a liquid. If we cannot say anything about the values of micro-observables when we are not making measurements except *merely* that they exist, then we will not be able to say that the swarm of particles constituting, let us say, a glass of water, will continue to be a liquid when no measurement is made. Or in the Schrödinger's cat case, we will be able to say only that the swarm of particles making up the cat and the apparatus exists and that it will certainly take the form "live cat" or the form "dead cat" if a measurement is made, i.e., if somebody looks. We will not be able to say that the cat is either alive or dead, or for that matter that the cat is even a cat, as long as no one is looking. If we go back to the idea that micro-observables have numerical values at all times, then if we are to be able to handle the Schrödinger's cat case in a satisfactory way, we will have to say *more* about these numerical values of micro-observables at times when no measurement is made than merely that they exist. In short, we will be back in the search for a hidden variable theory.

In conclusion, then, *no* satisfactory interpretation of quantum mechanics exists today. The questions posed by the confrontation between the Copenhagen interpretation and the hidden variable theorists go to the very foundations of microphysics, but the answers given by hidden variable theorists and Copenhagenists are alike unsatisfactory. Human curiosity will not rest until those questions are answered, but whether they will be answered by conceptual innovations within the framework of the present theory or only within the framework of an as yet unforeseen theory is unknown. The first step toward answering them has been attempted here. It is the modest but essential step of becoming clear on the nature and magnitude of the difficulties.

Notes

* This work was supported in part by the United States Army, the Air Force Office of Scientific Research, and the Office of Naval Research.

1. Spaces of yet another kind—"Hilbert spaces"—are also used in quantum mechanics, but these will be avoided here.

2. The formulation of complementarity is due to Paul Oppenheim and Hugo Bedau.

3. Recalling that $p = h/\lambda$ (the De Broglie relation)—i.e., the momentum p of a particle is just the reciprocal of the wave length of the corresponding wave, in units of h.

4. Cf. $[P_2]$.

5. This has sometimes been denied on the ground that we cannot know the exact state of a macrosystem. However, our *incomplete* knowledge of the state of a macrosystem can be represented in quantum mechanics exactly as it was in classical physics—by giving a "statistical mixture" of the various possible states of the system weighted by their probabilities (in the light of our information).

6. Cf. $[P_1, P_2]$.

7. This point needs emphasis. Suppose that the mixture representing the possible states of the cat at 12:00 P.M. is $M_1 = \{s_1, s_2, \ldots\}$. (Each s_i should be weighted by a probability, of course.) Then the mixture representing the state of the cat at 1:00 P.M. will be $M_2 = \{U(s_1), U(s_2), \ldots\}$, where U is a certain transformation function. Each $U(s_i)$ has the form "½ (Live cat) + ½ (Dead Cat)," putting it roughly. Thus M_2 does *not* contain the information that "the cat is either alive or dead, but we don't know which until we look." *That* information would rather be given by a third mixture $M_3 = \{t_1, t_2, \ldots\}$, where each t_i describes a pure case in which the cat is alive or a pure case in which the cat is dead.

It can sometimes happen that two mixtures $\{A_1, A_2, \ldots\}$ and $\{B_1, B_2, \ldots\}$ have the same expectation values for *all* observables (same "density matrix"). Such mixtures might be regarded as physical equivalent (although this would involve giving up the ignorance interpretation, and hence involve finding a new interpretation of *selection* from a mixture). However, the mixtures $M_2 = \{U(s_1), U(s_2) \ldots \}$ and $M_3 = \{t_1, t_2, \ldots \}$ do not have the same density matrix, no matter what probabilities may be assigned to the t_i (I plan to show this in a future paper). Thus this is not relevant here.

Lastly, it might be argued that M_2 and M_3 lead to *approximately* the same expectation values for all *macro*-observables, and hence (sic) describe the same physical states of affairs. It is easily seen that this is unsatisfactory (since the special status of *macro*-observables is presupposed, and not explained), and that otherwise this reduces to Ludwig's proposal, discussed above.

References

Ludwig, G., *Die Grundlagen der Quantenmechanik*. Berlin: Springer Verlag, 1954.

Margenau, H. and E. Wigner, "Comments on Professor Putnam's Comments," *Philosophy of Science,* Vol. XXIX (July, 1962), 292-293.

Putnam, H., "Comments on Comments on Comments—A Reply to Margenau and Wigner," *ibid.,* forthcoming.

————, "Comments on the Paper of David Sharp," *ibid.,* Vol. XXVIII (July, 1961), 234-237.

Sharp, D., "The Einstein-Podolsky-Rosen Paradox Re-Examined," *ibid.,* Vol. XXVIII (July, 1961), 225-233.

DAVID HAWKINS
University of Colorado

The Thermodynamics of Purpose

The statement that the natural laws are at the bottom, not only of the more or less permanent structures occurring in nature, but also of all processes of temporal development, must be qualified by the remark that chance factors are never missing in a concrete development. Classical physics considers the initial state as accidental. Thus "common origin" may serve to explain features that do not follow from the laws of nature alone. Statistical thermodynamics combined with quantum physics grants chance a much wider scope but shows at the same time how chance is by no means incompatible with "almost" perfect macroscopic regularity of phenomena. Evolution is not the foundation but the keystone in the edifice of scientific knowledge. Cosmogony deals with the evolution of the universe, geology with that of the earth and its minerals, paleontology and phylogenetics with the evolution of living organisms.

—HERMANN WEYL, *Philosophy of Mathematics and Natural Science*

In this paper I am going to put forth some natural teleologies. The scientific community generally and the tough-minded philosophers have long regarded this metaphysical nut with something close to horror. I want to crack and section it very carefully, therefore, to get at the sweet meat within. Aristotle is the main villain, although most people who read him find him rather disarming in his matter-of-fact, pedestrian way.

In fact, the revolt against Aristotelian traditions, so essential to the growth of modern physics, did scant justice to the older philosophy. This is not a complaint. The loss of insight was matched by greater gains, not only in physics but even in philosophy. There was a loss of innocence, as well as a gain of intellectual power, in the coming of the world-machine cosmology. In describing and measuring the loss of insight, I shall try to avoid the party of metaphysical Luddites, and may, therefore, be repudiated all around. On the other hand, the sources of my argument lie mainly in contemporary science, which, as I shall show, has grown around and covered up some of the old anti-teleological roadsigns.

One of the most worthy purposes of a philosophy is to discern, and coherently represent, the primitive and irreducible categories, "the nature of things." The new mechanistic philosophy was a claim of ascendancy in behalf of some ways of explaining things, of some categorial modes, over others to which Aristotle had held them subordinate. Although this new philosophy was lacking in philosophic adequacy, it had point; and it penetrated deeply, guiding not only physics, but also the development of ethics and politics and economics. The major conflict was over the relation between two of Aristotle's famous list of causes: the final cause, or *telos*, and the material cause. The new philosophy was called materialism because it claimed the ascendancy of material causes, reducing final causes to a derivative position. In its popular impact, the new philosophy was a denial of the reality of purpose in nature and a pushing back of all the older teleological questions into the supernatural order (if anywhere) and their replacement, here below, by mechanistic reformulations.[1]

The job of getting rid of final causes must be understood correctly, however. The form-matter dichotomy is relative, and what is matter to be formed, like the bricks for a house, is itself formed matter, like the clay that is to be molded and sunbaked. One never gets far enough down in this taxonomy to reach the prime matter, which can be formed into anything and is nothing formed. Prime matter was a theoretical requirement for Aristotle, but he did not think one would be able to lay hands on it *qua* prime matter. One may find a very generic matter, but it already has form; it has a nature and, therefore, a *telos*, a final cause. It was this generic nature that was taken over, in the mechanistic philosophy, for explication in terms of the laws of mechanics. It was this generic nature that would be manifest equally to the elliptical path of the moon, Newton's apple, or the boxed works of Aristotle suitably placed in orbit. The mechanistic claim was that once this *telos* of matter, the grand laws of interaction and motion, was adequately worked out, only the material components and their states of motion and rest needed to be discriminated, in order to distinguish any material system from any other.[2] From such discrimination alone every other form and quality of things would follow. To explain the way specific things go did not require a specific final cause in addition to the laws of matter, but only what we today call boundary conditions, data to be plugged into the statement forms of generic law.

There is a passage in Aristotle in which he notices that materialism in the above sense would be all right for those processes that are time reversible, like celestial orbits.[3] If he had been told about the grand laws of seventeenth-century physics, he might therefore have been put on the spot. What he would then have needed were the thermodynamic categories, accepted as irreducible modes of explanation. He was just as far along in that direction as the seventeenth century was, however, and I am getting ahead of myself.

So materialism did not get rid of final causes, but only of specific final causes. Materialism finds no need for ends or natures in the plural, but only for a kind of generic *telos* in nature, renamed "law." When we appeal to laws of nature in explaining anything, we are doing just the kind of thing Aristotle talked about under the heading of ends. Never mind the anthropomorphic connotations; they were just an accident of Aristotle's technical vocabulary. Unlike Kant and the rest of us moderns, he did not have a dead language handy, to rummage around in.[4] And never mind the Humean truism that laws are descriptive, not prescriptive. There is an obvious systematic ambiguity in the term "law," as between formulation and thing formulated. Law as thing formulated is prescriptive—not in the sense of a command given, of course, but in the sense of the metaphor of imperativeness—not to be avoided or evaded, unconditional, categorical. Philosophers worry a lot about this, but for the wrong reasons, I think. Of course one does not *know* that there are any unconditional or imperative laws. But one does know something else: that within the framework of any sufficiently ample body of knowledge there will be a distinction between what is necessary and what is contingent. The connection between force and acceleration is not conditional in Newtonian theory, but necessary. Since forces can always be detected independently of the accelerations coupled to them, the connection is not analytic. Since there is no way of describing conditions under which accelerations would be unaccompanied by corresponding forces, the connection is not contingent. It is not that the propositions of mechanics are necessarily true, but that, *true or false,* they have a *modality different from contingent propositions, namely, they are unconditional.*[5]

What is interesting in this connection is not the barren debate over the meaning of causal necessity, but the possibility, suggested by recent high-energy physics, that one could formulate dynamical laws in such a way that they *were* contingent, i.e., valid only for some bounded range of the dynamical variables. Such a theory would have other kinds of unconditional statements, but they would not be the dynamical ones.[6]

Time reversibility and categoricity, or imperativeness, are two characteristics of the classical laws. A third is completeness, and this makes quite a package. Completeness means that if a physical system is described in terms of its material composition, its internal states, and its interactions, there is a larger system to which these interactions are in turn internal and which can be described by the same laws. Momentum, or energy (in terms of which we now formulate the old laws), or any other conserved quantity may change in a given system, but only because it is not isolated, only by exchange with the environing system. Thus nothing can act that is not acted upon (this is good Aristotle) in accordance with the *same* laws (and this is not). The only unmoved mover possible is just the generic *telos*, the laws of mechanics themselves.

The classical laws are time reversible, categorical, and complete. But they are laws of transformation, mapping past and future into each other. This was seen explicitly only in the eighteenth century, but it was implicit before.[7] There is no classical imperative that requires or forbids a single event, or that requires or forbids any history of events internally consistent with the laws. On the other hand, any event that happens is in principle predictable from a sufficient knowledge of antecedent conditions. This is dynamical determinism, a fourth characteristic of the classical laws. Determinism means that every event is contingent and could have been different a fortiori if the whole history of nature were different.

The vivid way of saying what determinism means is that if the *telos* of matter is time reversible, unconditional, complete, *and* deterministic, then there is no choice left except the choice among possible worlds, alternative universal deterministic histories. But these are now suddenly abundant. They are indistinguishable with respect to the system of laws, and therefore with respect to matter, space, and time, but as varied as one pleases in any other respect. So the generic *telos* is not so potent after all. There is only a genre, so to speak, not a specific play that it determines. Why, then, just *this* world? The seventeenth century gave deistic answers. The question is important for my purposes because it shows the sense in which the classical laws were *not* complete. The eighteenth century recast the formulations in a more elegant way; the nineteenth century extended them to include the electromagnetic field; Einstein found the restriction to a maximal velocity change together with the implied revision of the space-time metric. But the metaphysics, so to say, remained unaltered. Field theories, to be sure, make determinism less useful sounding, especially when one sees the implication that news can be propagated no faster than some grosser disturbances of which one might want it to be news. The failure of predictability does not invalidate the classical determinism, but it does subvert it.

Even within this mechanistic metaphysical framework, however, there is room for the evolution of physics in novel directions. Within the domain of the *telos* of classical physics there is room for an enormous variety of possible worlds. Are there ways of narrowing the set of all possible worlds through the exercise of scientific inquiry before we throw the question open to theological discourse? Spinoza and Leibniz asserted it was the *only* possible world, and I have heard this metaphysical pronouncement from some contemporary physicists.[8] Is the *telos* of matter more powerful, more restrictive, than we have realized?

At any rate, one does not have to go all the way. There can be species and subspecies of possible worlds. One could exclude contracting universes, for example, and allow only those that expand. But this would be singularly *ad hoc,* since expansion or contraction is classically just a matter of boundary conditions. Some revision of dynamical laws might lead to such an exclusion, however.

A more familiar sort of limitation is that imposed by a probability metric, and this goes back into the nineteenth century. Ours does not appear to be a heads-much-more-often-than-tails world, for example, or the reverse, and this as well as many other statistical properties can be explained by the fact that almost all worlds are nearly symmetrical with respect to coin tosses. To say "almost all" in this way implies a metric, of course, because there could be a probability metric for which this "almost all" was transformed into "almost none." But the metric is not *ad hoc;* it is one that identifies equal probability with equal volume in phase space, subject to certain necessary constraints.[9]

The first important formulation of this kind in physics was the Maxwell-Boltzmann distribution, which has the constraints of given volume, particle number, and energy. We treat the sample of gas, as defined, only up to its parametrized properties, and we treat the rest as a set of equally probable alternatives.

It would be an odd metaphysical step to identify nature, i.e., reality, not with a uniquely actual member of the set of all possible mechanistic worlds, but with the ensemble of possible worlds consistent with our observation and measurement. On such a view, any measurement, any discrimination, may be thought of as a process of projection or preparation that annihilates worlds incompatible with the given discrimination.[10] The trouble with this, metaphysically, is that it undermines the distinction between actual and possible; for by symmetry, measurers in alternative worlds annihilate each other. Another way of taking it is to opt for idealism, making reality itself depend on observation, and this undermines the distinction between measurement and choice.

Nevertheless some such notion is implicit metaphysics for the classical statistical mechanics. It may be objected that the Gibbsian ensemble is only a fiction, a way of saying that as far as we *know,* the actual system may be any one of the ensemble. I would answer that the implicit metaphysics is not altered by a decision not to take it seriously, not even when the decision is sensible. A popular alternative is to replace the ensemble by a sequence of states of a single system, strung together *nacheinander* by deterministic laws instead of *nebeneinander* as alternatives. The probability metric then follows as a time average; instead of not knowing which world it is, one just does not know *when* it is. The trouble with this approach, the ergodic approach, is that it turns out to require something like an a priori probability metric anyway.[11] And it is true in many cases that a phase space, over which we want to have a probability metric, will have potential barriers in it that a system cannot climb over.

This implicit metaphysics of classical statistical mechanics has a very important utility. It loosens up the mechanistic framework of nature and thus allows laws of a new kind, probabilistic laws, a place, without at the same time, *altering* the classical laws. I have argued elsewhere that this loosening up can be achieved without the real ensemble and without relying solely on ergodic assumptions.[12] The qualitative notion of chance and the metric of probability can be defined in terms of interacting systems antecedently independent. The notion of causal independence gives physical meaning to the symmetry principle of indifference without the fiction of infinite ensembles or collectives. In this I am again on the side of Aristotle, who defined chance in terms of interacting systems with no joint *telos*. The existence of such a metric does not hold of all possible mechanistic worlds, for there are some such worlds where coins behave as badly as one pleases. The existence of a probability metric is thus a principle, a kind of conservation law, assumed for *our* world. It is not a dynamical law.[13] The point is that the multiplicity of mechanistically possible worlds does leave the *telos* of matter *per se,* the classical law system, incapable of explaining many sorts of things about *this* world; and there is room for radically new science even without loosening the logical constraints of the classical system.

But quantum mechanics does loosen these constraints. For each kind of microphysical system it defines a specific *telos* stipulating a set of sharply distinguished possible states within which the system can exist. Transitions between these states are therefore not through a continuum of intermediate states. Quantum mechanics is still an emendation of classical physics and not a complete system to replace the latter. The classical conservation laws, in particular, still hold, and their conjunction with the quantum principle implies that, in transition, the states of two interacting systems are not well defined.[14] It may prove possible to transcend the quantum theory in the future by a theory that re-establishes some kind of temporal determinism at a subquantum level—but not in a way to obviate the uncertainty principle for the states as classically defined. The Zeno-like incompatibility of quantum states with conservative transitions may prove to be interpretable as a kind of noise at the subquantum level analogous to Brownian motion, but it is no *less* real than the Brownian noise. Controversy over these questions leaves both the validity of the quantum principle and the reality of chance unaffected. And this is all I need for my purposes.

In my opinion, it is the combination of supermechanical order with submechanical noise, implied by the quantum principle, that defines the lasting philosophical impact of quantum physics. The former permits a world far more elegantly and complexly structured than classical physics could possibly allow, and the latter allows the classical conservation laws to hold sway essentially unaltered. The old *telos* of matter is still on the scene, but in a more radically pluralistic world where there is still a noise level even at absolute zero. And there are specific *teloi* possible only because of that noise level; they are determined by generic laws, but these are laws that give rise to a taxonomy of natural kinds—laws of state rather than laws of transition. Very Aristotelian!

The result of these laws of state is a map of the hierarchical structure of things, the ladder of nature,[15] and the relativizing of the concept of "atom." It is within that world of random noise and crystal order that my next steps will be taken.

Changes occur throughout the range of size in nature, but because of the quantum principle they occur at radically different rates. The hills are not eternal, but neither, for quite different reasons, are the chemical atoms. The temperatures that prevail in our world are enough to excite the superficial changes in atoms that we call chemistry. The deeper changes that transform them are now rare because these changes require energies no longer commonly available. The quantum changes of the larger structures, say, of crystals, extend down in the energy scale with increasing size. The low-energy end of this scale is submerged completely in the sea of thermal noise. Big things are held together by their relatively high-energy bonds. But they are internally active, noisy; they have specific heat. Only at absolute zero would they be true atoms.

The unevenness of change is what gives meaning to the idea of evolution, and in this very general sense, it is the quantum principle that allows us, in good scientific conscience, to replace the abstract world of classical physics with something much closer to the real world of nature. But there is one more law or, more properly, one more general boundary condition to be taken account of. It is the vastly uneven concentration of energy implied in the existence of hot stars, cold planets, and even colder space. Regardless of what genus of possible worlds we wish to consider as really possible, those that are significantly evolutionary belong to a very special class. They are worlds characterized by a radical disequilibrium in the thermodynamical sense. Energy is packed into some phases and almost absent from others, with energy densities varying over very many powers of ten.[17] It is this fact which constitutes the free energy that makes the world of nature a world and not a drift of atoms. Our own earth is only a special case, in which the white sun and black sky together feed the green fuse.

It is this prevalence of disequilibrium that gives relevance to the second law of thermodynamics. The world is a heat engine, in the account of which we come to a third kind of teleology.

I spoke earlier about the generic teleology of mechanistic laws and the specific teleology of quantum states. This was not done to inflame any metaphysical passions. I do not think I have imported any new mysteries by the label, which was applied for the sake of historical perspective. Now I may produce some inflammation. But first I propose to discuss a recent extension and clarification of the second law, called "information theory."

A heat engine does work, and it is driven by what is called "free energy," or "thermodynamic potential." One kind of work, the classic textbook example, is that of raising a weight. The kind of work that interests us especially is that involved in other and less obvious sorts of processes, by which thermodynamic potential is transported or transferred from one part of a system to another. The thermodynamical meaning of work is usually, but need not be, stated anthropomorphically. Nature is full of heat engines untouched by human hands.

To take one example, solar photons may lift electrons far out of their normal orbits. They fall back along a different path perhaps, and create some high energy chemical bond that would not have occurred in the purely earth-temperature milieu. Quite surely, most of the precursor energy-rich molecules of biology were formed in this way before there was any life. The temperature disequilibrium of hot sun and cold earth implies a thermodynamic potential, and some tiny bit of this has been trapped in a molecule that is now itself potent with a greater-than-average store of energy. In this way, the earth probably laid in a store of food and clothing long before it had any "occupants" in the more special sense of biology—complex molecular species that represented concentrations of energy considerably above the average, but with potential barriers high enough to protect them against thermal erosion.

I am not going to recapitulate current ideas about primitive evolution, but I do wish to observe that none of this portentous cookery is evolutionary in a biological sense. It presupposes the terrestrial heat engine and the potential variety of molecular species written in the quantum laws. I suppose one can say that there is random variation going on, and selection of the more stable resulting forms. But the same thing would be true in complete thermal equilibrium, only the concentrations would be very different. What is surprising is that among the array of molecular forms appearing on the earth, some should appear that would prosper only because of their capacity to extract the free energy already stored around them. They prosper because they are self-reproducing. When I say that it is surprising, I do not mean that it is miraculous. The molecular combinatorics are now very rich, and I assume it to be a statistical certainty that self-copying molecules appear in the fullness of time. To say that the emergence of life was "improbable" seems to me just begging for miracles. The combinatorics of the quantum ladder is miraculous enough. Life would be a miracle if it appeared in a world of thermal equilibrium. To take such a world as the norm of naturalness is itself, as I have said, a completely *ad hoc* starting point. Nature is more interesting than that.

The important point is that self-copying molecules should appear and thereby prosper. They are heat engines of a new kind. Their further evolution copies in them the order of the surrounding chemical world. Empathizing, as it were, with ancestors, one might think of the surrounding environment as existing for their sake. That would be a crude form of the teleological fallacy. What I want to emphasize about them is just that a description of such molecules, in terms of their detailed structure, would still be incomplete if it did not say also that this structure *did* constitute an effective copy of the surrounding milieu. It would be just as incomplete as a cataloguing of black grains on a photograph which failed to note that it was a picture of the moon.

The self-reproducing molecule is a heat engine whose characteristic work is that of reproducing its own peculiar order. Its population grows beyond all previous bounds, and now Darwinian principles of variation and selection hold sway. Thermodynamic potential is sucked from the chemical environment and transferred to the order of sheer species multiplicity. The food supply rebuilds slowly from the sun. Some variant self-reproducers come also to catalyze useful reactions in the neighborhood. To encapsulate and concentrate, a wall gets built. Cells appear and become selective. Through one-way passages they admit some kinds of molecular species and discharge others. They are, in fact, Maxwell demons.[18] "Copy" ceases to be merely a noun and becomes an active verb. Part of the work that these heat engines do is discrimination, self-information from the milieu. Finally, the solar cooking itself gets encapsulated, and we have photosynthesis. As the original environment gets depleted, the cells are discovered to each other and there emerge patterns of piracy and coexistence.

By this stage, clearly, we have a new sort of teleology, that of self-reproduction and, in later evolution, of self-maintenance, of function,[19] the latter implying a copying of environmental or internal variation as a means of self-maintenance.

Here is design. Who or what is the designer? Conceivably the once-in-a-million-years gardener who comes to weed and seed. Or perhaps (as we now mostly believe) the mother heat engine, the green earth. But is the latter kind of answer, appealing to variation and selection, teleological? Teleological mechanisms are produced, all right, but are they not produced by nature blindly running? Is not the survival of what survives, over a long time, just a tautology? This is not the meaning of Darwinism, however. The meaning of Darwinism is that what survives has, in fact, survived through a constantly increased informational coupling—natural selection—within itself and with the environment.[20] The time scale involved is of a cosmological order of magnitude. It is a sheer and momentous fact about this one of the possible worlds. The earth must have possessed a pretty stable environment for all of that time, and this stability is now subject to our disturbance or destruction. I believe that it is, or should be, a spur to counterselection against such possibilities that we have so ancient and aristocratic a lineage. The story of evolution has that kind of merit for our own human teleology. But the question persists: is our kind of teleology an extension of of nature's, or should we go into a fit of alienation about it? (Some psychologists think we should avoid the trouble altogether by eschewing anthropomorphism *even* in the study of man.) One may say that if the cell did *not* get evolved with the order of the world copied in it, it would not exist; and, in fact, it does exist—but, for heaven's sake, not by *design*. I think that the negative is wrong, because the design did exist before the cell as the very form of the environment. It is not variation but selection that produces new order in evolution; it is the relatively stable order of the environment that selects, and the free energy of the environment that fuels, the selection.

Still, the cell has no *design* to copy its environment, to rebuild itself for a steadier life. This brings us to the more complex organisms, in which the maternal heat engine has built components that do in turn design to copy the environment. Consider, for example, the antigen antibody mechanisms, which certify some proteins as friends and all others as enemies. Here we see a new sort of teleology emerging. Any suitable complex system will be informed by its environment when subject to selection pressures, but only special systems will contain heat engines creating their own selection pressures. The difference between evolution and learning is that in evolution the work of adaptation is performed on the system from the outside. In learning, the work, or part of it, is done within the system. Evolution has now become functionalized; life has its own organs for seeking information.

To see the implications of this I want to go into one technicality of information theory that is a source of frequent puzzlement. It has to do with the difference between seeking food, for example, and seeking information. Let me start with a contrast between the information in a thousand copies of the same book and that in a thousand different books. Naïvely we would say that the second set contains a thousand times as much information as the former, whereas information theory says they contain the same amount. The meaning of information in the latter case is a little wider than in the former; what the naïve answer overlooks is that the thousand identical books contain more information *about each other,* and that makes up for the difference. Let us now look at the relation between a reader and the two sets of books. He will get more information if he reads the thousand different books, but on the other hand he does not have to read more than one of the other set. The other 999 are *redundant.* This means that he *saves* an enormous amount of information-gathering work by taking note of the redundancy. This elimination of redundancy does not mean, however, that he *has* more information than from one book; it means instead that he *needs* less. If, as we sometimes say, a raised eyebrow communicates volumes, this means only that the receiver of the signal already knows most of those volumes, minus a few essential bits.

Since information is equivalent to negative entropy, and thus to thermodynamic potential (at the rate of .7 kT ergs per bit[21]), we may also consider the situation thermodynamically. In reading and recording the thousand books, the reader stores more thermodynamic potential than from reading one of the thousand identical books, and a suitably informed (i.e., equally well-read) heat engine could extract this from him as work, ideally a thousand times as much in the one case as in the other. The mutual thermodynamic potential of two systems can be imputed to either of them, but this does not in general apply to three or more.[22] Thus for the case of the thousand identical books, the reader would yield up only one volume equivalent; the rest would come not from him, but from the other 999 volumes. I hasten to say that these examples are for analysis only, since the energies involved are quite trivial. In the microbiological realm, however, the triviality vanishes.

Let us now apply the analysis to different *kinds* of books. Some books reduce the redundancy of the world far more than others. And to a particular reader, some books are far more redundant than others, depending on the antecedent coupling between the reader and the world. There are obvious aesthetic criteria here.

Let us apply the analysis, finally, to yeast cells and one or two other things. A well-informed heat engine can extract more work from yeast cells as a result of possessing a yeast-cell description. If yeast cells are abundant, the gain will far outweigh the cost. On the other hand, it will not outweigh nature's cost, which has already been paid in the copying process, which explains the abundance of these otherwise rare entities. In this case, therefore, redundancy corresponds to a supply of thermodynamic potential.

But now let us suppose that something else is copied, such as the structure of a well-formed diamond crystal. There may now be an enormous reduction of the information needed to define the crystal, but this yields no gain of thermodynamic potential to the heat engine in burning the diamond.[23] The structural entropy of the crystal is already zero, so there can be no gain from having that structure copied in the heat engine. Other things in nature can be copied with *no* meaning to be attached to a thermodynamic potential available. The bee, by using polar coordinates, has achieved a nonredundant use of spatial order, and in that sense has copied the geometry of space. But the latter is not, to put it quaintly, combustible. In nature there are many sorts of order that, being law-like rather than products of history, are exempt from any meaningful thermodynamic bookkeeping. Things so copied yield no thermodynamic potential; they are, so to speak, at the temperature of the Platonic world—absolute zero.

My point in all this exemplification is to set forth a sense in which the form of things can become a cause in the behavior of what is living, and in a sense different from the meaning of "cause" in the normal language of physics. The meaning, as well as the measure, of this information does not lie merely in the transfer of negative entropy to what evolves or learns, but in the reduction of redundancy. The reduction of redundancy is very much like the thing we talk about, however, when in human learning we identify *abstractions*.[24] When I said that the design of living things is in the order of the surrounding world, I meant to impute to living things something of this character of having and operating with abstractions—meaningful functions or habits. To *seek* information, finally, means to seek to reduce redundancy—to have in that sense better abstractions and to be guided by them.

In the evolution of mechanisms that have the function of seeking novelty and abstracting order, I think I am putting forth a pretty honest sort of teleology. I like the moral implications, which I think Aristotle missed but which other philosophers, without benefit of evolutionary exegesis, have with some frequency arrived at. A twentieth-century addition is Santayana: "We exist through form, and the love of form is our whole inspiration."[25] But ethics is another matter, which I am not now prepared to face. Instead, I shall come back to my major point with a simple example. I want to consider a case where redundancy is reduced, but where the system in question is not a physical system at all. It is the number system. We can talk about the mathematician in thermodynamic terms, but not the Platonic realm of number. Yet if we know that the mathematician has set himself to write down primes in succession, we can predict what sequence of marks he will make. This prediction is possible, in turn, only if *we* know the sequence of primes, or find them out concurrently. Relative to certain facts of the Platonic order, then, the mathematician's organism has a greater free energy than otherwise. Thus one uncovers a final cause. The mathematician has not reduced the entropy of the Platonic realm, because it had none to begin with. But *it* has reduced his entropy; in Platonic language he has become one of its participants.

You object that he did not become informed by inspection of the Platonic realm, but by recursive operations carried out with pencil and paper; these were the causes that informed him, all good and physical. I agree. But you would certainly be missing a good deal if you did not notice that he happened to be producing digits of the prime sequence, and not just random doodles. You might have explained the motions in an irrelevant fashion, but you would not have to explain how he happened to be writing the prime sequence. It takes a mathematician to catch a mathematician.

Let me push a little harder. One day we start receiving signals from a nearby star, and they come in clusters of pulses: first the natural numbers, then the triangulars and squares, and soon the primes. Next, by pedagogic induction, we get complementations, disjunctions, blanks, and the logical operators. Pretty soon we have a complete sort of mathematical language. Never mind what comes after that. We work at decipherment and get pretty excited, because it all seems to make sense. Test cases appear, and one of them translates as "every natural number can be represented in one way only as a product of powers of primes." All we know is that the pulses came from a spot in the sky, emitted perhaps by a large bag of cold hydrogen. If that kind of emitter is too unlikely, the *reason* for this unlikeliness is that we already recognize the fundamental theorem of arithmetic. The *fact* of the final cause is the *ground* for our inference that "they" exist. It seems to be a necessary condition for that particular code sequence to appear that every natural number can be written in only one way as a product of powers of primes.[26] Never mind what else we can infer, for example, about "them." Of course we might get some pulse sequences that decoded into false statements or nonsense. But this could happen only in a minority of cases, for otherwise we would not have been able to construct the code book in the first place. So my claim that a fact of number theory is a necessary condition for the occurrence of a set of otherwise very unlikely physical events has to be accepted on good, but merely statistical, evidence, like any scientific claim.

I may appear to be in trouble here with my logic. If an arithmetical truth p is said to be a necessary condition for its utterance "p," the same follows for any necessary proposition—it is trivially a necessary condition for any event that, in fact, happens. But this is obviously not the meaning of "necessary condition" here. The meaning is that a high preponderance of statements made in the sort of context assumed are true statements.

You may say that the whole sequence (assuming it occurs) has a perfectly good physical explanation of some kind. I would quickly agree, because I said at the outset that the teleology I am articulating emerges just out of the scientific view of nature, and imports no extras of the sort that vitalisms and idealisms have wanted to have need of. Even without extras, it seems to be the case that physical effects are not always adequately explained by physical causes (unless you want to expand physics to include arithmetic, and then, in turn, quite a few other disciplines. There might be some point in that, but it would hardly be an antiteleological point).

Notes

1. To expand the mechanical philosophy into a full-fledged metaphysics was not a choice accepted by many. The great figure is that of Thomas Hobbes, who laid the foundations of a science of human nature and of society in the framework of this metaphysics. Hobbes has the honor of being refuted by every im-

portant figure in the subsequent history of moral philosophy except Spinoza, who, without reservation, maps the Hobbesian philosophy of nature and human nature into his own ampler scheme of things. The central document is "On the Origin and Nature of the Affects," Book III of the *Ethics.*

2. The sense of this *telos* is strong in Hobbes's first law of human nature and Spinoza's *conatus,* analogical extensions of Galileo's conservation law of horizontal motion. There was, of course, nothing novel about the assertion that self-perpetuation or self-preservation is motivationally important. What is new is the context, in which self-preservation acquires the status of an explanatory law of all motivation rather than as itself a motive among others.

3. The passage is in *De Generatione et Corruptione,* 338ᵃ-338ᵇ. The discussion of physical necessity in Aristotle links it to the material cause. In cyclical motion, what is a necessary condition is also sufficient.

4. Uniform circular motion for heavenly bodies, radial motion for terrestrial bodies, growth and reproduction for living things, perception and motion for animals, exercise of reason for men are *teloi* in Aristotle's sense. A *telos* is not an end, either in the sense of a terminus or in the sense of a vital force tugging or pushing. Aristotle's final causes are morphic, but they are no more anthropomorphic than they are zoomorphic or astromorphic. The *kind* of *telos* is just the kind of state in which a thing tends to remain or the kind of behavior it tends to exhibit.

5. I agree with the position of Hilary Putnam, "It ain't necessarily so," *Journal of Philosophy,* Vol. LIX, No. 22 (October, 1962). Putnam is principally arguing that, relative to *no* body of knowledge, the entire distinction between the necessary and the contingent becomes unusable.

6. See, for example, the discussion of a possible, subquantum physics, sub-subquantum physics, etc., in David Bohm, *Causality and Chance in Modern Physics* (New York: Harper & Row, Publishers, Inc., 1961).

7. The famous statement is, of course, that of Laplace. See the first part of his *Philosophical Essay on Probabilities,* F. W. Truscott and F. L. Emory, trans. (New York: Dover Publications, Inc., 1952).

8. Spinoza's discussion of finite modes and their causality in Book III of the *Ethics* maps the implicit framework of mechanics into his metaphysics by the statement that God is the cause of a finite mode not unconditionally, but only insofar as God is affected by *another* finite mode. But again the metaphysics goes beyond the science, because God is the cause of the whole infinite chain of finite modes, insofar as He is affected by *other* infinite modes, etc. The only thing that God causes *unconditionally* is Himself as *causa sui.* Thus nothing is left open to accidental determination, and nothing except God is necessary *per se.*

9. Constraints are of the essence, for we cannot define a dynamical measure on an absolutely unrestricted phase space. Will we accept a finite universe? Will we allow universes of arbitrarily large energy per unit volume? For any manageable system of constraints there will be families of possible worlds without any probability comparison between them.

10. For a discussion of "projection" in quantum mechanics, see P. K. Feyerabend, "Problems of Microphysics," in R. G. Colodny, ed., *Frontiers of Science and Philosophy* (Pittsburgh: University of Pittsburgh Press, 1962). The proposition that measurements select is hard to avoid in quantum physics. All I am saying here is that although it is easy to avoid it in classical physics, one could *choose* to accept it there as well.

11. One has only to assign a suitable measure to make nonmixing initial conditions very probable.

12. D. Hawkins, *The Language of Nature*, Chap. 7 (San Francisco: W. H. Freeman & Company, 1963).

13. It might be considered as a sort of generic boundary condition. Cf. Adolf Grünbaum, "The Nature of Time," in R. G. Colodny, ed., *op. cit.*

14. Cf. Feyerabend, in *ibid.*, but also a concluding remark in Putnam, *op. cit.*

15. Quantum mechanics exploits another potential loophole in the scheme of classical necessities, which assumed corpuscles as given. It stipulates a hierarchy of corpuscles, purely atomic in character for interactions below a given threshold of energy, and essentially complex for interactions above that energy. Cf. V. F. Weisskopf, *Knowledge and Wonder* (New York: Doubleday & Company, Inc., 1962). The argument follows: by dimensional considerations from the uncertainty principle: $(\triangle p)^2 \cdot (\triangle x)^2 = \hbar$, where $\triangle p$ is momentum change and $\triangle x$ displacement. Then since $(\triangle p)^2 = 2m\triangle E$, the minimum energy change is related to the mass m and the radius $\triangle x = r$ of the system by $2m\triangle E = \hbar/r^2$.

16. See Bohm, *op. cit.* It was Bohm who first proved the logical possibility of a deterministic model.

17. One may treat this nonuniform distribution of energy as a mere fluctuation from a larger thermodynamical equilibrium. On the other hand, such an a priori framework is infinitely gratuitous. Why not start with the general, and nonequilibrium, conditions that in fact prevail? See Grünbaum, *op. cit.*, for a more careful argument.

18. Maxwell's demon discriminated between fast and slow molecules in a gas, thus raising the temperature on one side of a membrane at the expense of the other. This gave the demon face energy to live by, and constituted him a perpetual motion machine of the second kind. Maxwell did not know that it would cost the demon as much or more to make the discriminations as he regained in consequence.

19. Cf. Ernst Caspari, "On the Conceptual Basis of the Biological Sciences," in Colodny, *op. cit.*, for a careful statement of what "organ" and "function" mean.

20. The phrase "survival of the fittest" was the coinage of Herbert Spencer. Darwin thought the phrase very apt if its true meaning were understood. He also suggested that, if Spencer "had trained himself to observe more, even if at the expense, by the law of balancement, of some loss of thinking power, he would have been a wonderful man." *The Life and Letters of Charles Darwin*, Francis Darwin, ed., Vol. II (New York, 1887), pp. 229-230, 239.

21. The Boltzmann constant k, measured in ergs per degree Kelvin, is the constant relating energy and temperature, $1/2\ kT$ being the thermal energy of a system per degree of freedom. At a given temperature T, the free energy gain from one binary discrimination is $kT \log 2$, or $.7\ kT$.

22. If $H\ (ab)$ is the information needed to define the joint state of two systems a and b, and $H_a\ (b)$ the information needed to define the state of b when that of a is known, then we may write: $H\ (ab) = H\ (a) + H_a(b) = H(b) + H_b(a)$.

The decrease of needed information, as compared to two independent systems, is: $H(a) + H(b) - H(ab) = H(a) - H_b(a) = H(b) - H_a(b)$.

For three systems the gain is: $H(a) + H(b) + H(c) - H(abc) = H(b) + H(c) - H_a(b) - H_{ab}(c)$, etc.

23. If $H(a) = 0$, then $H_b(a) = 0$ too. On the other hand, of two diamond-

informed heat engines, either could extract more free energy from *the other* as a result of this informational bond between them than if one were uninformed. I believe that the phrase "well-informed heat engine" was first used by Jerome Rothstein.

24. Cf. W. Pitts and W. S. McCullock, "How We Know Auditory and Visual Forms," *Bulletin of Mathematical Biophysics,* Vol. IX (September, 1947), 127-147.

25. George Santayana, *Reason in Science* (New York: Collier Books, 1962), p. 63.

26. According to some conventionalist philosophers, my whole story is silly, for there is no reason to suppose that "they" would construct an arithmetic such as ours. If my story comes true, is that a refutation of conventionalism? If so, we might manage with intercultural rather than interstellar examples. If not, then what is it that conventionalism says?

PHILIP MORRISON
Cornell University

The Physics of the Large

Only on the basis of the spectroscopic investigation of stars and modern atomic physics, and only after well-founded opinions about the spatial order of the stellar universe had been derived by analysis of vast observational material, could the astronomers undertake to draw a picture, first of the inner constitution and then also of the temporal development, of stars. Cosmogony still remains a rather problematic enterprise.

—HERMANN WEYL, *Philosophy of Mathematics and Natural Science*

I. The Physics of the Large

It was the princely Gauss who first recognized that a measure of length might be hidden in physical but "empty" space. Euclid's geometry, taken over without question as the basis for the physics of the eighteenth century, was known to possess no such scale. Along with the smallest, even the largest of triangles must contain exactly two right angles. No scale need be marked on Euclidean diagrams. Yet the surface of our own earth is plainly non-Euclidean; the great triangle marked off by arcs from, say, Boston to Rome to the North Pole, of course, exceeds the Euclidean limit. Only the plausible act of imbedment of that spherical surface into the bland three-dimensional space of Euclid and Newton saves our geometrical naïveté.

But physical systems, distinct from mere space, bear always the mark of size. It is easy to see that a stationary teacup ten thousand miles across can never be built; its own self-gravity will crush it, whatever material be used. Nor can we conceive of a durable object of atomic scale, but of teacup shape; the quantum laws of motion will not admit it. We must, therefore, before entering upon the task of a scientific cosmology, first learn the physics of large scale. From this scrutiny not even time and space can be exempted a priori.

Astronomy is the paradigm of modern science. It was built upon the all but unquestioned—to Kant, even unquestionable—basis of Euclidean space and Newtonian time. The laboratory study of space over all the history of our science has moved us at most across one part in 10^8 of the region we know requires study; our knowledge of time from the clocks of the laboratory is still less complete. On this inadequate baseline do we presume to triangulate the possibly infinite depths of the space-time continuum. But now we do so in a far richer context. Astronomy is no longer so purely Newtonian; it is now seen to be contaminated at base by atoms and by their nuclei, by radioactive decay and by magnetic interactions, by relativistic transformation and even by biological evolution. It seems to me indispensable for any understanding of the issues of cosmology first to survey the physical circumstances of large volumes of space-time.

All that we knew until now of the wide universe must be carried to our senses or to their instrumental extensions across the barrier of the vacuum that surrounds the fringes of the atmosphere, well above the orbiting astronaut. It is not hard to prepare a short but exhaustive list of the channels by which such knowledge is gained:

electromagnetic radiation: The eye, the telescope, the radio dish, and all their rocket-borne cousins feed on this, which is by far the most abundant source of knowledge.

gravitational forces: Even buried deep in a mine, or inhabiting a perpetually foggy clime, the physicist could learn of the moon, the sun, their motions, and, in principle, the whole of the solar system and beyond, by watching with care the complex vibrations of a small weight hanging on a weak spring. These are the forces that raise the tides; their effects on laboratory instruments have, in fact, been deciphered, as a kind of exercise, to exhibit sun and moon. In the future it seems hopeful to exploit this channel very much more heavily; there may be gravitational waves of cosmic importance.

magneto-acoustics: Sound can hardly travel the near-vacuum of space. But space is not truly empty, and the interlinked motion of the electrons and ions of space, feeling each other out over great distances by virtue of magnetic and electrical fields, does constitute a kind of electromagnetic sound. It is not electromagnetic radiation itself, for matter must be present to allow this propagation; it is not ordinary sound, for there magnetism plays no role. Many geomagnetic phenomena, detected below the ionosphere by compass needles or electron amplifiers, reveal such waves. Some are believed to come from the sun; one day we may trace them even from the stars.

particles: The great meteorites that made the buried craters of our earth, and those visible on the moon, head the tally. The scale drops, through the meteorites of the museum, to dust, to gas atoms caught in the upper atmosphere, to the ions and electrons of the aurora-causing solar streams, to the whole range of the ions of the cosmic rays, up to single atoms that carry as much kinetic energy as a whole raindrop. Each scale tends to characterize a quite different set of processes. Finally, we may expect eventually an astronomy of cosmic neutrinos, those most penetrating and subtle of all particles, which are born in the depths of stars and move in undeviating rays to all of space.

From the data that these channels bring, and from the rich texture of interrelationships that these data bear to each other and to the corpus of theory and experiment of the terrestrial laboratory, all our knowledge of the large stems. Entry into space by instruments on rocket probes merely extends the use of these channels, and will give us samples of moon and planets. Manned voyages cannot promise much more. Only knowledge of life on another planet seems a major prize that space probes might win for us but that our present channels have so far not yielded. This is not to say that the new opportunities for using the familiar channels that space instruments provide will be only of small importance; they may be epochal, but they will exploit the same channels.

What is the map of this new world like? We live in a great wheel, the galaxy, whose cross section shines in the night sky like a knotted band. Galileo himself proved it to be starry. Today we know that it is in fact a giant disc of stars, about 10^{11} stars, of which prodigal number a billion or so are very like our own life-giving star. The stars whirl in near-Keplerian orbits through galactic space—the space itself holds gas, ions, dust, and is laced with magnetic fields. Beyond the disc there stretches a near-sphere of gas, with a few hundred clusters (each of a million stars or so) here and there in an otherwise nonluminous sphere, so that what the eye sees is the star-rich disc alone. The luminous stars form a complex population, but all of them are phenomena of gas. These great engines are made wholly of gases; no solids or liquids can withstand the high temperatures that produce the diagnostic glow. In such an engine, the pressure of hot gas holds the spinning sphere apart, against the untiring gravitational pull of all the mass seeking the center. The heat leaks out from center to surface, and the star shines. For masses like the sun, the internal heat must be dissipated in a geologic epoch or two, and the star would collapse. But instead the sun taps a new source of energy, thermonuclear energy, and the nuclear rearrangements, which convert simple hydrogen into more complex helium, keep the central gas hot and the star for the while stable. Thus the star evolves in time, as its thermonuclear furnace converts simple elements into more stable but more complex ones. This "element cookery" proceeds through many stages to span the periodic table. The stars and the chemical elements in this way have evolved together.

One simple but not self-evident fact of ordinary Newtonian dynamics —the virial theorem—governs the physiology of the star. A star, in which the inward pull of gravitation finds its match in the outward pressure of intensely hot gas, becomes hotter as it contracts. We are used to systems that contract on cooling, and the reverse is unfamiliar. To be sure, the star

Temperature in Degrees C.°	Nuclear Reaction	Typical Time
Below a few million	none; all star energy arises from slow gravitational contraction	up to 10^8 years
10-20 million	hydrogen burns, forming helium by direct proton fusion	about 10^{10} years
15-80 million	hydrogen forms helium by way of a catalytic carbon cycle	10^9 years
100-200 million	helium burning to form carbon, oxygen, etc.	millions of years
200-600 million	middle-range elements, especially the abundant ones near iron; equilibrium tendency dominant	under a million years
toward or above 10^3 million	the heavy elements, all the way up to uranium, by processes involving neutron capture	explosively fast (years to minutes)

FIG. 1.   A summary of the origin of elements. Nonthermal processes on stellar surfaces produce the lightest elements, which disappear rapidly by fusion within the star. This includes deuterium, lithium, and so on.

can contract only by giving off energy; so much remains true. But the lost energy, which *ceteris paribus* can come only at the expense of the heat energy, so that its loss would cool the star, is instead resupplied from the gravitational energy released as the star becomes denser. Contraction means, of course, that the mutual gravitational attraction of the aliquots of mass that make up the star becomes stronger, for on the average the masses are closer in the smaller state of the body. This implies a release of energy in "falling" in; it is this energy which compensates for that which is given off. Indeed, the virial theorem guarantees that the lost energy is always more than compensated. The denser a star becomes, the more energy it gives off, the hotter it gets; the cooler the star, the more extended and tenuous it would have to be. Only the entry of new forces, caused neither by thermal motion nor by gravitation, brings an end to this fundamental tendency to heat upon contraction.

A star is born, then, in slow condensation out of tenuous gas and dust. The gas must radiate away its internal heat energy if it is to condense; perhaps that emission is the role of the dust that we see associated with forming stars. The process is far from fully understood. Once condensed into a large, rather cool sphere, it continues to lose energy by radiation, to contract, and to heat centrally. As higher and higher central temperatures are reached, the nuclear fuels present in the central region burn element by element. The heavier the nucleus, the greater the Coulomb repulsion, which all nuclei bear toward each other because of their positive electrical charge. Low temperatures mean low thermal energies, and only the most perishable of nuclei, like deuterium, the H-bomb ingredient, can burn at temperatures below some ten millions of degrees C. Above that temperature, ordinary hydrogen—protons—can burn by a series of processes that become faster and faster as the temperature rises. At about one hundred million degrees, helium, the second most abundant element, begins to burn, and so on up the list of elements. By the time such central temperatures are reached, the rates of reaction are enormously heightened, and the time scale of stars that burn any fuel heavier than hydrogen is measured, not in billions of years, as is our sun, but by millions or even less. Such stars are therefore rare but spectacular, always bright, and often variable or explosive.

A star of modest mass will contract until it burns hydrogen. After a long while—one history of life on earth—the central fuel may be depleted. The slow contraction can begin again. But it stops when a new force comes into play, quantum forces that augment the ordinary gas pressure by adding a new kind of internal motion, a new resistance to collapse. This is nothing other than the apparent repulsion that keeps the electrons of the atoms themselves from collapsing into their positive cores; it is the consequence, perhaps the most important consequence, of the profound identity of all electrons in the quantum treatment named for Pauli. It turns out that there is a maximum load that even the quantum pressure can sustain, for the electrons which contribute most to the quantum pressure begin to move so rapidly that relativistic energies are reached, and then the pressure is inadequate to sustain the star, because of the slow linear rise of energy with momentum in the relativistic domain. The mass limit of a stable fuel-depleted sphere is reached with a mass of about Nm_H, with m_H the mass of the proton. Here N, a famous number first computed by Chandrasekhar, can be expressed as a combination of fundamental constants:

$$N = \left(\frac{\hbar c}{G\,m_H^2}\right)^{3/2} = 1.5 \times 10^{57}$$

where G is the constant of gravitation, $\hbar = h$ Planck's quantum of action divided by 2π, and the velocity of light. The mass Nm_H is about double the mass of the sun. Stars with only a few per cent of this mass, say, will merely condense and cool without ever reaching temperatures high enough to burn any nuclear fuel. Theirs is a brief, uneventful life. Stars with mass a little below the limiting mass will run through a few elements until the temperatures they can reach no longer permit burning any new type of fuel, and then they must settle down to the final state of changelessness, mere huge, gravity bound atoms. Stars above the limiting mass cannot enter the nirvana of finality without shedding mass; here there are many possibilities, but no sure knowledge. Some probably shed mass uniformly, spinning it off in huge, fiery pinwheels, while some may explode it off in one great cataclysm, the supernovae. We now know that matter which is too hot, beyond a few billions of degrees C., is indeed unstable, and must cool by the emission of neutrinos. Such cooling in the hot center of a star removes the support of the outside, and by radiating energy, and with it mass, also lessens the pull of the innermost core for the rest. The outer layers will expand, and if they contain burnt nuclear fuel, may explode. Stars with still higher masses, say above 100 times the limit of Chandrasekhar, are generally so unstable as never to condense permanently at all, but remain as tenuous clouds, here and there fitfully glowing and shifting, until the cloud has shed mass enough to begin a normal period of stellar evolution.

Surely no one will misunderstand the account of evolution of stars and of atoms given above. It is an account of the complex and wonderful

FIG. 2.   The Pleiades, familiar cluster of stars in the winter sky. Here they are seen through the one hundred-inch telescope, and the close interaction between these young, bright, blue stars and the gas around them is evident. These stars are in our own galactic neighborhood, about five hundred light years distant. They formed a hundred million years ago. (*Photograph from the Mount Wilson and Palomar Observatories.*)

tale no more complete than the story of the fourth, fifth, and sixth days of creation given in Genesis, though our story is closer to the sequence science can now establish.

Our story is still too atomic. Stars are generally discrete, but they are not lonely. The clouds from which they formed nucleated not merely lone stars but commonly the formation of double and multiple stars, of associations and open clusters of stars, like those that make up the half-dozen blue stars of Orion (probably all born out of a single expanding gas cloud whose motion they perpetuate) and the cluster of new stars we call the Pleiades (also close kin through gas). Larger clusters, like the globular cluster of Hercules—a couple hundred of the class are known in our galaxy —each contain millions of stars, all surely condensed from a vast sphere of gas very long ago.

All these objects are grouped in the interacting wheel and ball that is the galaxy. The age and nature of the stars are strongly correlated with their positions in the galaxy, fixed by the stage in galactic history when the particular star formed. Thus the very bright blue stars of today, which we know from the nature of their nuclear fuels could not possibly be older than mere millions of years, are all found in the disc of the galaxy itself, for today only there are found gas and dust in adequate density to allow stars to form. But old stars were formed everywhere, out to the fuzzy edge of the great sphere a hundred thousand light years across, condensing into globular clusters out of the great, bland, near-spherical mass of gas that our galaxy must have been in that dim past before its stars were born. So the galaxy, too, has evolved, though to trace the interplay among turbu-

Fig. 3.   The great globular cluster in the constellation Hercules. This spherical mass of some million giant stars lies somewhat outside the central region of our galaxy, some seventy-five times more distant than the Pleiades. These stars, in a region now devoid of gas and dust, formed some ten billions of years ago. (*Photograph from the Mount Wilson and Palomar Observatories.*)

Fig. 4.   NGC 4565 in Coma Berenices. A galaxy of the same general type as our own, viewed edgewise from the outside, as we view the Milky Way from within. It lies around fifteen million light years distant, hundreds of times further than any portion of our own galaxy. (*Photograph from the Mount Wilson and Palomar Observatories.*)

lent gas and magnetic fields, dark, cooling dust, the original endowment of spinning motion, and the over-all effects of gravitational attraction is more than we can now master. The progress of the atoms from hydrogen into the heavy metals has plainly gone on in the depths of stars. Nor has it been a single progression. The original material of the galaxy must have been all, or largely, pure hydrogen, for that simplest of atoms still outnumbers all the rest. The solid crust of the earth is built of atoms synthesized from hydrogen step after step in the alchemical furnace of long-dead stars that exploded to spew their produce out into the galaxy, from which once again it condensed into our sun by way of a nebular protosolar cloud from which the earth, too, was assembled.

The galaxy is an island in space, but it is not alone. Indeed, our galaxy belongs to an archipelago. Our twin lies there in the Andromeda direction, a great, spinning, spiral disc of stars, single and clustered, of dust, fields, and gas, very like our own home. A dozen or so islands make up the

FIG. 5. Our neighbor, the spiral galaxy M31 in the direction of Andromeda. Visible to the unaided eye in the summer sky, this object is judged to be very like our own galaxy in most properties. This photograph, taken with the forty-eight-inch camera on Palomar, shows the two small companion galaxies to M31, suggesting the intrinsic range of galaxies in size and in brightness. With all three of these, and with a dozen or so nearby galaxies, our galaxy forms a small cluster, the Local Group. M31 is around 1.5 million light years distant from our galaxy. (*Photograph from the Mount Wilson and Palomar Observatories.*)

FIG. 6. NGC 4486 in Virgo. A galaxy of a type very distinct from ours, a great, near-spherical mass of stars. The fuzzy spots that surround it are themselves globular clusters, similar to that shown in Fig. 3. This object is two or three times more distant than the flat spiral of Fig. 4. (*Photograph from the Mount Wilson and Palomar Observatories.*)

local group, galaxies differing from one another as do the three visible in the photo of M31 (Figure 5). And our archipelago stands in the ocean of space, not lonely at all, but surrounded, as far as we have seen, by cluster after cluster of galaxies, with single galaxies strewn in the field, sparsely arranged in that vast ocean so that clusters of galaxies stand about ten times as far apart as their own diameters, with some few single field galaxies in between. Clusters may hold from a dozen up to thousands of galaxies, always in the wide variety of forms that we can classify well. As for the evolution of galaxies and clusters of galaxies—for surely they, too, evolve—we have little to say. Here we remain still in the stage of taxonomy; the dynamics is too complex to have found a solution. Some galaxies are near-spheres, all halo; some, all disc; some, irregular mixtures of these two broadly defined elements. All are groupings of stars whose single lives we can go a long way to describe. As far as we can see, with the telescope or with our still more penetrating theories, the universe has a profound, but complex, uniformity. We see nowhere any region where the physics of the

FIG. 7.   A cluster of galaxies in the direction of the constellation Hercules. Many different types of galaxies can be seen in this one cluster. The scale is obviously enormous; these scattered patches are full-sized galaxies, distant from us by hundreds of millions of light years. (*Photograph from the Mount Wilson and Palomar Observatories.*)

large conflicts flatly with the physics of our laboratory. The structure of atoms and nuclei, the dynamics of Euler and of special relativity, the radiation theory of Dirac, the gravitation of Newton, the familiar motion of charges in magnetic and electric fields, these define the physics of the large. Many processes are still unexplained, but none seems inexplicable in familiar laboratory terms. The world is one great Cavendish.

There is one exception. We know now some hundreds of galaxies, distant and nearby, in which a gigantic explosion has taken place, an explosion of a sort that bears to the catastrophe of the collapse of a single star into a supernova about the relationship that the hydrogen bomb bears to a 500-pound bomb of Billy Mitchell's day. These prodigious events are not

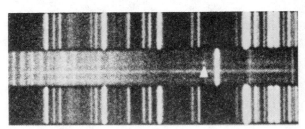

Fig. 8. A cluster of galaxies in Bootes. One of these—the marked object—is a strong radio source, and its unusual properties induced a successful effort to measure its distance from the spectrum shown. The Doppler shift amounts to 46 per cent, evidence that this is the most distant object that we have definitely identified, distant by five to eight billions of light years. (*Photograph from the Mount Wilson and Palomar Observatories.*)

understood. But it appears possible that they may have something to do with the collapse of a mass of gas millions of times a star's mass into domains of density in which Newton's laws of gravitation no longer describe phenomena well. It may be that on this unprecedented scale we are witnessing the first phenomenon not merely affected by, but dominated by, Einstein's view of gravity, a drastic modification of the local nature of

Fig. 9.   The two extreme types of star populations. Their nature and relations are the subject matter of present theories of the evolution of galaxies. Many intermediate populations also occur. *Left:* Andromeda nebula shows giant and supergiant stars of Population I in the spiral arms. The hazy patch at the upper left is composed of unresolved Population II stars. *Right:* NGC 205, companion of the Andromeda nebula, shows stars of Population II. The very bright, uniformly distributed stars in both pictures are foreground stars belonging in our own Milky Way system. In Figs. 2 and 3 are shown examples of these populations within our own galaxy. (*Photographs from the Mount Wilson and Palomar Observatories.*)

space and time. It is too soon to be sure, but it is valuable to recognize that once the negative gravitational potential energy of any mass becomes about equal in magnitude to its original inertial mass when spread at low density over much space, space-time must be re-examined, and no longer assumed Euclidean. The relationship is to be written:

$$\frac{GM^2}{R} = Mc^2$$

For $M =$ a stellar mass, the radius r is only a kilometer or two. Such densities have long been anticipated deep in the core of an exploding star, but there the physical environment has become so bizarre—rather like the interior of a compressed nucleus—that we could find no easy chain of argument. But if we imagine a really great mass, with M some 10^8 solar masses merely brought to a familiar density like that of water, where no strange nuclear phenomena can intervene, we learn that the above relation is satisfied. Such a ball of fluid would fill the orbital circle of Jupiter. It is possible that the events described will yet give us a means to study the still uncertain predictions of general relativity.

With the proviso just implied, physics in the large turns out to be only an appropriate extension of the laboratory. The familiars of the lab, the atoms, the constants G, c, h, m_H, and so on control the scene. As far as we can look, we see the universe about as it is closer to home, various, complex, diverse, and yet in its essence uniform; it is changing and evolving, but along paths whose beginnings we know, and whose routes we can hope one day to follow. There is no edge, no end, no sea change.

There are larger issues, issues that lie behind the easy physics of the laboratory, that seem to involve us in a new way in the depths of space and time, issues that we cannot find well studied in the lab. It is not hard to list three:

1. The universe as we see it, here or afar, is electrically neutral. The total charge, which as far as we know must remain unchanged in all processes, is zero, or close to it. But the number of protons and neutrons, the so-called baryon number, which stands on the same footing as charge in our theories of the fundamental particles, is, of course, far from zero. We can imagine systems in which it is close to zero. In such systems, the number of nucleons might be as we find it, but for each nucleon there would be somewhere an antinucleon, just as in the charge analogue we find both positive protons and negative electrons. The same question could be put: where is the antimatter? Is there none of it? Or is it in some remote place? We can be sure that it must be segregated, for otherwise we could not stably exist; annihilation would claim all the world. Nor do we see the products of annihilation. Here we are brought against the unstated problem of the initial circumstances of the universe.

2. A similar question arises when we seek to understand the anisotropy of the course of events: the decay of radium, the cooling of a star, the ex-

pansion of a gas. All these happen far more often than do their inverse processes, yet the laws of laboratory physics admit both directions equally. To be sure, the statistical interpretation of the laws of thermodynamics has explained the matter in part. But there remains the prior fact that the universe is not in equilibrium, that it was stocked with free energy long ago, and that this asymmetrical way of beginning still affects all processes. The great world has subtly but deeply biased the laboratory results.

3. Finally, there remains even in classical mechanics, free of particle physics and even of statistics, the effect of the remote world, for there must be an inertial frame in which the laws of mechanics hold. Long since we learned that there is no ether to mark space, and that earth, sun, galaxy cannot provide us with a sure inertial frame. Yet mechanics works well; the Foucault pendulum reveals the selfsame rotation speed as the transit telescopes describe. It was Mach who sought to found this result on the direct material presence of the remote universe; not absolute space, he said, but matter must define the inertial frame. If he is right, and the principle he uses is persuasive, how does this happen? Here too we are unsure; general relativity seems to contain the answer, though the issue is not fully settled. But certainly we have not solved it within the laboratory.

We will close with a very much simplified account of the way in which it appears—for rather deep technical reasons, it is not quite sure—that the insight of Mach is fulfilled by the subtle gravitational effects of the most remote matter. We shall assume that gravitational forces are genuine field phenomena, described by Einstein's famous tensor equations. The inertial force on any mass, which is its response to local acceleration relative to the bulk of the matter in the universe, is induced by the summed gravitational field of all the matter.

The inertial force arises, of course, when the sample nearby is set into accelerated motion with respect to the distant galaxies. Since in the view of Mach only relative motion can have meaning, this is equivalent to watching the motion of our sample were all the galaxies to begin a sudden acceleration as we watch. The gravitational field they produce here seizes the body; it thus resists its acceleration viewed from the other frame of reference. The portion of the gravitational field here effective is not, of course, the familiar static inverse-square field of Newton; rather, it is that portion of the field of a gravitating mass that is proportional to the acceleration of the mass. Such a portion is indeed predicted by the field equations, in close analogy to the more familiar electromagnetic radiation described by Maxwell's equations. The tensor equations are more complex than the vector fields of Maxwell, but in the present problem this added complication turns out to provide exactly the correct behavior to agree with the laws of Newton for low velocities.

Just as in the case of electromagnetism, the relevant quantity is the time derivative of the potential. We shall approximate crudely throughout, ignoring numerical factors not much different from unity. We begin by quoting the result of solving the field equations, namely, that a body of inertial mass M_I set into accelerated motion with acceleration a produces, at a distance of r centimeters, a field of force given closely enough by:

$$F = \frac{GM_I}{rc^2}\, a$$

where G is the familiar gravitational constant that occurs in the inverse-square law formula of Newton, which serves essentially to measure how much gravitational "charge" is associated with a unit inertial mass. Here c is as usual the velocity of light, which for gravitation too is the ultimate propagation speed. If a body that feels such a field strength has mass m_I, the force it experiences is just

$$F = m_I f = m_I \frac{GM_I}{rc^2}\, a$$

In our approximation, we need now to add up the contribution from every mass in all the universe, to form the grand sum

$$\sum \frac{M_I}{r}$$

where the sum includes all masses everywhere. Setting this equal to the force of inertia required in Newton's second law of motion, we have:

$$F = m_I a = m_I a \sum \frac{GM_I}{rc^2}$$

or

$$\sum \frac{M_I}{r} = \frac{c^2}{G} = \frac{(3 \cdot 10^{10})^2}{7 \cdot 10^{-8}}\, \frac{\text{g}}{\text{cm}} \cong 10^{29}\, \frac{\text{g}}{\text{cm}}$$

It is most instructive to see how much of the sum is contributed by masses close at hand. We tabulate the effect of various known masses, estimating how large a mass might be readily introduced or removed in a laboratory experiment.

Relative Contributions to Inertia

	A laboratory mass	The earth	The sun	The galaxy	The universe
Mass in grams	10^9	10^{26}	10^{33}	10^{44}	—
Distance in cm	10^2	10^9	10^{13}	10^{22}	—
Term M_I/r in g/cm	10^7	10^{17}	10^{20}	10^{22}	10^{29}
Relative contribution	10^{-22}	10^{-12}	10^{-9}	10^{-7}	1

This is the startling result: no measurable effect can be expected from anything but the distant galaxies, whose distribution we can never change, whose number is so great that their real graininess is most probably smoothed out beyond detection.

It is this result which explains how it can be that the inertial force, which appears to us fixed, intrinsic, local, is indeed the summed effect of the remotest matter. For it is beyond our powers—or at least fiercely challenging—to induce a measurable change in the value of the sum. Yet only some sort of external change can distinguish the intrinsic from the externally conditioned.

We observe the galaxies scattered like chaff to the limits of our present telescopes. Their mean matter density can be estimated, but we cannot easily detect matter that does not glow. Cold gas or dust, possibly chunks of rock, even neutrinos, would escape the census. Thus the mean density of space in the large can now be given only a lower limit. With that limit, perhaps 10^{-30} or 10^{-31} grams/cc, the visible world within the grasp of our best instruments contributes a small, but appreciable, fraction to the sum, say about a part in a thousand at least. Plausible upper limits suggest that we can now at most see that part of the universe which contributes a tenth of the total inertia-inducing field. It is these arguments which bolster the view that we do know at least a fair sample of the whole.

Finally, if the universe were infinite, the sum would, of course, diverge, for we can sum by using spheres of density ρ_{ave}. Then clearly successive shells grow in mass like $\rho_{ave}R^2$, while their contribution per unit mass falls only by $1/R$. Such divergence is characteristic of infinite universes. The Draconic solution is to bound the universe, or even to let its mean density, paradoxically, fall to zero, a suggestion, generations old, of Charlier. Today we have another way out, far more acceptable; the distant matter moves so fast in the universal expansion that its field contributions tend to vanish. The field, so to speak, does not reach us in time, since all effects move only with the speed c, which is opposed by comparable speeds of recession. We can describe this by saying that the equivalent contributing universe is bounded of radius R, where R is cT, and T is the characteristic time of universal expansion. How conformable this is to the great facts of the world is the topic of the second half of this essay.

II. First and Last Things

The extension of physics into the large has moved from scale to scale, leaving us finally with an incredibly rich and yet somehow unsurprising region of space and time to survey. We are left with those questions that demand more than any set of observations, that demand at a minimum the imposition of boundary conditions, either because we discover true boundaries in space or in time, or in lieu of an infinite set of observations. This subject, which transcends laboratory physics and its extrapolations by at least one sweeping assumption, is called cosmology.

Most of what we will say is held in the sway of a far-reaching but irresistible assumption, the very assumption of Copernicus: that the world we see, however grand in extent, is a representative sample of the whole, conceivably exhausting it. This is often called the cosmological principle: the uniformity of the universe in space. The hierarchical ordering of the stars into clusters, populations, galaxies, and clusters of galaxies at least gives us the sign that no sample which does not include clusters of galaxies can have plumbed the full range of phenomena. For half a century there has been the skeptical cosmology of Charlier, which holds that there is an infinite material hierarchy, which might even have a zero net density, space being almost everywhere empty, in the set-theoretic sense. Superclusters of clusters have indeed been suggested, but they do not appear well marked, if in fact they exist. But an unending hierarchy of matter in space so plainly escapes our theories that this scheme is always tentatively rejected. (Observed expansion after all takes away its motive: to avoid the divergences of light from distant sources.)

Between 1920 and 1935 or so there became empirically established a regularity of the greatest importance, large-scale motions in the world of the galaxies of a peculiarly simple sort, given the name "expanding universe." The nature of this ordered motion, with the relativistic physics of the laboratory, lends reassurance that we have indeed seen a sample large enough to generalize. First, a remark is needed on the measurements of distance that we can make in the realm of the astronomically large. Here are the methods now used, more or less independent:

1. Radar. This, the only method not wholly passive, is in principle powerful enough for all purposes, and has the clearest conceptual base. In practice, it will for the foreseeable future span only the solar system.
2. Triangulation. Leaning more on geometry, using just optics as radar does, but now passively and statically, this method reaches a few thousand stars.
3. Apparent angular motion, plus a physical measure of true velocity. This scheme enables the measure of distance to any changing phenomenon whose changes we can see geometrically and whose nature we can surmise. The Doppler shift in optical frequencies, the orbital velocity of gravitationally bound masses, and perhaps other velocities, can be measured or inferred. Then if apparent motions be seen as well, true distance can be computed. This has been applied widely in the galaxy and even beyond, to the great distant explosions. By its nature, it requires rather special objects.
4. Brightness. Here again optics and geometry combine to allow distance estimates to any object whose intrinsic light output we can estimate by analogy or by understanding of its true local nature. This works even at the very greatest distances, but never very securely.
5. Size. Again, a known object can be ranged by its apparent size, subject to similar assumptions about optics and geometry.

Applying these means, really mainly (4) and (5), to the galaxies, first Slipher, then Hubble found that the more distant a galaxy, the faster its Doppler-measured outward radial velocity. In Figure 10 is plotted the typical result of this work, a dominating correlation between distance (always as judged by some indirect method, of course) and "red shift." The effect is by no means weak. Of course nearby objects reveal no such correlation, for the motions implied by gravitational interaction on the scale of galaxies and of small clusters of galaxies, like the one we dwell in, amount to some one or two hundred kilometers per second. The slope of the mean curve amounts (by 1962 results) to some twenty-five or thirty kilometers/ second per million light years of distance. The Virgo cluster, nearest of the large clusters of galaxies, shows a typical red shift of about 0.004, which implies $v_r \cong 1100$ km/sec, and distance near 40×10^6 light years. Red shifts have been seen up to 0.46, which implies a distance of 5×10^9 light years, and a recessional speed of 140,000 km/sec. Much effort has been spent on improving the knowledge of the red shift function, to establish the nature of the deviation away from the simplest approximation of a straight-line relationship, but no agreement on the facts is yet at hand. Of course there is a presumption of a plateau at $v \rightarrow c$. Other features of the shape are predicted by various cosmological theories.

The accidents of history and of phrase have somewhat confused the correlation between distance and speed, which we call the red shift. Its existence was first predicted by the earlier studies made in the 1920's of the consequences of Einstein's theory of gravitation applied to the universe as a whole. Its discovery was therefore viewed as the most spectacular of the triumphs of that theory, which dominated a couple of decades of cosmological thought. But it was shown very beautifully in 1932 by Milne, McCrea, and McVittie that the "expanding universe" is contained in any effort to describe the cosmological problem in gravitational terms, even using pure Newtonian theory. (Some writers in the philosophical literature have confused the red shift we are here describing with another red shift predicted by Einstein, that of light reaching us from a stellar atmosphere, red shifted by the change in gravitational potential. This latter effect, too, can be derived from much simpler theories than the full field equations. It has by now received direct laboratory confirmation.)

The question of the reality of the recessional speeds is often raised. If the issue is that of a physical reddening of light in space, without geometrical meaning, then one can say only that no proposal which agrees with laboratory experience, and which neither blurs the reddened images by scattering nor deviates from the well-established frequency independence of the Doppler ratio, has been suggested. It is, of course, not inconceivable that such a process exists, but we do not know of any. If the issue is that of the naïve kinematic interpretation of the motions, the matter is more subtle. The theories of non-Euclidean geometry do not truly imply the naïve Euclidean limit, but they come operationally to the same thing. For example, if one waits long enough, a red shifted galaxy will, in certain circumstances, not only redden more deeply but shrink in size. Of course in regions of stronger space curvature its image might begin to grow again in size, but the non-Euclidean implications would by then be clear.

The most obvious source of the non-Newtonian properties of the expansion is the delay in time between emission and detection of light or radio signal. If we see a galaxy that we regard as billions of light years away, we must admit that the light we observe was emitted into free space billions of years ago. We see a world *picture* at a time that we regard as the local present; a world *map* could be drawn, with the time coordinate at every point given a fixed equal value. This last step demands that there be in fact a single universal time stream to which every galaxy can refer; its existence, put into the formal language of field theory, is called the Weyl postulate, and is by no means self-evident. Again naïvely, we can discuss the ages of objects that we infer from various physical models, and compare those values with the present red shift observations.

We can list a few measured ages all within our own galaxy. It ought to be clear that these measures are always very much dependent on physical theories of system change, on models of stars and of stellar evolution:

	Age in billions of years
The radioactive elements of the solar system	5-10
The sun	4-15
Stars near the sun	12-15
Stars in certain old clusters	up to 25

It is a moot point whether these ages scatter because of deficiencies in the measures and models or for intrinsic reasons. None of them is extremely secure, but taken as a whole they suggest that some ten billion years must be allotted to the history of our galaxy. Now the red shift can be written:

$$\frac{-\triangle v}{v} = \frac{v_r}{c} = +\frac{Hr}{c} = +\frac{r}{cT}; H = \frac{1}{T}$$

where H is the Hubble constant, and T can be given an obvious, rather too dramatic name, "the age of the universe." Naïve extrapolation of the red shift law would indeed tell us that looking at objects with a red shift $\triangle v/v \rightarrow 1$ would mean looking back to the beginning. No theory is, of course, quite so crude as to make such a strong extrapolation of the simple law. But the approximate agreement between the Hubble constant and the measured age has always meant to the workers in the field that the distance-motion correlation was indeed a general phenomenon of deep meaning, and not a mere happenstance of the local materials. The galaxies occupy a sphere whose radius roughly doubles in each Hubble time, and the age, either of all the galaxies, as some think, or at least of a typical one—our own—is of the same magnitude as the time T. The characteristic time of the universe is T, and its characteristic space dimension (which might perhaps be a radius of curvature) is some ten billion light years.

The universe has, therefore, a large-scale structure, a correlation between position and motion, which was surely unexpected. It is this correlation that lies at the bottom of all theories of cosmology, that is, of the universe as a whole. It is evident that this phenomenon, the grandest in scale of any we know, runs quite contrary to the notion of the equivalence of inertial frames, for an observer who moves, not with the earth and sun, but instead with a cosmic ray particle near the earth, will see no broadly isotropic and uniform surroundings as he looks out, but rather, looking forward, a world that seems hot and blue and, looking behind him, cold and red. There is indeed a preferred inertial frame: it is that frame at each position in space in which the large-scale motions appear isotropic. It is the averaged frame of the galaxies and their clusters.

The very idea of homogeneity, the extended Copernican assumption, taken with isotropy, fits well into what we see. This is true even in the approximation of Newtonian physics. Consider an observer who looks out at the distant world, which we will imagine smoothed out so that the mean motion of the galaxies is all he sees. Now he can label all the moving points he sees by a vector field (Figure 10). Suppose he looks from a point

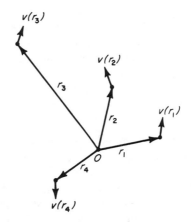

FIG. 10. An arbitrary observer at O looks out into the smoothed field of the galaxies. At every point marked by a distance vector r, he sees a particular velocity $v(r)$.

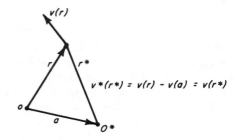

Fig. 11.   Construction showing the form imposed on the velocity-distance relationship by the requirement of uniformity. Two observers, at points O and O^*, moving with the local matter, must agree on the world picture of the motions they see.

O to see the motion of the world fluid at r. Another observer at a remote point O^* also observes the same physical object. We will assume Euclidean geometry for vector addition, Newtonian "absolute" time, and the key assumption, the uniformity of the world in the large. Then the view from O and from O^* must lead to the same over-all picture. Take the vector from O to O^* as a. We have:

$$r = a + r^*$$

and if v^* is the velocity seen at O^*, then

$$v^*(r^*) = v^*(r-a) = v(r) - v(a)$$

For uniformity of conclusion, the function v^* must depend on its argument r^* exactly as the function v depends upon its argument r. Only so can the two observers come to the same world picture. Then it follows that:

$$v^*(r^*) = v(r-a) = v(r) - v(a)$$

for all r and a. Then the function v *must* be a linear vector function of its argument. If we require also that the world be isotropic, the only multiplier of r can be a scalar F. This scalar can, of course, depend upon the epoch of the universe, the time t. Finally, then:

$$v(r) = F(t)r$$

and we are left with the red shift correlation in its simplest form. Only the assumptions of uniformity and isotropy have entered in an essential way. Relativity, while necessary because of our knowledge of the failure of Newtonian spacetime, is not necessary for the red shift law. That law is a consequence of the idea of Copernicus: we are not privileged observers.

The simplest solution still is to place $F(t)$ equal to zero. Then the universe is without a correlated motion. It need not be quite static; it might resemble a gas in equilibrium, with a wide but isotropic velocity spread, the same at each point. That is perhaps the simplest motion to be expected, short of universal rest. It is not the case.

The expansion that we observe—of course without any necessity to impute a center—is not quite without physical harbingers. One might put it that the function F must be different from zero, unless indeed the universe is very young, that is to say, unless the laws were greatly different at a time not much more than T years back. For if the universe were much older than T, and if there were no expansion, then the distant parts of space would contribute much radiation to our position. The night sky could not be dark. This argument, called Olber's paradox, and as old as the eighteenth century, is weighty. The point is that if time is infinite (or large compared with T), if the material universe is infinite (or large compared with cT), there is no way to prevent the remote parts of the world from having a dominant influence at every point. One must find a cutoff; the paths of modifying the laws of physics, or of a finite (or short) time scale, are not only disagreeable, but are unnecessary, since the expansion provides an empirically founded cutoff. Distant masses move so fast that their radiations become Doppler shifted out of any observable range, and their local events appear to us, at least in relativity theories, to pass so slowly as to lead to no physical information here; they can affect us, if at all, only in the subtle ways that make possible the mechanical definition of an inertial frame, for their aggregate masses would appear great, even dominant. And as we saw at the end of Sec. I, their contribution to the inertial force must be made finite.

It is within this qualitative framework that the technical literature of cosmological theory has grown up. It proceeds from some basis of mechanical and optical law, Newtonian, relativistic, or even other and new versions, adds the assumptions of uniformity in some degree, and derives classes of models to compare with experiment.

The elementary Newtonian argument just given allows us to characterize a cosmology by the single scalar function $F(t)$, which plainly describes the large-scale correlations of a universe that can move as it will. It can expand, contract, oscillate, mix all those motions, or even stand still. Such motions are imputed in a normal Euclidean space to a quasi-fluid substratum, which in the real world would have to be identified with galaxies and their clusters. Plainly this theory is still only kinematical. To adjoin physics to this geometry, one ought to add the Newtonian laws of motion, and examine the effects of the only large-scale force we know to exist, Newtonian gravity. This has been done, and the functions $F(t)$ determined as solutions of the appropriate equations of motion. A variety of runs of the function $F(t)$, which describes the universal motions, appears, choice among them being dependent upon the assumed mean density of gravitating matter and upon the initial conditions of the universal motion. Once the question has been put about the function F, the problem is on all fours with that of any other Newtonian motion. In such a model, the universe is a great gravitating gas, with a uniform simple motion either of an indefinitely expanding or of a bounded, and hence oscillatory, sort. The work of Einstein need play no part.

The history of the subject is very different. Exactly this problem was put and solved by Einstein toward the end of the First World War, using not Newton's laws, but his own new relativistic theory of gravitation and its relation to geometry. The expanding universe soon appeared as a natural possibility, and the marvelous verification of that prediction by Hubble and the others in the 'twenties seemed to confirm the grandest consequences of Einstein's difficult theory. But the simple classical approach of Milne and McCrea in 1934, which we have just sketched, demonstrates that the notion of the expanding universe is not at all a property of general relativity and its more mysterious curved spaces.

Still, physics is not Newtonian any more. When we can see Doppler shifts of nearly 50 per cent, we cannot rest content with the low-velocity form of dynamics. Exactly similar programs can be carried out, though with much greater conceptual and mathematical generality—and complexity—in the framework of relativity, with the limiting velocity of light in free space as the key idea.

A similar kinematical study can be made, using as fundamental measuring means a conceptually clear, if impractical, method of gaining knowledge. Instead of imputing to the observers at O and O^* the knowledge of absolute time and distance, it is more plausible to let them work, say, by means of radar pulses that enable them to gauge their surroundings in a physically reasonable way. It was first Milne, and later Robertson and, independently, Walker, who carried out such a scheme. The result of their work, which leans upon powerful symmetry arguments of a group theoretic kind, is to present a quite general characterization of the geometry of *any* cosmological theory that remains consistent with the main ideas of relativity, but not necessarily with its theory of gravity. If observers, who move with the local matter, see events in a smooth and one-valued, four-dimensional world to which they assign local comoving space and time coordinates r and t, they will find themselves with a local interval ds between events, such that the minimum interval between cause and effect, carried say by light signals, has $ds = 0$. The only form of the interval that satisfies the conditions of over-all uniformity and isotropy, and for which a postulate of uniqueness of velocity at any point obtains, is the Robertson-Walker form:

$$ds^2 = c^2 dt^2 - \frac{R^2(t)dr^2}{(1 + kr^2/4r_0^2)^2}$$

where $R(t)$ is the relativistic counterpart of the function $F(t)$, describing the large-scale motions. The denominator represents the possibility, absent, of course, in Euclidean space, of uniformly curved space of arbitrary "radius of curvature" r_0, which may be closed like a sphere, open like a hyperboloid, or flat. These cases are characterized by, respectively, $k = +1$, $- 1$, or 0. (These possibilities are physically the same as the bounded, unbounded, or limiting motions of the external matter in the Newtonian version described above.) A postulate, called after Weyl, is needed. It assumes that the motions stream smoothly, through one singularity at most in the past, so that t defines a single cosmic time, valid everywhere.

The task remains, as before, to give a dynamical theory of the function $R(t)$. Here at last the equations of Einstein's gravitational field theory, sometimes with various modifications, are applied. A whole branch of theory now exists for this study. Motions and curvatures of all sorts, and even topological modifications—partly disjoint universes of one or another kind—have been examined, and a few can be compared with the motion and statistics of the galaxies. There is still no strong evidence for or against any of a wide range of plausible schemes. It should be noted that curved and closed spaces, which do sound strange, are in some physical sense not difficult to compass under intuition. For the straight line, the geodesic is defined by the ray of light. A closed space is merely one in which the matter distribution acts gravitationally on light in such a way that the light ray can never leave some finite volume, but may circulate for a long time, making multiple images, in principle, for some observers. There is, of course, no boundary. The course of various light signals in time and space, while complicated in some cases, is not beyond a plausible qualitative description, unfortunately often omitted in the rather austerely mathematical tradition of writers in the field.

There are other cosmologies. Perhaps the most original is the so-called steady state cosmology, which goes Copernicus one better. In it one assumes that our vantage point is without distinction not only in space, but also in time. In our time the universe looks no different, it is postulated, than it looked a hundred Hubble times ago, or will a hundred from now. The function $\dot{R}(t)/R(t)$ is an observable (just Hubble's constant) and must in this theory be set equal to a constant, therefore independent of time. The line element can then be only

$$ds^2 = c^2 dt^2 - \exp \frac{2t}{T} \, dr^2$$

The dynamics of this theory are unknown, and require some more or less serious modification of Einstein's field equations.

It is physically clear that this steady state theory demands the creation of matter, for matter must continually expand out of sight, and yet the observable mean density must be constant in time. Creation of matter by the formation of new nucleons in space, either out of nothing or out of some field invented *ad hoc*, is thus indispensable. Since either (1) creation of matter, or (2) infinite cyclical passage through bizarre and singular conditions, or (3) single passage after an infinite initial delay, or (4) a deep-going modification of the idea of time is required by *any* cosmology that must deal with the problem of matter of finite age in a time that is putatively infinite, the steady state proponents (of whom I am one) are not greatly embarrassed by having made explicit and continuous a process that is either implicit or catastrophic in other theories. The theory is plainly vulnerable to experimental attack, for according to such a theory, matter can be found of all ages; while the older matter is widely diffused, newer matter still remains nearby. Increased knowledge of evolution of galaxies can settle the issue. It is possible that counts of distant objects, now made best by radio means, will show that remote objects look very different from those nearby because they are seen at much earlier stages. This is sure in most evolving cosmologies, such as the conventional relativistic ones, but cannot be true in the steady state theory. Present radio indications are against the steady state, but the lack of any certain physical identification of the nature of the weakest sources (perhaps they are local?) prevents one from regarding this decision as final.

More drastic modifications of physics than even the spontaneous appearance of nucleons or of atoms have been introduced by some authors. These theories, starting with the visible importance of very large numbers in cosmology, such as the baryon number of the visible world, or with the postulated Mach relation between inertia and distant mass, consider a time variation of important physical constants, such as the intrinsic strength of the several kinds of forces. Variations that require a Hubble time or two to reach a measurable threshold are probably not excluded by anything we know, though so far no demonstration of a secular change in the laws of physics has been successful. Attempts to explain geophysical phenomena, such as continual drift, by fundamental changes in natural law, as in the ratio of electrical to gravitational forces, seem strained. The natural systems—earth, sun, stars—are sufficiently complex to resist full understanding even if one assumes that we have merely to apply the correct and complete physical laws now found in the laboratory. This is not to deny the possibility that such phenomena might occur; it is rather to show the difficulty of establishing new and drastic physics without either precise measures or phenomena of qualitatively unexpected type (like the general red shift).

There is one approximation dear to cosmologists that is surely in error and whose effect has been very little studied. That is the truly atomic nature of matter, or even closer to the point, the quasi-atomic nature of the galaxies, which are surely no uniform matter fluid. In the kinematical studies, for example, the radar return from distant parts of the universe will not at all be single. Many geodesic paths will exist, as the signal returns through the fluctuating gravitational fields of the many intervening galaxies. Distant contact must be a diffuse one. This remains true in principle whatever the signal carrier, for even the penetrating neutrinos will diffuse in the lumpy gravitational field. Whether this fact seriously affects conclusions about the universe as a whole is quite unclear. Perhaps it is another example of the total gap that lies between contemporary micro- and macrophysics. How can the particles of matter be described by the smooth fields of Einstein's theories? This fundamental problem has not escaped notice, but its solution remains wholly dark.

A related but simpler question, one that serves to illuminate rather than to raise doubts about cosmological theories, is that of horizons. If one looks far enough out, even into a smoothed over model, his view may reach a horizon: a surface beyond which some degree of unobservability must arise. Plainly, no signal can come from a galaxy moving at light speed. The asymptotic approach to such horizons differs widely according to the model. In some models, whole new galaxies become visible as one looks farther and farther out; new sources cross the horizon. They are seen at "creation." In others, galaxies disappear. In still others, galaxies once seen remain forever, but their light grows dimmer and redder, so that the events that occur within them become asymptotically unknowable. A close study of these implications for each model is always of interest. It is worth remarking that looking far enough in some closed models should reveal the moment of the "big bang," and the observer would see the same single "central" glowing sphere in whatever direction he turned his telescope. Like the horizons, the expected appearance of the earliest stages is a characteristic of a model.

All theories so far face the tasks set for them by physics with little success. Neither the Mach principle, nor the nature of time's arrow, nor the remarkable size of the baryon number has received any very convincing statement in any model so far worked out. Perhaps the pretensions of cosmology are still presumptuous; first and last things may lie beyond our reach. Certainly the years still bring us new and apparently important phenomena not yet subsumed in any cosmology; this suggests that cosmologies without an indwelling particle theory, without neutrinos, without great radio sources, without whatever is new tomorrow, cannot exhaust the subject. Yet they have made progress. We are learning at least the preconditions of a theory. The theory appears on an unwritten page.

Finally, I shall dare to enter upon metaphysical terrain. What is it that fixes the very laws of nature which we express in our cosmological theories? I can see only a few ways to seek such an answer. It is easiest to say *ignoribimus*. It is possible that subtle analysis can show that only one universe is free of logical contradiction; this is the best, the worst, the only possible world. That is challenging, but far from present thought. Or it may be that, like everything else which is not mere logic, like gas and galaxies, stars and men, the laws of nature themselves evolve. Their present state, perhaps their state over an enormous time in the past, may be only a durable state. One can picture a kind of chaos, without fixed particles, without observers or laws or even spatio-temporal substrata. Such a chaos may have flowed without specific causes, by chance, through the possible forms of disorder. When, again by chance, it reached a state possessing compatible large- and small-scale laws that allowed their own self-perpetuation, its state was forever after fixed. Thus the cosmos maintains itself; its order is the order of inheritance. Anything else would become modified beyond recognition. It is not clear that there is in these remarks anything beyond a form of metaphor; but if there is meaning, it is the meaning that the state of the world is the survival of the most stable.

It would at least be fitting if the natural order mirrored in this grand vein the protean quality of life by which we know that man—surely not the only cosmologist, but the only local one—was formed by the Master Potter, the test of survival.

References

For a general account of modern astronomy, see L. V. Berkner and H. Odishaw, eds., *Science in Space* (New York: McGraw-Hill Book Company, 1961).

For an introduction to the ideas of cosmology, an excellent and nonmathematical, but not simple, work is D. Sciama, *The Unity of the Universe* (New York: Doubleday & Company, Inc., 1961).

A more complete and difficult, but equally excellent, book is H. Bondi, *Cosmology*, 2nd ed. (London: Cambridge University Press, 1961).

The Newtonian cosmology was presented first by E. A. Milne and W. H. McCrea, *Quarterly Journal of Mathematics*, Vol. V, No. 64 (1934), 73.

A valuable account of age measures is found in R. Dicke, *Reviews of Modern Physics*, Vol. XXXIV (1962), 110.

The great radio sources and their possible interpretation are discussed, for example, in G. R. Burbidge, E. M. Burbidge, and A. R. Sandage, *ibid.*, Vol. XXXV (1963), 947.

Horizons in cosmology are first treated by W. Rindler, *Monthly Notes of the Royal Astronomical Society*, Vol. CXVI (1956), 662.

The red shift, its observational basis, and its interpretation are reviewed in a rich and useful paper by A. Sandage, *Astrophysical Journal*, Vol. CXXXIII (1961), 335, the basis of Figure 10.

PAUL K. FEYERABEND
University of California

Problems of Empiricism

Many issues are decided by many people on a basis of party spirit, not of detailed examination of the problems involved. In particular, whatever presents itself as empiricism is sure of widespread acceptance, not on its merits, but because empiricism is the fashion.

—BERTRAND RUSSELL, *The Philosophy of Bertrand Russell*

I. Introduction

Today empiricism is the professed philosophy of a good many intellectual enterprises. It is the core of the sciences. It has been adopted by influential schools in aesthetics, ethics, theology. And within philosophy proper the empirical point of view has by now been elaborated in great detail and with considerable precision.

This predilection for observation and experiment has been, and still is, due to the assumption that only a thoroughly observational procedure can exclude fruitless hypothesizing and empty talk; to the hope that empiricism is most likely to prevent stagnation and to further the progress of knowledge; and to the conviction that the sciences, and especially the physical sciences, owe their existence and spectacular successes to the fact that they have adopted the empiricist creed. Herschel's enthusiastic description of the rise of the new physics is commonly regarded as the correct historical account and as an indication that lesser subjects, too, might profit from the elimination of metaphysics and the concentration upon undiluted data of observation.[1]

The popularity of empiricism would make one expect that this doctrine has been developed in some definite form and that there exist *arguments* indicating why observations are to be preferred to, say, predictions on the basis of dreams. Now it is to be admitted

145

that quite a few empiricists have tried to give an account of the thesis that our ideas derive from the senses (aided or unaided by experimental equipment) and of the corresponding belief that the truth of statements containing such ideas essentially can be ascertained by observation. They have in this way explained how the doctrine of empiricism was understood by *them*. The trouble is that there are as many such accounts as there are writers.[2] Nor has the development of the sciences brought about unanimity in these matters. Thinkers who accept different views concerning the nature of experience and the way in which theories[3] are to be related to experience[4] have proceeded in different ways and thereby given rise to different kinds of scientific knowledge.[5] The unity of doctrine insinuated by the scientist's appeal to experimentation and by his hostility toward "hypotheses" must therefore be viewed with extreme caution.[6] But this is not all. Although the *propaganda* for observation and experiment and the general *clamor* concerning the importance of careful attention to fact may be a fairly recent feature of the history of thought,[7] the *use* of observational results and the careful attention to details is not. At all times people have found it important and instructive to make observations. Even the most abstruse myth is formulated in a fashion that provides detailed descriptions and explanations of a great variety of phenomena and brings coherence into otherwise unintelligible occurrences.[8] This could hardly be otherwise. A myth is often held by people who are engaged in a desperate struggle for survival and who have no time to indulge in the luxury of considering possible hypotheses. Its function is to provide a guide through life, and this means through a series of occurrences conceived in the most concrete manner possible (e.g., reaping of crop, treatment of cattle, illness, storms, the seasons, etc.). Considering this function of a myth, it is utterly unrealistic to assume that it will resemble a dreamed up superstructure that is but loosely connected with the hard facts of life.[9] Quite the contrary. We shall have to expect a much *closer* relation to the facts of experience than will be met in a scientific theory whose inventor may be a gentleman "free and unconfined," to use Bishop Sprat's characterization of the early members of the Royal Society.[10] Even the most determined metaphysicians who intend to discuss Being in Itself, and who are upset by the most insignificant sensual element, have a hard time, and it is likely that they will never succeed in keeping experience completely out of their account. Thus it is not at all difficult to *consider* experience, to *incorporate* it into one's system of thought, to *bring about* a "closer alliance of . . . the experimental and the rational faculty."[11] But it is almost impossible to omit from one's discourse all reference to detailed observation.[12] The variety of practices and doctrines that could be called "empirical" is therefore much larger than one would be inclined to think. Does this mean that the demand for observation and experiment that has played such an important role in recent centuries, and has been raised with renewed vigor by the followers of Niels Bohr, signifies nothing? Or does it perhaps mean that those who thought that it was the emphasis upon observation which led to their discoveries *misplaced* the element that distinguishes them from their predecessors? Does it mean that observation played only an accidental part in the reform of the sciences, and that the importance of the particular observations which *were* made was due to circumstances of an altogether different nature? What feature of knowledge *did* these thinkers have in mind when demanding observations? It is hoped that we shall be able to provide some sort of answer to this problem, which we might well call the "problem of the *omnipresence of experience.*"

Turning now to the *arguments* for the empirical point of view, we notice first that the most basic assumption, namely the superiority of observations over dreams (for example), is only rarely supported by cogent reasons,[13] and second, that the related question as to what is more fundamental—sensation or thought, observation or theory—is generally assumed to be independent of scientific research. In the domain of physics, of chemistry, of biology, the empiricists implore us to wait, not to anticipate observation, not to observe hastily. Yet the presumably much more subtle and difficult domain of the mind is supposed to bare its secrets to the most primitive kind of investigation.[14] Even worse, it is assumed that the relation between mental events (inside the human observer) and physical events (outside the human observer) can be decided by *philosophical* argument, by the abstract consideration of possibilities, without any help from the exigencies of empirical research. This means, of course, that the stern admonition to avoid hypotheses, not to consider matters in the abstract, but always to use all the facts available can be supported only by considerations of the very kind that the empiricist wants to avoid. Russell's remark that empiricism "is sure to be of widespread acceptance, not on its merits, but because empiricism is the fashion," does not seem to be too far removed from the truth.

It would be extremely interesting to find out the reasons for this paradoxical situation. Such an examination, which cannot be carried out here, would reveal that the various forms of empiricism which we know today were originally parts of wider theories and could be supported by reference to the principles of these theories. The general aversion against speculation and metaphysics that characterizes a good deal of the thought of the seventeenth and eighteenth centuries, as well as the tendency to secularize the science of cosmology, eliminated most of these principles. The causal theories of perception of Locke and Descartes, which could have provided a new framework, were dropped under the influence of the criticism of Berkeley and Hume. Much of contemporary empiricism is the unreasonable end result of this philosophical spring cleaning. It is a *fragment* that cannot stand on its own feet. Once, when metaphysics was still a respectable enterprise, the reference to observation made excellent sense. Today it is hardly more than an article of faith.[15]

Now it is interesting to note that a single step further on this road away from metaphysics could bring reason back into empiricism. This step consists in dropping the idea of *certainty,* or of *directness,* or the weaker idea of *support,* which has been and still is an essential part of almost all philosophizing and is mainly responsible for the rejection of the causal theories. Why were the ideas of matter or of substance eliminated by Berkeley? Because it was impossible to obtain well-founded knowledge of substances or of material objects *unless* these objects were interpreted as consisting only of sensations. Why was the distinction between primary and secondary substances dropped? Because it could not be justified on the basis of immediate observation, which is an observation that leads to certainty (or at least to results that are highly confirmed). Abandoning this demand for certainty, we may introduce a *hypothesis* concerning the relation between experience and the external world. Such a procedure removes the assertion of the basic character of experience from the domain of philosophical consideration to the domain of scientific research. It also restores to empiricism that amount of reasonableness which the elimination of the assumption of the *veracitas dei* and of similar metaphysical assumptions had taken away from it. Sections XV and XVI contain a sketch of an empiricism that has been revised and made reasonable in the manner indicated. I hope to be able to give a more detailed account both of the historical development and of the desirable end product in the not too distant future.

It is my intention to keep the discussion as concrete as possible. This has the advantage that more recent developments, which are fairly detailed, can be taken into account. It has the disadvantage that only *some* facets of the very complex phenomenon "empiricism" can be investigated. I have tried to choose these facets in a manner that does not exclude the transition to a more general discussion and that makes it possible to deal profitably with many philosophical questions. Indeed, I believe that the criticism of the facets chosen will not affect just a few technical problems but will lead to considerable changes, both in the sciences and in philosophy. What I have in mind is one particular thesis that has played a considerable role in the history of empiricism. The thesis is of much wider application than its first and rather technical formulation might suggest. Briefly, it is this:

Assume that we possess a theory in a certain domain which has been highly confirmed. Then this theory must be retained until it is refuted, or at least until some new facts indicate its limitations. The construction and development of alternative theories in the same domain must be postponed until such refutation or such limitation has taken place. Any doctrine containing the thesis just outlined will be called a *radical empiricism*.

Radical empiricism is a *monistic* doctrine. It demands that at any time only a single set of mutually consistent theories be used. The *simultaneous* use of mutually inconsistent theories or, as we might call it, a *theoretical pluralism* is forbidden.

It will be argued that the demand for a theoretical monism is liable to lead to the elimination of evidence that might be critical for the defended theory; it lowers the empirical content of this theory and may even turn it into a dogmatic metaphysical system.

This result has important tactical consequences. It forces us to admit that the fight for tolerance in scientific matters and the fight for scientific progress that was so important a part of the lives of the early scientists *must still be carried on.* There are still attempts made to arrest progress and establish a doctrine. What *has* changed is the denomination of the defenders of such doctrines. They were priests, or "school philosophers," a few decades ago. Today they call themselves "philosophers of science," or "logical empiricists." There are also a good many scientists who work in the same direction. Almost all the members of the so-called "Copenhagen school" must be mentioned here. But whereas the traditional champions of obsolete ideas did their defending openly, and could therefore be discerned quite easily, their modern successors proceed under the flag of progressivism and empiricism and thereby deceive a good many of their followers. Hence although their presence is noticeable enough, they may almost be compared to a fifth column, the aim of which must be exposed in order that its detrimental effect be duly appreciated. It is our purpose to contribute to such an exposure.

The attempt will also be made to formulate a methodology that can still claim to be *empirical* but that is no longer beset by the problems of a *radical* empiricism. Put in a nutshell, the methodology requires a theoretical pluralism instead of the theoretical monism that is the ideal of much of contemporary empiricism, of philosophy in general, as well as of almost all "primitive" thought. This plurality of theories must not be regarded as a preliminary stage of knowledge that will at some time in the future be replaced by the "one true theory." Theoretical pluralism is assumed to be an *essential* feature of all knowledge that claims to be objective. Nor can one rest content with a plurality that is merely abstract and created by arbitrarily denying now this and now that component of the dominant point of view, as is the plurality created by the various attempts of modern artists to free themselves from the conventions of their predecessors. Alternatives must rather be developed in such detail that problems already "solved" by the accepted theory can again be treated in a new *and perhaps also more detailed manner.* Such development will, of course, take time, and it will not at once be possible, for example, to construct alternatives to the present quantum theory that will compare with it in richness and sophistication. Still it would be very unwise to bring the process to a standstill in the very beginning by the remark that the new ideas are undeveloped, general, metaphysical. *It takes time to build a good theory;*[16] and it also takes time to develop an alternative to a good theory. The *function* of such concrete alternatives is, however, this: they provide means of criticizing the accepted theory in a manner that goes beyond the criticism provided by a comparison of that theory with the "facts." However closely a theory seems to reflect the facts, however universal its use, and however necessary its existence seems to be to those speaking the corresponding idiom, its factual adequacy can be asserted only *after* it has been confronted with alternatives *whose invention and detailed development must therefore precede any final assertion of practical success and factual adequacy.* This, then, is the methodological justification of a plurality of theories: such a plurality allows for a much sharper criticism of accepted ideas than does the comparison with a domain of facts that is supposed to be given independently of theoretical considerations. The function of unusual *metaphysical* ideas is defined accordingly: they play a decisive role in the criticism and in the development of what is generally believed and "highly confirmed," and they must therefore be present at *any* stage of the development of our knowledge.[17] A science that is free from all *metaphysics* is on the way to becoming a *dogmatic* metaphysical system.

The results just mentioned are not restricted to the philosophy of science. They have important applications to philosophy proper. Indeed, it is my contention that once the scientific method has been freed from certain dogmatic elements that still reflect its past involvement with the philosophical tradition, it will provide a basis for the discussion and the solution of *all* philosophical problems dealing with matters of fact. After all, philosophical doctrines—be they now formulated explicitly, in the form of a philosophical *system,* or implicitly, in the form of linguistic *rules* governing a certain language—are quite comparable to general theories of physics. They propose an ontology; they defend this ontology either by reference to observations of a general kind or by theoretical arguments; and they reject alternatives in much the same way as a radical empiricist would reject them, by pointing out that the accepted doctrine is not yet in difficulty and will perhaps never be in difficulty. The last assertion can be easily believed. The observations referred to figure in the arguments only to the extent to which they have received a theoretical interpretation on the basis of the system or the linguistic rules chosen; and the theoretical arguments consist in going back and forth between the accepted principles and their consequences. Philosophical arguments (with the possible exception of dialectical arguments) are therefore invariably *circular.* They show what is implied in taking for granted a certain point of view,[18] and do not provide the slightest foothold for a possible criticism. This being the case, it is quite impossible to evaluate them in the traditional manner. We must choose a point *outside* the system or the language defended in order to be able to get an idea of what a criticism would look like. We must use an alternative.

Such a transition to a theoretical pluralism would also seem to increase our chances of solving some philosophical problems that are regarded as essentially insoluble or are perhaps even denied the status of legitimate problems. Such insolubility is usually due to the fact that the system on which a philosopher wants to base his arguments is a mixture of incommensurable theories.[19] The demand for a theoretical monism (and the related demand of meaning invariance that will be formulated below), working in reverse, (1) assumes it to be coherent and (2) forbids any fundamental change. The combination of (1) and (2) then blocks the solution of all those problems whose origin is the internal incoherence of the chosen system. The mind-body problem and the problem of the reality of the external world are excellent examples of the situation just described. The customary solution of these two problems and the arguments adduced in their favor and against rival theories exhibit beautifully the circularity of philosophical argumentation to which we have alluded. A brief discussion of some aspects of the former (and to some extent also of the latter) will add detail to the general remarks we have just made. It will also prepare the development of a theory of experience that is more satisfactory than the customary accounts.

In these accounts it is taken for granted that observational results can be stated and verified independently, at least independently of the theories investigated. This is nothing but an expression, in the formal mode of speech, of the common belief that experience contains a factual core that is independent of theories. Such a core must exist, or else we can never be sure that our ideas have any relation to fact. Theoretical pluralism is inconsistent with the idea of a core. The reason is quite simple. Experience is one of the processes occurring in the world. It is up to detailed research to tell us what its nature is, for surely we cannot be allowed to decide about the most fundamental thing without careful research. Carrying out such research, we must devise and examine critically various hypotheses concerning the nature of experience and its relation to external fact. A critical examination is an examination in the light of alternatives. No aspect must be left unscrutinized, which means that no feature of experience must be left undenied. The idea of a core is incompatible with this methodological principle.[20]

Observation statements, then, are not *semantically* different from other contingent statements. They do not possess a special content (any conceivable theory of experience must lead to observation statements) or a special core of content. If there is a difference between them and other statements, then this difference is provided by the psychological, or physiological, or physical circumstances of their production. This position, which we shall call the *pragmatic theory of observation,* was formulated some time ago by Popper, and it was accepted by Carnap and others.[21] Since then it has been all but forgotten.

The pragmatic theory denies that there is an asymmetry between theory and observation (observation statements eliminate theories; theories do not eliminate observation statements) or between metaphysics and observation (observation statements are vastly superior to metaphysical statements; if a clash occurs, then it is the metaphysical statement that has to go). We invent theories in order to criticize observational results. A clash between an observational statement and a metaphysical point of view is therefore a most welcome indication that the chosen metaphysics is fit for criticism and is in this way an important argument for, and not against, its retention. The occurrence of observations in a system and their compatibility with the system are not necessarily a distinguishing mark of excellence.

This leads back to the first question: what is the role of experience in the Scientific Revolution of the sixteenth and seventeenth centuries? The reply must be: the transition from Aristotelianism to the new physics is *not* a transition to a radical empiricism (although it is presented as such a transition even by some of those who brought it about). Quite the contrary, we have here the attempt to use a new point of view for the criticism both of the older *doctrine* and of the *observations* supporting this doctrine. Considering the omnipresence of the older doctrine—even the most common occurrences were presented in its terms—these supporting observations were of no particular interest. Their support was a very cheap affair. There existed a dependence between observation and theory that made even empirical arguments circular. What *was* of interest was the question whether a different and more daring interpretation of the very same observable processes could be devised, and whether such an unusual interpretation would lead to a coherent system of thought. Equally interesting was the question whether there were not processes, known to all, that had either received little attention from the Aristotelians—that were, in a sense, situated at the periphery of their enterprise—or that had not yet been incorporated in a simple and straightforward manner, and whose occurrence nevertheless lent very immediate support to the new doctrine. Such processes were indeed found.[22] Considering the expectation involved, the possibility of a disagreement between the independently framed hypotheses, the very real possibility of a *failure,* one felt that for the first time one could make discoveries leading beyond what was already known and obtain worthwhile support from observation. It is evident that this situation depended on the invention, *prior to experiment,* of a new point of view admitting of independent tests. The impact of the new observations also depended on the existence of Aristotelianism. Without the Aristotelian system, it could always have been objected that a different interpretation might be possible and that the new theory had not proved its mettle. *With* Aristotelianism, it became clear that the new system was at least better than the Aristotelian point of view and thus that some progress had been made. Conversely, it is clear that the Aristotelian philosophy could not possibly have been refuted by a direct appeal to the facts. The new observations that were made assumed the decisive character ascribed to them only for the believer in the new astronomy; they were irrelevant, beside the point, or perhaps even unnoticed by the Aristotelians. Moreover, the Aristotelians were quite right in emphasizing their irrelevance; they could not be convinced by new facts (which most likely they were able to accommodate),[23] but only by proof of the comparative superiority of the new theory, which therefore had to be developed *before* factual arguments could become relevant.

The doctrine that all our knowledge comes from the senses and rests securely upon the information provided by them has led to a complete misrepresentation of this situation. The new observations, which were revolutionary only because they were part of a new point of view, and whose interest (including the fact that they were made at all) depended quite decisively on the existence of this new point of view as well as of the Aristotelian doctrine, were separated from all these components and presented as isolated data that had been willfully neglected by the bad Aristotelians but were taken up by conscientious Galileo and made the inductive basis of a new and truly empirical theory. This is how an unexamined philosophical doctrine (empiricism) has led to an historical myth.[24]

This study is based on results that I explained in the following earlier essays: "Explanation, Reduction, and Empiricism"; "Problems of Microphysics"; "A Note on Linguistic Philosophy and the Mind-Body Problem"; "An Attempt at a Realistic Interpretation of Experience"; "Complementarity."[25] All the detailed acknowledgements can be found there. My general outlook derives from the work of David Bohm and K. R. Popper and from my discussions with both. The idea that a theoretical pluralism should be the basis of knowledge can be found both in the dialectical philosophy of Bohm and in Popper's critical rationalism. However, it seems to me that it is only within the framework of the latter that it can be developed without undue restrictions. Such development is bound to eliminate whatever remainder of an empirical "core" may still be contained in Popper's point of view.[26] My discussions with T. S. Kuhn and his skillful defense of a scientific conservatism (which does not work with alternatives, or which uses them only in exceptional cases, and then rather unwillingly) triggered two papers dealing with the problems in the sections to follow, especially Sec. VI. To D. Rynin and J. J. C. Smart, as well as to J. W. N. Watkins I owe much good advice that has been used in the final version of this essay. The most personal motive, however, for writing the paper derives from my discussions with Herbert Feigl. It is to him that many arguments are addressed in the hope that they will be satisfactory replies to the objections he untiringly raised against the preliminary versions of some of the opinions contained herein.

II. Historical Sketch: Different Types of Empiricism

Before starting with the critical discussion of radical empiricism, I want to give a more detailed, though still most insufficient account of a point that was touched upon in the last section—the great variety of doctrines hiding behind the label "empiricism." This account might help to identify the historical positions to which the criticism must be applied, and it might also help to see some arguments in contemporary physics in the proper light.

We may distinguish three periods in the history of empiricism. The first is Aristotelianism. The second is the "classical" empiricism of the seventeenth, eighteenth, and nineteenth centuries. The third period, which shows some surprising similarities with the first (and *not* with the second, or with the beginning of the second period, as has been insinuated by Heisenberg[27]), is the empiricism of modern physics, more especially, the empirical doctrine of the so-called Copenhagen school of the quantum theory. The first and third periods are characterized by a fairly close agreement between the house philosophy of the scientists and the way in which they build up their theories. The second period is characterized by a kind of schizophrenia. What is propagated and declared to be the basis of all science is a radical empiricism. What is *done* is something different. This difference between the professed philosophy and the actual practice is covered up both by a manner of presentation which makes it appear that the theories are indeed nothing but true reports of fact and by a tradition of history writing whose function has been described as being purely "ritualistic,"[28] and which creates the impression that after an initial revolution in the Renaissance, science has been steadily progressing through the accumulation of more and more facts.[29] The developments occurring under this guise constitute one of the most interesting chapters in the history of thought. The false idea is adhered to as closely as possible in words; the actual results deviate radically from it. The crisis of physics brought about by relativity and by the discovery of the quantum of action terminates this period of schizophrenia. The majority of physicists, with the sole exception of Einstein and his followers, now adopt a practice that is much more closely connected with a radical empiricism than is the practice of the preceding classical period. The consequence is a partial return to Aristotelian empiricism.

Such are the rough outlines; I shall now present some details. However, I must warn the reader that the discussion will still be fairly general, and that many developments will have to be disregarded. This is due partly to lack of space, partly to my ignorance. Yet if the reader is prompted by the sparse evidence I provide to delve into the still unused mass of historical material, then the inclusion of these remarks will be justified.

Aristotelian physics is quite explicitly observational; to be observable is part of the definition of physical nature it uses.[30] The constitutive forms of physical objects are required to be observable,[31] and the laws of motion, too, are "the most commonplace experiences [about] motion [formulated] as universal scientific propositions."[32] The empiricist epistemology of the Aristotelians and their physics are in harmony.[33] This harmony disappears after the Scientific Revolution in the seventeenth century.[34] The epistemology *remains* empiricistic—and a very militant empiricism it is. It influences not only the philosophy of the physicists, but also the way in which they present their theories, even their style.[35] From now on we find only very few scientists who violate explicitly the inductivist prescription to start from observations, to keep one's theories in close contact with them, and to refrain from hypotheses. Newton introduces his law of gravitation in a manner which suggests that it was a direct consequence of his own inductivist rules of procedure.[36] His work was regarded as a shining example that only few could properly imitate.[37] So convincing was his presentation that many later thinkers, including Drude[38] and Born,[39] believed the law of gravitation to be nothing but a mathematical transcription of Kepler's laws and that Hegel denied to Newton the discovery of anything over and above Kepler.[40] Since Duhem's investigations,[41] and especially since the general theory of relativity and the publication of Einstein's papers on the theory of knowledge,[42] it has become clear that this is a travesty of the actual situation. Not only does Newton's theory transcend the domain of observation; it also contradicts the observational laws that were available when the theory was first suggested.[43] It is therefore quite impossible to obtain it by inductive generalization, which leaves the "facts" unchanged; and if it *did* seem possible to obtain it in this fashion, then this was due to the omission from the argument of some essential premises.[44] Only by such a procedure of suppression and unwitting concealment could the gulf be bridged that existed between the theory and the philosophy allegedly responsible for its invention.[45]

Duhem was not the first to point to the discrepancy between the philosophical ideal and the rules that were actually followed. Newton's own colleagues in the Royal Society had felt, for example, that the method of his earlier papers on optics was not truly empirical. It is quite illuminating to study the various points of view that the Society included in its early history and that were all identified with the new Baconian philosophy. We may distinguish three, perhaps four, versions of empiricism. They are (1) a crude empiricism that encourages the uncritical collection and preservation of facts; (2) Hooke's empiricism, which emphasizes the importance of factual research without prejudice, and values multiplication of experiment, but is not averse to the consideration of hypotheses, provided they are used in their proper place and are clearly characterized as such; (3) Newton's empiricism, which starts not from some raw facts of everyday life, nor from an experience that has been cleaned of all prejudicial elements, but rather from an intimate synthesis of experience and mathematical representation, and which regards a derivation from an experience thus conceived as unique and as free from arbitrary elements; finally, we might consider (4) Bacon's own empiricism, with its vague demand for an alliance of thought and experiment, with its insistence upon the removal of prejudice, and with its emphasis upon the importance of negative instances. The subdivision is somewhat schematical, but it corresponds to important tendencies which existed at the time. A more detailed study of these tendencies and of their later history would seem to be a most valuable contribution to the better understanding of the historical impact of empiricism.

Version (1) found expression in the very curious museum that was set up by the society and "into which flowed a strange mixture of objects of real value with others of only passing interest; and the passion for collecting biological freaks brought many worthless and even fraudulent objects into the case."[46] Among the curiosities of the museum there were to be found "an ostrich whose young were always born alive; an herb which grew in the stomach of a thrush; . . . the skin of a moor, tanned, with the beard and hair white. . . ."[47] It also found expression in many experiments that seemed to serve no purpose whatever, except perhaps the purpose of arriving at an inconclusive opinion concerning a common superstition. An example is this: "1661, July 24: a circle was made with powder of unicorn's horn, and a spider set in the middle of it, but it immediately ran out several times repeated. The spider once made some stay upon the powder."[48] Weld feels obliged to apologize for this side of the society. "Let not the reader therefore, when he smiles, as he assuredly will, at many of the seemingly absurd and ridiculous experiments tried by the society . . . criticize them as mere folly. . . . They were necessary for the welfare of science."[49] No one will criticize a thinker trying to test a hypothesis. What *is* to be criticized is the attempt *without* much guidance from thought to collect as many useless facts as possible from as many domains as possible, and to expect that science will one day miraculously profit from the collection thus assembled.

A very surprising indication of tendency (1) is the sometimes quite extraordinary credulity extended to the reports of eyewitnesses.[50] Reports from foreign countries about conditions not known in England were most welcome to the society and accepted without much discrimination, except perhaps that the reports of sailors (allegedly not prejudiced and therefore especially good observers) were received slightly more favorably by some of the members. This may account for the following strange story (not the only one of its kind) to be found in the first volume of the register book of the Royal Society:[51]

> An English mariner was wounded at Venice in four severall places soe mortally, that the murderer took sanctuary; the wounded bled three days without intermission; fell into frequent convulsions and swounings, the chirurgeons, despairing of his recovery, forsook him. His comrade came to me [i.e., Sir G. Talbot, who held high offices under the crown] and desired me to demand justice from the duke upon the murderers (as supposing him already dead). I sent for his bloud and dressed it, and bad his comrades haste back, and swathe up his wounds with cleane linnen. He lay a mile distant from my house, yet before he could gett to him, all his wounds were closed, and he began visibly to be comforted. The second day the mariner came to me, and told me his friend was perfectly well, but his spirits soe exhausted, he durst not adventure soe long a walke. The third day the patient came himself to give me thanks, but appeared like a ghost—no bloud left in his body.

I cannot help seeing the grin of the "mariner" who had found such an eager scientific customer for his yarn.

(2) Hooke, in his evaluation of Newton's first paper, pointed out that "having, by many hundreds of trials, found them so" he was prepared to accept Newton's experimental results.[52] He had promised to "furnish the society every day they met [once a week] with three or four considerable experiments, expecting no recompense till the society get a stock enabling them to give it."[53] He seems to have adopted a combination of (1) and some vague elements of the Baconian philosophy. However, the philosophical radicalism inherent in such a combination is tempered by his close association with craftsmen and instrument makers and by his familiarity with all the "useful arts." Experience is for him factual knowledge improved by experimentation and controlled by common sense and by a keen critical intellect.[54] It is on *this* basis, and not on the basis of a clearcut philosophical doctrine, that he thought he had looked through Newton's pretense to have given nothing but an accurate description of phenomena and to have avoided all speculation: "Nor would I be understood to have said all this against his theory, as it is an hypothesis; for I do most readily agree with them in every part thereof and esteem it very subtil and ingenious, and capable of solving all the phenomena of colours; but I cannot think it to be the *only* hypothesis, nor so certain as mathematical demonstrations."[55]

(3) Newton himself, while paying lip service to the empiricism of the Royal Society and defending his own ideas above all by reference to observation, was quite averse to a senseless multiplication of experiments.

> Experience is necessary but yet there is the same difference between a mere practical mechanic, and a rational one; or between a mere practical surveyor or gauger, and a good geometer; or between an empiric in physics, and a learned and rational physician. . . . I would add that if instead of sending the observations of seamen to able mathematicians on land, the land would send able mathematicians to sea, it would signify much more to the improvement of navigation and safety of men's lives. . . .[56]

Indeed, so great was the disparity between his philosophy and the general attitude of the Society, that Goethe regards the acceptance of his theory of colors by the Society as a most interesting historical problem: "The society had hardly come into existence as Newton was received into it, in his thirtieth year of age. Yet how he was able to introduce his theory into a circle of men who were most definitely averse to theories, this is an investigation well worthy of a historian."[57] A look at one detail shows the character of the problem.

Newton had described his spectrum as "terminated at the sides with straight lines" and capped on both ends with "semicircular lines."[58] Linus, Professor of Mathematics and Hebrew at Liège, objects that "these semicircular ends are never seen in a clear day" and that "when the image be made so much longer than broad, then the one end thereof will run out into a sharp cone or pyramid like the flame of a candle and on the other side into a cone somewhat more blunt. . . ."[59] As Kuhn remarks, this is the appearance to be expected.[60] Newton never responded to this criticism of Linus's. Can there be a more striking deviation from the principles of a radical empiricism as defended by Newton? As the empiricist he professes to be, he must report the observational facts as they are. This is not what he does. And yet his reports are eventually accepted as correct and are defended against those who take the demand for undiluted observation seriously. How can this development be explained? This is the problem posed by Goethe.

Before venturing an explanation, let me make a few comments. An observer who is also a realist has a twofold task. He must report what he *observes,* and he must make sure that his report is *relevant,* that is, that it concerns the particular entity, state of affairs, process that he happens to be interested in. The first demand requires him to keep as closely as possible to "what he sees." The second demand requires him to interpret his report as concerning an objective state of affairs, i.e., to say *more* than just what he sees. The situation becomes especially transparent in those cases where a piece of physical apparatus is involved, for in this case the assertion made at the end of the experiment, and regarded as expressing the result of the experimental investigation, will describe many things that a direct look does not reveal. Extrapolations are made not only quantitatively, but also qualitatively. These extrapolations are an essential part of the observational report. Without them it would not be a report concerning some feature of an objective physical world. We may, of course, demand that the extrapolations be checked independently. Still, their occurrence in an observation statement does not render that statement careless or rash. It is only an extreme empiricist, a thinker who refuses any kind of theoretical interpretation, who will regard such a procedure with suspicion. And it is only such a thinker who will object to Newton's somewhat idealized description of his experiments. But then such a thinker will not be a realist either; realism is incompatible with an extreme empiricism. The problem is that Newton and his colleagues in the Royal Society were both realists and somewhat extreme empiricists. How did they manage to achieve such a combination? How was it possible for Newton to succeed *as an empiricist?*[61]

The situation we are dealing with here shows the great elasticity, not to say vagueness, of the doctrine of empiricism and the superb talent of empiricists to pretend that they are doing one thing when in fact they are doing something very different.[62] The success of empiricism and its great staying power are not least due to this possibility to interpret it in various ways and to use it as surrogate for almost any idea that for some reason seems to be worthy of acceptance. In the present case, a new and more abstract account of experience was given. The initial resistance against this procedure disappeared when physicists became more familiar with the theoretical ideas contained in the new experience. Being able to produce these ideas effortlessly, they soon thought them to be a direct expression of what was "given" in observation. Interpreting them in this way, they refused to accept different accounts as being incompatible with "experience." Thus a valuable advance (from the narrow-minded insistence upon undiluted observational reports to a more abstract notion of experience) is made possible only by being connected with a dogmatic appraisal of its result, and it is neutralized thereby.[63]

Newton's "phenomena," which are the elements of the new "experience," are not everyday facts pure and simple; nor are they an experience that has been cleared from prejudicial elements and left that way. They are rather an intimate *synthesis of laws,* possessing instances in the domain of the senses and certain mathematical ideas. The synthesis is so close that the division into the elements "law," "instances," "mathematical ideas" can be carried out only abstractly and only by comparison with quite different points of view. One might imagine it to be the result of two steps.[64] First, *dispositions* are defined (such as refrangibility, density, refraction coefficient) that admit of continuous variation. This step amounts to expressing certain *regularities* characteristic either of a substance or of light with the help of singular statements.[65] Second, experimental results expressed in terms of these dispositions are formulated mathematically. Considering this origin of the "phenomena," it is clear that their descriptions are condensed and generalized reports of numerous experimental results that have been purged of peculiarities due to the specific nature of the apparatus used. Actual experiment, which always depends on a large variety of irrelevant variables, may therefore *illustrate* the phenomenon; it cannot *establish* it (this, by the way, is a very obvious consequence of the so-called pragmatic theory of observation, which we have briefly mentioned in the introduction and which will be discussed in greater detail in Sec. XV below). Strictly speaking, describing a phenomenon means stating a law.[66] It is therefore not at all surprising that one can now obtain laws by a derivation "from the phenomena." Nor is it surprising that many of the laws obtained are regarded as irrevocable. After all, they are based upon premises that are part of "experience" and that are therefore beyond reproach.

The new experimental philosophy that is introduced by Newton (and that was prepared by Archimedes, Galileo, and Descartes) therefore contains a progressive and a conservative element. The progressive element is to be sought in the attempt to transcend the everyday experience that formed the basis of all previous thinking and played such an important role in the Aristotelian philosophy. The conservative element lies in the fact that the new experience is viewed in the same fashion as were all previous "foundations" of knowledge—it is regarded as sacrosanct.

We have already pointed out that this curious combination of progressive and conservative elements is due to the empiricist's insistence upon a sound basis of knowledge and to his vagueness concerning the nature of this basis. It is clear that as long as empiricism is structured in this fashion it will be able to accommodate almost any consistent point of view. On the other hand, such accommodation will not be achieved without some distortion. Accommodation is possible because the basis is left *indefinite*. Distortion occurs because part of the point of view accommodated is regarded as a *basis*. We have indicated how empiricism has led to an account of the nature of knowledge (and of myth, religion, superstition, etc.) that is at variance with actual fact and reasonable conjecture. In the present example, we can see even more distinctly how the doctrine transforms facts. Certain *ideas* (the ideas constituting the phenomena in Newton's sense) are accepted not for what they are (viz., ideas, *good* ideas, perhaps, but still *fallible* ideas); they are presented as an immediate expression of fact. The circumstances of their origin (that an immense feat of imagination was needed, an ability to think in as yet untried categories) are covered up. Praise is applied for the wrong reasons (Newton was an excellent *observer*). Hypotheses are retained for the wrong reasons. And history, which in any case is full of fairy tales, is enriched by the narration of a few more fictitious events.

In addition to changing the notion of experience, Newton also attempted to show that the laws he stated explicitly were a direct consequence of his phenomena. These two features characterize what we have called the *second period,* a period where theories are accepted because they are supposed to be facts, where facts are rejected because they do not fit the theories that are supposed to be facts, where mathematical demonstration is used to prove what cannot be proved. At the same time, we have here a period of the greatest scientific triumphs, a period that has to report more discoveries, that has given more understanding to otherwise disconnected occurrences than almost any other period in the history of thought. The predictive successes achieved, the extension of knowledge obtained, the boundless optimism of most of these early modern scientists, all this seemed to be a sufficient refutation of the objections that were occasionally raised by some philosophers and that suggested (as we know now, correctly) that the alleged *foundation* was nonexistent and that the successes were lucky accidents. Empiricism and the double talk created by its presence dominated the scene down to the nineteenth century.[67]

There were some changes, however, that did not significantly influence the philosophy of scientists. Attempts were made to reformulate the principles of empiricism in order to obtain agreement with actual scientific practice. They went in two different directions and foreshadowed the split of opinions characteristic of the empiricism of the twentieth century. Kant's theory of knowledge fully recognized the theoretical elements contained in classical physics (which he identified with Newton's celestial mechanics) and discussed their influence upon the most common observational results. This trend was continued by Whewell and culminated in Einstein's *On the Method of Theoretical Physics* and Popper's *Logic of Scientific Discovery.* Mach, on the other hand, attacked the exceptional position given to mechanics on the ground that it could not be justified by experience.[68] He envisaged a new, phenomenological physics and hoped it might be based upon electrodynamics.[69] But the criticisms of these philosophers count only little when compared with the tremendous effect that the final breakdown of the classical point of view had upon the attitude of physicists. This, after all, was the first real and undeniable crisis after one had pretended for more than 250 years that an empiricist who refrained from speculation would not have to take back anything he said, and that in fact nothing was ever taken back.[70] The two philosophical positions that we have just described are now further developed into two distinct methodologies offering two distinct sets of prescriptions for the future. It is admitted by both that classical physics was not fully observational. Very different lessons are drawn from this admission. Einstein pointed out that all physics shares the fallibility of the classical point of view and that every physical theory must therefore be treated with caution. Physical theories are "free mental creations" that can only occasionally be related to observations.[71] The younger physicists, notably those thinkers who tried to improve the theory of matter under the guidance of Niels Bohr, draw a very different conclusion. The breakdown of the classical physics indicates to them that it was speculation, and therefore not physics at all. The fact that despite the general upheaval some classical laws had remained strictly valid (laws of interference, conservation laws, fluctuations) suggests that it nonetheless contains a physical core. They set out to free this core from its metaphysical trappings, and to add to it by further investigation. This, they think, might lead to a point of view that is (a) correct and (b) fully empirical. Considering this origin of the envisaged future theory, the elements already obtained (and especially the elementary quantum theory) are called a "rational generalization of the classical mode of description."[72] Such a generalization cannot leave untouched the concepts of classical physics. These concepts contain theoretical elements, and their use implies the assertion of certain theoretical assumptions. A radical empiricist must restrict the validity of these assumptions and thereby reveal their limitations (A 4, 5, 8, 13, 16, 53, 108).[73] New rules for the use of the classical concepts have to be devised "in order to evade the quantum of action" (A 18). These rules must satisfy the following demands: (a) they must allow for the description of any conceivable experiment in classical terms (which are still taken to be the at least partly correct representatives of experience); (b) they must "provide room for new laws" (R 701; A 3, 8, 19, 53),[74] and especially for the quantum of action (A 18); (c) they must always lead to correct predictions. (a) is needed if we want to retain the idea that experience is to be described in classical terms; (b) is needed if we want to avoid any clash with the quantum of action; (c) is needed if this set of rules is to be as powerful as a physical theory in the usual sense. Any set of rules satisfying (a), (b), and (c) is called by Bohr a "natural generalization of the classical mode of description" (A 4, 56, 70, 92, 110; D 316; E 210, 239) or a "reinterpretation . . . of the classical electron theory" (A 14). "The aim of regarding the quantum theory as a rational generalization of the classical theory led to the formulation of the . . . correspondence principle" (A 37, 70, 110). The correspondence principle is the tool by means of which the needed generalizations may be, and have been, accomplished. The rules of quantization allow us to replace the fairly intuitive considerations of the correspondence principle by a procedure that can be applied more mechanically. It was Bohr's belief that *any future quantum theory* could only be a rational generalization of classical physics in the sense just explained. Most physicists have followed him in this respect. They have even taken it for granted that theirs was the only method deserving serious consideration, and they have been quite emphatic and surprisingly intolerant in the rejection of the metaphysical "monsters" of their opponents.[75] Indeed, one can say that the ideal of a purely factual theory which formed the background of much that was done in the Royal Society was first realized by Bohr and his followers and not, as had been thought, by Galileo and by Newton himself. Bohr seems to have been aware of this feature of his investigations, and he therefore included even relativity in his notion of "classical physics." Classical physics is still metaphysics.[76] Physics starts with the research carried out by him, Kramers, Heisenberg, and the other members of the Copenhagen school of quantum theory.

It is not possible to examine here all the details of this development. Part of such an examination can be found in *Problems of Microphysics* and in *Complementarity*. For the present, the following remark must suffice: in *theory*, the structure consisting of a well-defined dualistic experience and a series of formalisms designed to make the correct predictions seems to be impeccable. In *practice*, there is so much dabbling and guessing that the success of the structure cannot at all be properly appreciated. Moreover, we cannot accept without criticism the epistemological arguments supporting this structure. The structure is said to be final and impeccable because it is firmly based upon experience: the parts of classical physics that went beyond experience have been eliminated; the successful parts have been retained and have been carefully described in such a manner that no further assumptions are involved. Now the investigation leading to this result consisted only in a *direct* comparison between the theory and "experience." Alternatives were not used. They were even excluded, and the argument was that the direct tests have not uncovered any difficulty. This assumes that a theory must be retained until it is refuted, or at least until some new facts indicate its limitations. It assumes that the construction and the development of alternative theories in the same domain must be postponed until such refutation or limitation has taken place. That is, it assumes a *radical empiricism*. It is now time to examine this more specific philosophical doctrine in some detail. I shall proceed in a piecemeal fashion and discuss, first, two principles that are connected with it, the principles of consistency and meaning invariance.

III. Two Conditions of Contemporary Empiricism; Historical Predecessors

In this section, I intend to give a more detailed outline of some assumptions of contemporary radical empiricism that have been widely accepted. It will be shown in the sections to follow that these apparently harmless assumptions, which have been explicitly formulated by some logical empiricists, which seem to guide the work of a good many physicists, but which have also played a most important role in the past history of physics and of philosophy, are bound to lead to exactly the results I have outlined in the introduction: dogmatic petrification and the establishment, on so-called "empirical" grounds, of a rigid metaphysics.

One of the cornerstones of contemporary philosophical empiricism, which is but a highly formalized version of radical empiricism, is its theory of explanation.[77] This theory is an elaboration of some simple and very plausible ideas that were first proposed by Popper,[78] and it may be introduced as follows: Let T and T' be two different scientific theories, T' the theory to be explained, or the *explanandum,* T the explaining theory, or the *explanans.* Explanation (of T') consists in the *derivation* of T' from T and initial conditions, which specify the domain D' in which T' is applicable. *Prima facie* this demand of derivability seems to be a very natural one to make, for "otherwise the explanans would not constitute adequate grounds for the explanation."[79] It implies two things: first, that the consequences of a satisfactory explanans T inside D' must be compatible with the explanandum T'; and second, that the main descriptive terms of these consequences must either coincide, with respect to their meanings and with the main descriptive terms of T', or at least be related to them via an empirical hypothesis. The latter result can also be formulated by saying that the meaning of T' must not be affected by the explanation. "It is of utmost importance that the expressions peculiar to a science will possess meanings that are fixed by its *own* procedures, and are therefore intelligible in terms of its own rules of usage, whether or not the science has been, or will be [explained in terms of] some other discipline."[80]

Now if we take it for granted that more general theories are always introduced with the purpose of explaining the existent successful theories, then every new theory will have to satisfy the two conditions just mentioned. Or, to state it in a more explicit manner,

> Only such theories are then admissible in a given domain which either *contain* the theories already used in this domain, or which are at least *consistent* with them inside the domain;[81] and meanings will have to be invariant with respect to scientific progress; that is, all future theories will have to be framed in such a manner that their use in explanations does not affect what is said by the theories, or factual reports to be explained.

These two conditions I shall call the *consistency condition,* and the *condition of meaning invariance,* respectively.[82]

Both conditions are *restrictive* and therefore bound profoundly to influence the growth of knowledge. They restrict theorizing to a single, internally consistent point of view. This point of view may be expanded or enriched in its details, but it must not be changed in any fundamental way. The conditions of consistency and of meaning invariance therefore encourage a *theoretical monism* and discourage a *theoretical pluralism.* They have a long and rather interesting history. They form an essential part of such philosophies as Platonism and Cartesianism. It is clear that Platonism, at least in the form in which it is usually presented, demands meaning invariance. Meanings are here eternal and unchanging entities. A theory will *be* satisfactory if it correctly represents the properties of, and the relations between, these entities. It will *remain* satisfactory if the

correct mode of representation is preserved in the course of further research, i.e., if the meanings of the key terms of the theory are kept
unchanged. The same demand reappears in empiricism, observational
meanings replacing the mathematical meanings that were the Platonic
representatives of knowledge. Thus the existence, within a certain tradition, of a variety of opinions (or of a variety of theories) has *almost always*
been regarded as proof of the unsoundness of the method adopted by the
members of this tradition, and even criticism of *character* has been applied to those who either would not agree with the basic principles of a
prevalent point of view or found reason to change their own philosophy
in the course of their lives.[83] Almost always it was assumed, as being
nearly self-evident, that the proper method, sincerely applied, must lead
to the truth, that the truth is one, and that the proper method must
therefore result in the establishment of a single theory and the perennial
elimination of all alternatives. Conversely, the existence of various points
of view and of a community where discussion of alternatives is regarded
as fundamental has always been regarded as a sign of confusion or as due
to a lack of modesty and subordination.[84] Curiously enough, this attitude
is found in thinkers who otherwise have very little in common. This can
be seen from an examination of various criticisms of the pre-Socratic
philosophers that have been developed in the course of history.[85]

As has been argued by Popper,[86] these early philosophers not only
were the inventors of a *theoretical science* (as opposed to a science that is
content with assembling empirical generalizations as was the physics,
mathematics, and astronomy of the Egyptians), but they also invented the
method characteristic of this kind of science, i.e., the method of test within
a class of mutually inconsistent, partly overlapping, and to that extent
empirically adequate theories. This was immediately attacked by recourse
to general prejudice. Thus the sophists are reported to have ridiculed the
Ionians by pointing out that their motto seemed to be "to every philosopher his own principle." Plato made full use of this popular sentiment
(*Sophistes* 242ff.), and so did the Church Fathers later on: "As of canonical
authors, God forbid that they should differ. . . . [But] let one look
amongst all the multitude of philosophers' writings, and if he finds two
that tell both one tale in all respects, it may be registered for a rarity."[87]
Soon after the rise of Baconian empiricism, with its apparently so very
different message, the variety of the theories discussed by the pre-Socratics
was used as an example of where one gets when leaving the solid ground,
now not of revelation or of philosophical intuition, but of *experience.*
The following quotation is very characteristic: "As to the particular tenets
of Thales, and his successors of the *Ionian* school, the sum of what we
learned from the imperfect accounts we have of them is that each over-
threw what his predecessor had advanced; and met with the same treatment himself from his successor . . . so early did the passion for systems

begin."[88] In fact, nearly all inventors of new methods in philosophy as well as in the sciences were inspired by the hope that they would be able to put an end to the quarrel of the schools and to establish *the one true body of knowledge*. (This inspiration is present even today in some of the defenders of the Copenhagen interpretation of the quantum theory.) It will be instructive to quote some examples that show to what extent the demand for a theoretical monism and the more specific demands of consistency and meaning invariance have influenced scientists.

Taking first an earlier example, we find that in his *Wärmelehre*, Ernst Mach makes the following remark: "Considering that there is, in a purely mechanical system of absolutely elastic atoms, no real analogue for the *increase of entropy*, one can hardly suppress the idea that a violation of the second law . . . should be possible if such a mechanical system were the *real* basis of thermodynamic processes."[89] And referring to the fact that the second law is a highly confirmed physical law, he insinuates that for this reason the mechanical hypothesis must not be taken too seriously.[90] There were many similar objections against the kinetic theory of heat.[91] More recently, Max Born has based his arguments against the possibility of a return to determinism upon the consistency condition and the assumption, which we shall here take for granted, that wave mechanics is incompatible with determinism. "If any future theory should be deterministic," he writes, expressing what is believed by many physicists, "it cannot be a modification of the present one, but must be entirely different. How this should be possible without sacrificing a whole treasure of well-established results (i.e., without contradicting highly confirmed physical laws and thereby violating the consistency condition) I leave the determinist to worry about."[92] Most members of the so-called Copenhagen school of quantum theory would argue in a similar manner. As has been indicated, they believe that the idea of complementarity and the formalism of quantization expressing this idea are free from any hypothetical element, and that both are "uniquely determined by the facts."[93] A theory that contradicts them is factually inadequate and must be removed. Conversely, an explanation of complementarity is acceptable only if it either *contains* this idea or is at least *consistent* with it. This is how the consistency condition is used in arguments against theories such as those of Bohm, De Broglie, and Vigier, and of other physicists who disagree with the Copenhagen point of view.[94]

The use of the consistency condition is not restricted to such general remarks, however. A decisive part of the existing quantum theory itself, viz., the projection postulate,[95] is the result of the attempt to give an account of the definiteness of macro-objects and macro-events that is in accordance with it. The influence of the condition of meaning invariance goes even further. "The Copenhagen interpretation of the quantum theory starts from a paradox. Any experiment in physics, whether it refers to the phenomena of daily life or to atomic events, is to be described in the terms of classical physics. . . . *We cannot and should not replace these concepts by any others.* Still, the application of these concepts is limited by the relations of uncertainty. We must keep in mind this limited range of applicability of the classical concepts while using them, but we cannot, and we should not try to improve them."[96] This means, of course, that the meaning of the classical terms must remain invariant with respect to any future explanation of the microphenomena. Microtheories have to be formulated in such a manner that this invariance is guaranteed. The principle of correspondence and the formalism of quantization connected with it were explicitly devised for satisfying this demand. Altogether, the quantum theory seems to be the first theory after the downfall of the Aristotelian physics that has been quite explicitly constructed, at least by some of the inventors, with an eye both on the consistency condition and on the condition of meaning invariance.[97] In this respect it is very different indeed from, say, relativity, which violates both consistency and meaning invariance with respect to earlier theories. Most of the arguments used for the defense of its customary interpretation also depend on the validity of these two conditions, and they will collapse with their removal. An examination of these conditions is therefore very topical and bound deeply to affect present controversies in microphysics.

The demand for a theoretical monism is not restricted to physics and to philosophy. It is also implied in the customary evaluation of the *history* of scientific ideas.[98] This evaluation proceeds from the *fact* (which, however, is sometimes suppressed by a history too much under the influence of a philosophy favoring a theoretical monism[99]) that the sciences show periods of normal development and periods of crisis. Normal science is dominated by a single theory; almost all research problems are determined by the attempt to apply this single theory to as many and as diverse facts as possible. No difficulty seems to exist; the confidence in the capabilities of the theory is unbounded. Theoretical monism is an historical fact. Crises, on the other hand, have a very different structure. The one theory that guided research in the preceding time of normal development has broken down. A great variety of attempts are made to find a new theory that is no longer beset by the difficulties of its predecessor and will be capable of playing a similarly singular role in the future. Theoretical pluralism is the most decisive feature of a crisis.

Now it is commonly assumed—and this is where the *evaluation* starts —that crises are, or at least should be, *transitory stages* in the history of thought, that they are periods of disorder and embarrassment which are void of knowledge and provide no suitable basis for methodological discussions. Only science at its best contains genuine knowledge, and science is at its best in those sometimes very long periods when a single point of view reigns supreme.[100] This evaluation quite clearly flows from the belief that genuine knowledge has to consist of a single, consistent point of view; it flows from a preference for theoretical monism.

It is now time to investigate in detail the validity of this common belief in uniformity. I shall start the investigation by showing that some of the most interesting developments of physical theory in the past have violated both the consistency condition and the condition of meaning invariance.

IV. These Conditions Not Invariably Accepted by Actual Science

The case of the consistency condition can be dealt with in a few words; it is well known (and has also been shown in great detail by Duhem[101]) that Newton's theory is inconsistent with Galileo's law of the free fall and with Kepler's laws; that statistical thermodynamics is inconsistent with the second law of the phenomenological theory; that wave optics is inconsistent with geometrical optics; and so on. Note that what is being asserted here is *logical* inconsistency: it may well be that the differences of prediction are too small to be detectable by experiment. Note also that what is being asserted is not the inconsistency of, say, Newton's theory and Galileo's law, but rather the inconsistency of *some consequences* of Galileo's law and of Newton's theory in the domain of validity of that law. In the former case, the situation is especially clear. Galileo's law asserts that the acceleration of the free fall is a constant, whereas application of Newton's theory to the surface of the earth gives an acceleration that is not a constant but *decreases* (although imperceptibly) with the distance from the center of the earth. Conclusion: if actual scientific procedure is to be the measure of method, then the consistency condition is inadequate.

The case of meaning invariance requires a little more argument, not because it is intrinsically more difficult, but because it seems to be much more closely connected with deep-rooted prejudices. Assume that an explanation is required, in terms of the special theory of relativity, of the classical conservation of mass in all reactions in a closed system S. If m', m'', m''', . . . , m^i, . . . are the masses of the parts P', P'', . . . , P^i, . . . , of S, then what we want is an explanation of

$$\Sigma \, m^i = \text{const} \qquad\qquad (1)$$

for all reactions inside S. We see at once that the consistency condition cannot be fulfilled. According to special relativity, Σm^i will vary with the velocities of the parts relative to the coordinate system in which the observations are carried out, and the total mass of S will also depend on the relative potential energies of the parts. However, if the velocities and the mutual forces are not too large, then the variation of Σm^i predicted by relativity will be so small as to be undetectable by experiment. Now let us turn to the meanings of the terms in the relativistic law and in the corresponding classical law. The first indication of a possible change of meaning may be seen in the fact that in the classical case, the mass of an aggregate of parts equals the sum of the masses of the parts: $M(\Sigma P^i) = \Sigma M(P^i)$. This is not valid in the case of relativity, where the relative velocities and potential energies contribute to the mass balance. That the relativistic concept and the classical concept of mass are very different indeed becomes clear if we also consider that the former is a *relation,* involving relative velocities, between an object and a coordinate system, whereas the latter is a *property* of the object itself and independent of its behavior in coordinate systems. True, there have been attempts to give a relational analysis even of the classical concept (Mach). None of these attempts, however, leads to the relativistic idea, which involves quite specific transformations and must therefore be added even to a *relational* account of classical mass. The attempt to identify the classical mass with the relativistic rest mass is of no avail either, for although both may have the same *numerical value,* they cannot be represented by the same concept. The relativistic rest mass is still dependent on the coordinate system chosen (in which it is at rest and has that specific value), whereas the classical mass is not so dependent. We have to conclude, then, that $(m)_c$ and $(m)_r$ mean different things and that $(\Sigma m^i)_c = \text{const}$ and $(\Sigma m^i)_r = \text{const}$ are different assertions. This being the case, the derivation either of (1) or of a law that makes slightly different quantitative predictions from relativity, with m^i used in the classical manner, will be possible only if a further premise is added that establishes a relation between the $(m)_c$ and the $(m)_r$. Such a "bridge law"—and this is a major point in Nagel's theory of reduction and explanation[102]—is a hypothesis "according to which the occurrence of the properties designated by some expression in the premises of the [explanans] is a sufficient, or a necessary and sufficient, condition for the occurrence of the properties designated by the expressions of the [explanandum].[103] Applied to the present case, this means the following: under certain conditions the occurrence of relativistic mass of a given magnitude is accompanied by the occurrence of classical mass of a corresponding magnitude. This assertion is inconsistent with another part of the explanans; it is inconsistent with the theory of relativity. After all, the theory of relativity asserts that there are no absolute (classical) masses, and it therefore asserts that "$(m)_c$" does not express real features of physical systems. We inevitably arrive at the conclusion that classical mass conservation cannot be explained in terms of relativity, or reduced to relativity, without a violation of meaning invariance. And if one retorts, as has been done by some critics of the ideas expressed here,[104] that meaning invariance is an *essential* part of both reduction and explanation, then the answer will simply be that (1) can neither be reduced to, nor explained in terms of, relativity in this new, and more strict, sense of "to reduce" or "to explain." Whatever the *words* used for describing the situation, the *fact* remains that actual science does not always observe the requirement of meaning invariance.

It can easily be seen why this is so. After all, we very often discover that entities we thought existed did, in fact, not exist. Realizing this, we must *eliminate* and *replace* the terms designating these entities from our factual descriptions (which, at least in the sciences, are supposed to be about *existing* things, processes, relations, etc.), and this may, of course, lead to a violation of the demand for meaning invariance.

This argument is quite general and independent of whether the terms whose meaning is under investigation are observational or not. (Even objects of alleged direct observation may turn out not to exist.) It is therefore stronger than may seem at first sight. Most empiricists would admit that the meaning of *theoretical* terms may be changed in the course of scientific progress. However, hardly anyone nowadays is prepared to extend meaning *invariance* to *observational* terms too. In Sec. XV we shall investigate some of the ideas responsible for such reluctance and shall then also discuss in some detail the problem of observation. Anticipating results obtained there, one may connect the demand for observational meaning invariance with the belief that the meaning of observational terms is uniquely determined by the procedures of observation and/or by the phenomena appearing as their result. These procedures and phenomena very often remain unaffected by theoretical advance; hence observational terms, too, remain unaffected. What is overlooked here is that the "logic" of the observational terms is not exhausted by the procedures of their application "on the basis of observation." It also depends on the more general ideas that determine the "ontology" (in Quine's sense) of our discourse. These general ideas may change without any corresponding change of observational procedures. For example, we may change our ideas about the nature, or the ontological status (property, relation, object, process, etc.) of the color of a self-luminescent object without changing the methods used for ascertaining that color (looking, for example).[105] Clearly, such a change is bound profoundly to influence the meanings of our observational terms.

All this has a decisive bearing upon some contemporary ideas concerning the interpretation of scientific theories. According to these ideas, theoretical terms receive their meanings via correspondence rules which connect them with an observational language that has been *fixed in advance* and independently of the structure of the theory to be interpreted. The results we have just obtained would seem to show that if we interpret scientific theories in the manner accepted by the scientific community and *not* in the way suggested by the double language model, then most of these correspondence rules will be either false or nonsensical. They will be false if they *assert* the existence of entities denied by the new theory, and they will be nonsensical if they *presuppose* this existence. Turning the argument around, we can also say that the attempt to interpret the calculus of some theory that has been voided of the meaning assigned to it by the scientific community with the help of the double language system will lead to a very different theory. Let us again take the theory of relativity as an example. It can be safely assumed that the physical thing language of Carnap, and any similar language that has been suggested as an observation language, is not Lorentz invariant. The attempt to interpret the *calculus* of relativity on *its* basis cannot therefore lead to the *theory* of relativity as it was understood by Einstein. What we shall obtain will be at the very most *Lorentz's interpretation,* with its inherent asymmetries. This undesirable result cannot be evaded by the *demand* to use a different, and more adequate, observation language. The double language system assumes that theories which are not connected with some observation language do not possess an interpretation. The demand assumes that they do, and asks us to choose the observation language most suited to it. It reverses the relation between theory and experience that is characteristic for the double language system. This means that it gives up this system. Contemporary empiricism, therefore, has not led to any satisfactory account of the meaning of scientific theories.[106]

Nor is it correct to assume that the theories we possess today could have been obtained by strict adherence to a theoretical monism. True, the assumption of the superiority, from the point of view of genuine knowledge, of periods of normal development (in the sense explained at the end of Sec. III) frequently has some very conspicuous *practical* consequences. Attempts to develop alternative theories during a period of normal science are discouraged; they are regarded as a willful, childish, and very "unscientific" activity, which, if successful, would result only in another period of multiplicity, in another period of chaos. Strong terms and strong methods (refusal to publish) are used against those who favor alternative views (for examples, see Secs. VI and VII below). This defense of uniformity sometimes assumes an almost religious fervor. Still, the revolutions that terminate such periods of uniformity are almost always the result of the fact that alternatives *have been* developed and that their existence has precipitated further change. Thus the Newtonian point of view was revised because of the problems arising from its confrontation, in the light of empirical facts, with alternatives that had been developed in the nineteenth century. The different transformation properties of Maxwell's theory and of Newton's dynamics were an important motive for the development of the theory of relativity[107]—as a matter of fact, this was the way in which Einstein originally arrived at some of the basic ideas of the theory.[108] The confrontation of the kinetic theory of heat and of the second law of the phenomenological thermodynamics forces one to recognize the existence of objective probabilities *in addition* to the deterministic fields postulated by the Newtonian point of view.[109] And finally, the comparison, in the light of experimental evidence, of Young's version of the wave theory with a particle philosophy led to the idea of complementarity and to all the succeeding attempts to give a rational account of this idea. It is worth repeating that the alternatives leading to such progress were always introduced against great resistance from the scientific community. This is true of Maxwell's theory, of Boltzmann's investigations,[110] of Young's ideas, and even of the second law of the phenomenological theory.[111] Still, without these alternatives, contemporary science would not be what it is today.

The realization that alternatives precipitate progress suggests an evaluation of the relative merits of normal science and periods of crisis that differs radically from what was reported at the end of Sec. III. Normal science, extended over a considerable time, now assumes the character of stagnation, of a lack of new ideas; it seems to become a starting point for dogmatism and metaphysics. Crises, on the other hand, are now not accidental disturbances of a desirable peace; they are periods where science is at its best, exhibiting as they do the method of progressing through the consideration of alternatives. "Revolution in Permanence!"—this, so it seems, is the battle cry of science, which is not the petrified leftover of discoveries made long ago, but an evergrowing enterprise of empirical knowledge. Do these considerations by themselves establish the desirability of a theoretical pluralism for the sciences?

They do not. There are two reasons. First, we have indicated only that pluralism precipitated progress *as a matter of historical fact;* we have not shown that it is a necessary way of advancing knowledge (coffee stimulates thought, but this will not be treated in the theory of knowledge). Second, it is not at all clear, in advance of a more detailed investigation, that what we call progress (transition to theories inconsistent with previous theories) is something valuable and that stability is to be rejected. We have seen that almost every philosopher of note has argued the opposite, that change has universally been regarded as indicating absence of truth, and that possession of knowledge has been made dependent on stability. Moreover, we have seen that even those thinkers, those "philosophers of science," who pay lip service to progress and who loudly welcome any single scientific advance usually defend a philosophy, which makes it impossible even to *state* the most important elements of such advance.[112]

Our argument is therefore as yet incomplete. What we have shown is that the two conditions of Sec. III have been violated in the course of scientific practice, and especially at periods of scientific revolution, and that the implied monism of theory construction is not a universal feature of existing scientific knowledge. These are historical facts, and facts, taken by themselves, cannot justify methodological principles. Actual scientific practice, therefore, cannot be our last authority.[113] We have to find out whether consistency and meaning invariance are desirable conditions and whether a theoretical monism is a desirable state of affairs; and we have to find this out independently of who accepts them and praises them and how much money has been won with their help. Such an investigation will be carried out in the sections to follow.[114]

V. Inherent Unreasonableness of Consistency Condition

Prima facie the case of the consistency condition can be dealt with in very few words. Consider for that purpose a theory T' that successfully describes the situation in the domain D'. From this we can infer (1) that T' agrees with a *finite* number of observations (let their class be F); and (2) that it agrees with *these* observations inside a margin M of error only.[115] Any alternative that contradicts T' outside F and inside M is supported by exactly the same observations and is therefore acceptable if T' is acceptable (we shall assume that F are the only observations available). The consistency condition is much less tolerant. It eliminates a theory not because it is in disagreement with the *facts,* but because it is in disagreement with *another theory,* with a theory, moreover, whose confirming instances it shares. *It thereby makes the as yet untested part of that theory a measure of validity.* The only difference between such a measure and a more recent theory is age and familiarity. Had the younger theory been there first, then the consistency condition would have worked in its favor. In this respect, the effect of the consistency condition (and of the inductivist methodology that is sometimes used to support it) is rather similar to the effect of the more traditional methods of transcendental deduction, analysis of essences, phenomenological analysis, linguistic analysis. It contributes to the *preservation of the old and familiar,* not because of any inherent advantage in it—for example, not because it has a better foundation in observation than has the newly suggested alternative, or because it is more elegant—but just because it is old and familiar. This is not the only instance where, on closer inspection, a rather surprising similarity emerges between modern empiricism and some of the school philosophies it attacks.[116]

These observations, although leading to a very decisive *tactical* criticism of the consistency condition, do not go to the heart of the matter. The observations at the beginning of the present section show that an alternative to the accepted point of view which shares its confirming instances cannot be *eliminated* by factual reasoning. They do not show that such an alternative is *acceptable;* even less do they show that it *should be used.* It is bad enough, so a defender of the consistency condition might point out, that the accepted point of view does not possess full empirical support. Adding new theories of an equally unsatisfactory character will not improve the situation; nor is there much sense in trying to *replace* the accepted theories by some of their possible alternatives.

Such replacement will be no easy matter. A new formalism may have to be learned, and familiar problems may have to be calculated in a new way. Textbooks must be rewritten, university curricula readjusted, experimental results reinterpreted. And what will be the result of all the effort? Another theory that, from an empirical point of view, has no advantage whatever over the theory it replaces. The only real improvement, so the defender of the consistency condition will continue, derives from the *addition of new facts*. Such new facts will either support the current theories or will force us to modify them by indicating precisely where they go wrong. In both cases they will precipitate real progress and not just arbitrary change. The proper procedure must therefore consist in the confrontation of the accepted point of view with as many relevant facts as possible. The exclusion of alternatives is then required for reasons of expediency: their invention not only does not help, but it even hinders progress by absorbing time and manpower that could be devoted to better things. And the function of the consistency condition lies precisely in this: it eliminates such fruitless discussion and forces the scientist to concentrate on the facts, which, after all, are the only acceptable judges of a theory. This is how the practising scientist and his philosophical apologist will defend concentration on a single theory to the exclusion of all empirically possible alternatives.[117]

It is worthwhile repeating the reasonable core of this argument: theories should not be changed unless there are pressing reasons for doing so, and the only pressing reason for changing a theory is disagreement with facts. Discussion of incompatible facts will therefore lead to progress; discussion of incompatible theories will not.[118] Hence it is sound procedure to increase the number of relevant facts. It is not sound procedure to increase the number of factually adequate, but incompatible, alternative theories. One might wish to add that formal improvements, such as increase of elegance, simplicity, generality, and coherence, should not be excluded. But once these improvements have been carried out, the collection of facts for the purpose of test seems indeed to be the only thing left to the scientist.

VI. Relative Autonomy of Facts

And this it is—*provided these facts exist, and are available independently of whether or not one considers alternatives to the theory to be tested.* This assumption on which the validity of the argument in the last section depends in a most decisive manner I shall call the assumption of the relative autonomy of facts, or the *autonomy principle.* It is not asserted by this principle that the discovery and the description of facts are independent of *all* theorizing. But it *is* asserted that the facts which belong to the empirical content of some theory are available whether or not one considers alternatives to *this theory*. I am not aware that this very important assumption has ever been explicitly formulated as a separate postulate of the empirical method. Yet it is clearly implied in almost all investigations dealing with questions of confirmation and test. All these investigations use a model in which a single theory is compared with a class of facts (or observation statements) that are assumed somehow to be "given." I submit that this is much too simple a picture of the actual situation. Facts and theories are much more intimately connected than is admitted by the autonomy principle. Not only is the description of every single fact dependent on *some* theory (which may, of course, be very different from the theory to be tested), but there also exist facts that cannot be unearthed except with the help of alternatives to the theory to be tested and that become unavailable as soon as such alternatives are excluded. This suggests that the methodological unit to which we must refer when discussing questions of test and empirical content is constituted by a *whole set of partly overlapping, factually adequate, but mutually inconsistent theories;* in short, it suggests a theoretical pluralism as the basis of every test procedure. Only the barest outline of such a test model will be given here. However, before doing this, I want to discuss an example that shows very clearly the function of alternatives in the discovery of facts.

As is well known, the Brownian particle seen from a microscopic point of view is a perpetual motion machine of the second kind, and its existence refutes the phenomenological second law. It therefore belongs to the domain of relevant facts for this law. Now could this relation between the law and the Brownian particle have been discovered in a *direct* manner, i.e., by an investigation of the observational consequences of the phenomenological theory, without borrowing from an alternative account of heat? This question is readily divided into: (1) could the *relevance* of the Brownian particle have been discovered in this manner? (a psychological question) and (2) could it have been demonstrated that it actually *refutes* the second law? (an empirical question). The answer to the first question is that we do not know. After all, it is impossible to say what would have happened had the kinetic theory not been considered by some physicists. It is my guess, however, that in this case the Brownian particle would have been regarded as an oddity much in the same way in which some of Ehrenhaft's astounding effects[119] are regarded as an oddity, and that it would not have been given the decisive position it assumes in contemporary theory. The answer to the second question is, simply, no. Consider what the discovery of the inconsistency between the Brownian particle and the phenomenological second law would have required. It would have required (a) measurement of the *exact motion* of the particle in order to ascertain the charges of its kinetic energy plus the energy spent on overcoming the resistance of the fluid, and (b) precise measurements of temperature and heat transfer in the surrounding medium in order to as-

certain that any loss occurring here was indeed compensated by the increase of the energy of the moving particle and the work done against the
fluid. Such measurements are beyond experimental possibilities.[120] Neither
is it possible to make precise measurements of the heat transfer; nor can
the path of the particle be investigated with the desired precision. Hence
a "direct" refutation of the second law that considers only the phenomenological theory and the "fact" of Brownian motion is impossible. And,
as is well known, the actual refutation was brought about in a very different manner—via the kinetic theory and Einstein's utilization of it in the
calculation of the statistical properties of the Brownian motion.[121] In the
course of this procedure, the phenomenological theory (T') was incorporated into the wider context of statistical physics (T) in such a manner
that *the consistency condition was violated,* and *then* a crucial experiment
was staged (investigations of Svedberg and Perrin).

We may generalize this example as follows: assume that a theory T
has a consequence C and that the actual state of affairs in the world is correctly described by C', where C and C' are experimentally indistinguishable. Assume furthermore that C', but not C, triggers, or causes, a macroscopic process M that can be observed very easily and is perhaps well
known. In this case there exist observations, viz., the observations of M,
which are sufficient for refuting T, although there is no possibility whatever to find this out on the basis of T and of observations alone. What is
needed in order to discover the limitations of T implied by the existence
of M is another theory, T', which implies C', connects C' with M, can be
independently confirmed, and promises to be a satisfactory substitute for
T where this theory can still be said to be correct. Such a theory will have
to be inconsistent with T, and it will have to be introduced not because
T has been found to be in need of revision, but in order to discover
whether T *is* in need of revision.[122]

It seems to me that this situation is typical for the relation between
fairly general theories, or points of view, and "the facts." Both the relevance and the refuting character of many decisive facts can be established
only with the help of other theories that, although factually adequate, are
not in agreement with the view to be tested. This being the case, the production of genuine refutations may have to be preceded by the invention
and articulation of alternatives to that view. Empiricism demands that
the empirical content of whatever knowledge we possess be increased as
much as possible. Hence *the invention of alternatives in addition to the
view that stands in the center of discussion constitutes an essential part of
the empirical method.*[123] Conversely, the fact that the consistency condition eliminates alternatives now shows it to be in disagreement with empiricism, and not only with scientific practice. By excluding valuable tests,
it decreases the empirical content of the theories that are permitted to remain (and that, as we have indicated above, will usually be the theories

that have been there first); and it especially decreases the number of those facts that could show their limitations. The last result of a determined application of the consistency condition is of very topical interest: it may well be that the refutation (or an interesting confirmation) of the quantum mechanical uncertainties presupposes just such an incorporation of the present theory into a wider context that is no longer in accordance with the idea of complementarity and that therefore suggests new and decisive experiments.[124] And it may also be that the insistence of the majority of contemporary physicists on the consistency condition will, if successful, forever protect these uncertainties from refutation. This is how a radical empiricism may lead to a situation where a certain point of view petrifies into dogma by being, in the name of experience, completely removed from any fundamental criticism.

VII. The Self-Deception Involved in All Uniformity

It is worthwhile to examine this empirical defense of a dogmatic point of view in somewhat greater detail. Assume that physicists have adopted, either consciously or unconsciously, the idea of the uniqueness of complementarity, and that they therefore elaborate the orthodox point of view and refuse to consider alternatives. In the beginning, such a procedure may be quite harmless. After all, a man can do only so many things at a time, and it is better when he pursues a theory in which he is interested rather than a theory he finds boring. Now assume that the pursuit of the theory he chose has led to successes, and that the theory has explained in a satisfactory manner circumstances which had been unintelligible for quite some time. This gives empirical support to an idea that to start with seemed to possess only this advantage: it was interesting and intriguing. The concentration upon the theory will now be reinforced; the attitude toward alternatives will become less tolerant. Now if it is true, as was argued in the last section, that many facts become available only with the help of alternatives, then the refusal to consider them *will result in the elimination of potentially refuting facts*. More especially, it will eliminate facts whose discovery would show the complete and irreparable inadequacy of the theory.[125] Such facts having been made inaccessible, the theory will appear to be free from blemish,[126] and it will seem that "all evidence points with merciless definiteness in the . . . direction . . . [that] all the processes involving . . . unknown interactions conform to the fundamental quantum law."[127] This will further reinforce the belief in the uniqueness of the current theory and in the complete futility of any account that proceeds in a different manner. Being now very firmly convinced that there is only one good microphysics, the physicists will try to explain even adverse facts in its terms, and they will not mind when such explanations are sometimes a little clumsy. By now the success of the theory has also become public news. Popular science books (and this includes a good many books on the philosophy of science) will spread the basic postulates of the theory, and applications will be made in distant fields. More than ever the theory will appear to possess tremendous empirical support. The chances for the consideration of alternatives are now very slight indeed. The final success of the fundamental assumptions of the quantum theory and of the idea of complementarity will seem to be assured.

At the same time it is evident, on the basis of the considerations in the last section, that this appearance of success *cannot in the least be regarded as a sign of truth and correspondence with nature.* Quite the contrary, the suspicion arises that the absence of major difficulties is a result of the decrease of empirical content brought about by the elimination of alternatives, and of facts that can be discovered with the help of these alternatives only. In other words, *the suspicion arises that this alleged success is due to the fact that in the process of application to new domains, the theory has been turned into a metaphysical system.* Such a system will, of course, be very "successful," not, however, because it agrees so well with the facts, but because no facts have been specified that would constitute a test and because some such facts have even been removed. Its "success" is *entirely man-made:* it was decided to stick to some ideas, and the result was, quite naturally, the survival of these ideas. If now the initial decision is forgotten, or made only implicitly, then the survival will seem to constitute independent support; it will thereby reinforce the decision, or turn it into an explicit one, and in this way close the circle. This is how empirical "evidence" may be *created* by a procedure that quotes as its justification the very same evidence it has produced in the first place.

At this point, an "empirical" theory of the kind described (and let us always remember that the basic principles of the present quantum theory[128] are uncomfortably close to forming such a theory) becomes almost indistinguishable from a myth. In order to realize this, we need only consider that because of its all-pervasive character, a myth such as the myth of witchcraft and of demonic possession would possess a high degree of confirmation on the basis of observation.[129] Such a myth has been taught for a long time; its content is enforced by fear, prejudice, and ignorance, as well as by a jealous and cruel priesthood. It penetrates the most common idiom, infects all modes of thinking and many decisions that mean a great deal in human life. It provides models for the explanation of any conceivable event—conceivable, that is, for those who have accepted it.[130] This being the case, its key terms can be fixed in an unambiguous manner, and the idea (which may have led to such a procedure in the first place) that they are copies of unchanging entities and that the change of meaning, if it should happen, is due to human mistake will appear very plausible. Such plausibility reinforces all the maneuvers that are used for the preservation of the myth (elimination of opponents included). The conceptual apparatus of the theory and the emotions connected with its application having penetrated all means of communication, all actions, perceptions,[131] and, indeed, the whole life of the community, such methods as transcendental deduction, analysis of usage, phenomenological analysis, which are means for further solidifying the myth, are bound to be extremely successful. This again shows that all these methods which have been the trademark of various philosophical schools, old and new, have one thing in common: they tend to preserve the *status quo* of the intellectual life.[132] Observational results, too, will speak in favor of the theory, as they are formulated in its terms. It will seem that at last the truth has been arrived at. At the same time, it is evident that all contact with the world has been lost and that, the stability achieved, the semblance of absolute truth *is nothing but the result of an absolute conformism.*[133] For how can we possibly test, or improve upon, the truth of a theory if it is built in such a manner that any conceivable event can be described and explained in terms of its principles? The *only* way of investigating such all-embracing principles is to compare them with a different set of *equally* all-embracing principles—but this way has been excluded from the very beginning.[134] The myth is therefore of no objective relevance; it continues to exist solely as the result of the effort of the community of believers and of their leaders, be these now priests or Nobel prize winners. *Its "success" is entirely man-made.* This, I think, is the most decisive argument against any method that encourages uniformity, be it now termed "empirical" or not. Any such method is in the last resort a method of deception. It enforces an unenlightened conformism, and speaks of objective truth; it leads to a deterioration of intellectual capabilities, and speaks of deep insight; it destroys the most precious gift of the young, their tremendous power of imagination, and yet speaks of education.

To sum up: *unanimity of opinion may be fitting for a church, for the frightened or dazzled victims of some (ancient or modern) myth, or for the weak and willing followers of some tyrant; variety of opinion is a feature necessary for objective knowledge; and a method that encourages variety is also the only method that is compatible with a humanistic outlook.*[135] To the extent to which the consistency condition (and, as will emerge, the condition of meaning invariance) delimits variety, it contains a theological element (which lies, of course, in the worship of "facts" so characteristic of nearly all empiricism[136]). This finishes my criticism of the consistency condition.

VIII. Inherent Unreasonableness of Meaning Invariance

What we have achieved so far has immediate application to the question whether the meaning of certain key terms should be kept unchanged in the course of the development and improvement of our knowledge. After all, the meaning of every term we use depends upon the theoretical context in which it occurs. Words do not "mean" something in isolation; they obtain their meanings by being part of a theoretical system.[137] Hence if we consider two contexts with basic principles that either contradict each other or lead to inconsistent consequences in certain domains, it is to be expected that some terms of the first context will not occur in the second with exactly the same meaning. Moreover, if our methodology demands the use of mutually inconsistent, partly overlapping, and empirically adequate theories, then it thereby also demands the use of conceptual systems that are *mutually irreducible* (their primitives cannot be connected by bridge laws that are meaningful *and* factually correct), and it demands that the meanings of all terms be left elastic and that no binding commitment be made to a certain set of concepts.

It is very important to realize that such a tolerant attitude toward meanings, or such a change of meaning in cases where one of the competing conceptual systems has to be abandoned, need not be the result of directly accessible observational difficulties. The law of inertia of the so-called *impetus theory* of the later Middle Ages[138] and Newton's own law of inertia are in perfect quantitative agreement: both assert that an object which is not under the influence of external forces will proceed along a straight line with constant speed. Yet despite this fact, the adoption of Newton's theory entails a conceptual revision that makes it necessary to abandon the inertial law of the impetus theory, not because it is quantitatively incorrect, but *because it achieves the correct predictions with the help of inadequate concepts:* the law asserts that the *impetus* of an object that is beyond the reach of external forces remains constant.[139] Impetus is interpreted as an inner force that pushes the object along. Within the impetus theory, such a force is quite conceivable, as it is assumed here that forces determine velocities rather than accelerations. But this means that the concept of impetus is formed in accordance with a law (forces determine velocities), which is inconsistent with the laws of Newton's theory and must therefore be abandoned as soon as the latter law is adopted. This is how the progress of our knowledge may lead to conceptual revisions for which no direct observational reasons are available. The occurrence of such changes quite obviously refutes the contention of some philosophers that the invariance of *usage* in the trivial and uninteresting contexts of the private lives of not-too-intelligent and inquisitive people indicates invariance of *meaning* and the superficiality of all scientific changes. It is also a very decisive objection against any crudely operationalistic account of both observable terms and theoretical terms.

What we have said applies, of course, also to singular statements of observation. Statements that are empirically adequate and are the result of observation (such as "here is a table") may have to be reinterpreted, not because it has been found that they do not adequately express what is seen, heard, felt, but because of changes in sometimes very remote parts of the conceptual scheme to which they belong.[140] Witchcraft is again a very good example. Numerous eyewitnesses claim that they have actually *seen* the devil, or *experienced* demonic influence. There is no reason to suspect that they were lying. Nor is there any reason to assume that they were sloppy observers, for the phenomena leading to the belief in demonic influence are so obvious that a mistake is hardly possible (possession, split personality, loss of personality, hearing voices, etc.). These phenomena are well known today.[141] In the conceptual scheme that was the one generally accepted in the fifteenth and sixteenth centuries, the only way of describing them, or at least the way that seemed to express them most adequately, was by reference to demonic influences. Large parts of this conceptual scheme were changed for philosophical reasons, and also under the influence of the indirect evidence accumulated by the sciences. Descartes' materialism played a very decisive role in discrediting the belief in spatially localizable spirits. The language of demonic influences was no part of the new cosmology that was created in this manner. It was for this reason that a reformulation was needed, and a reinterpretation of even the most common "observational" statements. Combining this example with the remarks at the beginning of the present section, we now realize that according to the method of classes of alternative theories, a lenient attitude must be taken with respect to the meanings of *all* the terms we use. We must not attach too great an importance to "what we mean" by a phrase, and we must be prepared to change whatever little we have said concerning this meaning as soon as the need arises. Too great concern with meanings can lead only to dogmatism and sterility. Flexibility, and even sloppiness, in semantical matters is a prerequisite of scientific progress![142]

IX. Consequences of Point of View Adopted

Three consequences of the results obtained deserve a more detailed discussion. The first consequence is an evaluation of metaphysics that differs significantly from the standard empirical attitude. As is well known, there are empiricists who demand that science start from observable facts and proceed by generalization, and who refuse the admittance of metaphysical ideas or of "hypotheses" at any point of this procedure. For them only a system of thought that has been built up in a purely inductive fashion can claim to be genuine and undisturbed knowledge. Theories that are partly metaphysical, or that derive from metaphysical parents, or that have been brought to life by metaphysical midwives are suspect and are best not used at all. They might be contaminated, and they might give a false account of facts. They might introduce words where knowledge of facts is needed. Metaphysicians, says Bacon, "certainly have this in common with children, that they are prone to talking, and incapable of generation, their wisdom being loquacious and unproductive of effects."[143] A very similar attitude is expressed by Newton in his reply to Pardies' second letter concerning the theory of colors: ". . . if the possibility of hypotheses is to be the test of truth and reality of things, I see not how certainty can be obtained in any science; since numerous hypotheses may be devised, which shall seem to overcome new difficulties."[144]

This radical position, which clearly depends on the demand for a theoretical monism, is no longer as popular as it used to be. It is now granted that metaphysical considerations may be of importance when the task is to *invent* a new physical theory: such invention is a more or less irrational act containing the most diverse components. Some of these components are, and perhaps must be, metaphysical ideas. However, it is also pointed out that as soon as the theory has been developed in a formally satisfactory fashion and has received sufficient confirmation to be regarded as empirically successful, it is asserted that in the very same moment it can and must forget its metaphysical past; metaphysical speculation must now be replaced by empirical argument. "On the one side I would like to emphasize," writes Ernst Mach, "that *every and any* idea is admissible as a means for research, provided it is helpful; still, it must be pointed out, on the other side, that it is very necessary from time to time to free the presentation of the *results* of research from all inessential additions."[145]

This means, of course, that empirical considerations are still given the upper hand over metaphysical reasoning. Metaphysical reasoning has at most a psychological function. It is on the same level as strong coffee, absence of fatigue, satisfactory sex life. I have the impression that it is on these grounds that many students of the sciences now delve into the history of their subject. History is not studied for any theoretical purpose; it is regarded as a reservoir of ideas that might be used to advance knowledge. Taken by itself, this is not at all a deplorable attitude. Some of the best scientists have received stimulation from the past. However, this concern for the past affairs of science becomes dangerous when combined with certain more recent ideas concerning the task of the theory of knowledge.[146] It is for this reason that historians of the sciences seem to be in special need of a methodological backbone.

However, even if metaphysics should be given a theoretical function, then it is still assumed that its pronouncements must yield to the results of empirical investigation. Thus in the case of an inconsistency between metaphysics and some highly confirmed empirical theory or some results of observation, it will be decided, *as a matter of course,* that the theory, or the results of observation, must stay and that the metaphysics must go. An example that I have analyzed in *Problems of Microphysics* is the attempt to eliminate certain very general ideas concerning the nature of micro-entities on the basis of the remark that they are inconsistent with "an immense body of experience" and that "to object to a lesson of experience by appealing to metaphysical preconceptions is unscientific."[147] For a scientist, this seems to be a very natural attitude to take. What is more surprising is that philosophers very often proceed in exactly the same fashion. Even thinkers who favor a metaphysical point of view and who frown upon empiricism as much too simple-minded a theory of knowledge frequently use observational results for the purpose of discrediting a rival metaphysics. A very simple example, which will be discussed at length later, is the way in which materialism is being criticized by some of its opponents. "It is only too clear," writes a wrathful theologian,[148] "that such a primitive explanation of mental phenomena . . . cannot really do justice to their nature. *What we experience in us* is the loudest protest possible against such a shallow philosophy." Again experience is given the upper hand over speculation.

The methodology developed in the present paper leads to a very different evaluation of metaphysical speculation. Metaphysical systems are scientific theories in their most primitive stage. If they *contradict* a well-confirmed point of view, then this indicates their usefulness as an alternative. Alternatives are needed for the purpose of criticism. Hence metaphysical systems that contradict observational results or well-confirmed theories are *most welcome* starting points of such criticism. Far from being misfired attempts at anticipating, or circumventing, empirical research[149] that have been deservedly exposed by reference to experiment, they are the only means we possess for examining the assumptions implicit in our observational results.

A second consequence is that a new attitude has to be adopted with respect to the *problem of induction*. This problem consists in the question of what justification there is for asserting the truth of a statement S given the truth of another statement S' whose content is smaller than the content of S. It may be taken for granted that those who want to justify S would also assume that, after the justification, the truth of S will be *known*. Now knowledge that S is the case implies the *stability* of S (we must not change, or remove, what we know to be true). The method we are discussing at the present moment cannot allow such stability. It follows that the problem of induction, at least in some of its formulations, is a problem whose solution would lead to undesirable results. The problem may therefore be properly termed a pseudo-problem.

The third consequence concerns the manner in which *philosophical problems* and even scientific problems (especially in the social sciences) are habitually formulated, as well as some of the conditions that this formulation imposes upon the solution. Philosophical problems very often assume the form "What is X?" (for example: "what is pain?", "what is thought?", "what is space?"). This naturally implies that the answer must run something like "*X* is *Y*." Now this latter statement may be regarded as the result of an analysis or as an empirical hypothesis. In the first case, "*X*" must be synonymous with "*Y*"; in the second case, "*X*" and "*Y*" must possess at least the same extension, which means the intension of "*X*" must not contain components denied in the intension of "*Y*" or vice versa; moreover, "*X* is *Y*" must in this case also express a meaningful statement, which presupposes that formulating "*X* is *Y*" does not violate grammatical rules of the language of which "*X*" is part (it is, of course, assumed all the time that "*X*" has the same meaning in the answer that it had in the question). *In either case, the content of the answer is restricted by the form of the language of which* "X" *is part.* Now if it is taken for granted that all advance of knowledge consists in answering questions of the form "what is *X*?," then we thereby decide *forever to retain* certain parts of the language from which we started and in which we formulated our question. And we retain these parts not because they have been investigated and found to be without fault; we retain them because of the peculiar fashion in which we state our problems. Of course the question "what is *X*?" sounds harmless enough. Also, there does not seem to exist any way of replacing it by different questions without altogether giving up the quest for knowledge. Still, even the most evident procedure is unacceptable if it implies that certain parts of the knowledge or of the language we possess will forever remain unexamined (and will therefore appear to be true a priori). Moreover, it should be remembered that the parts thereby retained will be the most primitive ones (after all, we all start *historically* from some kind of ordinary language). Clearly, then, we shall have to give up either questions of the form "what is . . . ?" or the demand that "*X*" possess the same meaning in the question and in the answer. In view of our above considerations, the latter would seem to be the more natural procedure.

However, we can and must go still further. The ideas of test which we have developed in Secs. VI and VII are general enough not only to *reject* the demand for meaning invariance and the implied demand for either synonymy or partial agreement of intension, but even to *support* the stronger demand for irreducibility or semantical incompatibility. The reason is easily seen: the meaning of a term depends on the grammatical rules applied to it as well as on the theoretical assumptions in which it occurs essentially. The totality of these rules and assumptions forms a theory, or a point of view. This point of view has the peculiarity of not being wholly explicit; only part of it is expressed in the form of clear *theses* that can be examined and criticized; the rest "shows itself" in the way in which some of the terms of the point of view are being used, and particularly in dispositions to utter, or to refuse to utter, certain sentences. It is customary to say that such usage and such dispositions are governed by "grammatical rules." This verbal custom is somewhat misleading; it suggests that we are dealing with well-defined principles, whereas "grammatical rules" are hardly ever available in explicit form and often resist all attempts at an explicit formulation. This being the case, how can alternatives be constructed? We may be able to change those principles of the theory to be criticized that are explicitly stated. But in order to explore any and every aspect of it, we should also have to use alternatives that are inconsistent with the mostly unknown and perhaps even unknowable grammatical rules of the point of view. Is there any way out of the difficulty? There clearly is—we have only to insure that the main descriptive terms of the alternatives are very different from the main descriptive terms of the view to be examined; more especially, we must formulate the alternatives in such a fashion that they cannot provide any synonyms or co-extensionals.[150] An outer indication of such irreducibility that is quite striking in the case of commonly accepted ideas is the feeling of *absurdity:* we deem absurd what goes counter to well-established linguistic habits. The absence of synonymy relations connecting a newly introduced set of ideas with parts of the earlier point of view, and the feeling of absurdity connected with the new ideas indicate that these ideas are fit for the purpose of criticism, i.e., for leading either to a strong *confirmation* of the earlier theories or else to a very revolutionary *discovery:* absence of synonymy, clash of meanings, initial absurdity *are desirable;* presence of synonymy, intuitive plausibility, agreement with customary modes of speech, far from being philosophical virtues, indicate that not much progress has been or will be made. Such features are a sign that we are still moving safely within the boundaries of knowledge set by our ancestors, and that we have not even started examining whether the boundaries are correctly drawn or what goes on outside them.

It is most instructive in this connection to consider the quandaries of thinkers who want to put forth their ideas concerning some well-known entity X in the usual "X is Y" form, who also try to be philosophically conscientious and try to keep stable the meaning of "X," and who must then learn that the ideas they want to express cannot possibly be true or even meaningful. Just take the case of a materialist. Being a materialist he wants to assert the *material* character of mental processes. Being an empiricist he wants his assertion to be a testable statement about *mental* processes. Adopting meaning invariance, he tries to formulate part of his materialism with the help of the hypothesis H: X is a mental process of kind A if and only if X is a central process of kind a. But this hypothesis backfires. It may be declared to be outright nonsense (see Sec. X). But assume that it is admitted to be meaningful. It then not only implies, as it is intended to imply, that mental events have physical features; it also seems to imply (when read from right to left) that some physical events, viz., central processes, have nonphysical features. It thereby replaces the initial dualism of events (which the materialist wants to overcome) by a dualism of features. Many materialists have been convinced by an argument of this kind, or at least they have been severely shaken and have surrounded their belief with all sorts of precautionary clauses, thereby emptying it of a great deal of interest. Yet all that happened was that they were fooled by words: if materialism is correct, then hypothesis H is false. And if the hypothesis should be declared to be without sense, then the correctness of materialism implies that the mentalistic language is unfit for describing the world. Now the fitness, or unfitness, of a language for description in a certain domain surely is a matter that must be decided by factual research. If follows that *linguistic arguments are never strong enough to unsettle either an empirical hypothesis or a metaphysical system.*[151]

It seems to me that these considerations already clear the way for any kind of metaphysical speculation, provided the aim is to criticize some point of view or to have the speculation itself criticized, and not only to give learned expression to cosmic feelings. However, in philosophical matters, abstract argument is never enough. It may create annoyance, but it will not be accepted until the set of intuitions supporting a traditional point of view is replaced by a different set. It is for this reason that I shall now undertake a more detailed investigation of some aspects of the mind-body problem. This investigation will also prepare the way for the main task of this essay, which is a sketch of a new theory of experience. I start the investigation with a discussion of a very common argument against materialism.

X. Mind-Body Problem: Simple Argument Against Materialism Refuted

Materialism assumes that the only entities existing in the world are atoms and aggregates of atoms, and that the only properties and relations are the properties of and the relations between such aggregates. A simple atomism such as the theory of Democritus will be sufficient for our purpose. The refinements of the kinetic theory, or of the quantum theory, are outside the domain of discussion. And the question is: will such a cosmology give a correct account of human beings?

It is assumed by many philosophers that this question must be answered in the negative. Human beings, apart from being material, also have *experiences;* they *think;* they *feel pain;* etc. These processes cannot be analyzed in a materialistic fashion. The reason is that there are various things I can assert, say, about a particular sensation I am having now that I cannot assert about any material process, and this because such an assertion would be either *false* or *meaningless.* Conversely, there are various things I can assert about material processes that I cannot assert about sensations, and this again because such assertions would be either false or meaningless. Thus I can say of a sensation that it is confused or clear, whereas a material process is neither confused nor clear, but just what it is. Sensations do not exist unperceived, whereas material processes do exist unperceived. Conversely, a material process may consist of parts, it may contain elements that move on a curved path, whereas it would be utter nonsense (and not only false) to say that this sensation of RED that I am having now contains a part that is moving on a curved path (a simple thought such as Bühler's *Aha Erlebnis* would be an even more damaging counterexample, and so would be feelings, intentions, mathematical ideas, and the like). Does this argument show the absurdity of materialism?

In order to decide this question, let us consider meaninglessness first. Whether or not a statement is meaningful depends on the grammatical rules guiding the corresponding sentence. The argument appeals to such rules. It points out that the materialist, in stating his thesis, is violating them. Note that the particular *words* he uses are of no relevance here. Whatever the words employed by him, the resulting system of rules would have a structure incompatible with the structure of the idiom in which we usually describe pains and thoughts. This incompatibility is taken to refute the materialist.

It is evident that this argument is incomplete. An incompatibility between the materialistic language and the rules implicit in some other idiom will criticize the former only if the latter can be shown to possess certain advantages. Nor is it sufficient to point out that the idiom on which the comparison is based is in common use. This is an irrelevant historical accident. Is it really believed that a vigorous propaganda campaign that makes everyone speak materialese will turn materialism into an acceptable doctrine? The choice of the language that is supposed to be a basis of criticism must be supported by better reasons.

So far as I am aware, there is only one further reason that has been offered: it is the practical success of ordinary English that makes it a safe basis for arguments. "Our common stock of words," writes J. L. Austin, "embodies all the distinctions men have found worth drawing, and the connections they have found worth marking, in the lifetime of many generations: these surely are likely to be more numerous, more sound, since they have stood up to the long test of the survival of the fittest, and more subtle . . . than any that you or I are likely to think up. . . ."[152] This reason is almost identical with certain views in the philosophy of science that we discussed in Sec. VI. However, it must also be emphasized, in all fairness to the scientists, that the parallel does not go very far. Scientific theories are constructed in such a way that they can be *tested*. Every application of the theory is at the same time a most sensitive investigation of its validity. This being the case, there is indeed some reason to trust a theory that has been in use for a considerable time, and to look with suspicion at vague and new ideas. The suspicion is mistaken, of course. Still it is not completely foolish to have such an attitude. At least *prima facie* there seems to be a grain of reason in it.

The situation is very different with "common idioms." First of all, such idioms are adapted not to *facts,* but to *beliefs.* If these beliefs are widely accepted, if they are intimately connected with the fears and hopes of the community in which they occur, if they are defended and reinforced with the help of powerful institutions, if one's whole life is carried out in accordance with them, then the language representing them will indeed be regarded as most successful. At the same time, it is clear that the question of the truth of the beliefs has not even been touched.

A second reason why the success of a "common" idiom is not at all on the same level as the success of a scientific theory lies in the fact that the use of such an idiom, *even in concrete observational situations,* can hardly ever be regarded as a *test.* There is no attempt, as there is in the sciences, to conquer new fields and to try the language in them. And even on familiar ground one can never be sure whether certain features of the descriptive statements used are *confronted* with facts and are thereby *examined,* or whether they do not simply function as accompanying noises. It is clear, then, that the argument is without force unless it is supplemented by a detailed investigation of the structure of the language used. Until such an investigation is forthcoming, we ought to disregard the dictum of common speech.

However, even if the idiom to which reference is made were testable in the sense that its inadequacy could be detected in the course of speaking it, even then it would not be possible to reject materialism or to stop the attempts of the materialist to develop his particular account of human beings. The reason that we have explained in some detail in preceding sections is that a mere "comparison with facts" is not sufficient to establish the value of a certain point of view and of the language used to express it. Many facts are formulated in terms of this language and are therefore already prejudiced in its favor. Also, there may exist facts that are inaccessible, for empirical reasons, to a person speaking a certain idiom and that become accessible only if a different idiom is introduced. This being the case, the construction of alternative points of view and of alternative languages, far from precipitating confusion, is a necessary part of an examination of the customary usages. More concretely, if one wants to find out whether there are pains, thoughts, feelings, in the sense indicated by the common usage of these words, then one must become a materialist. Trying to eliminate materialism by reference to the common idiom, or to some other favored language, therefore means putting the cart before the horse.[153]

XI. Mind-Body Problem: Observable Facts

While the argument from meaninglessness is wholly based upon language, the argument from falsity is not. That a thought cannot be a material process is established, so it is believed, *by observation*. It is by observation that we discover the difference between the one and the other and refute materialism. We now turn to an examination of this argument.

To start with, we must admit that the difference does indeed exist. Introspection *does* indicate, in a most decisive fashion, that my present thought of Aldebaran is not localized, whereas Aldebaran is localized; that this thought has no color, whereas Aldebaran has a very definite color; that this thought has no parts, whereas Aldebaran consists of many parts exhibiting different physical properties. Is this character of the introspective result proof to the effect that thoughts cannot be material?

The answer is no, and the argument is the truism that what *appears* to be different does not on that account need to *be* different. Is not the seen table very different from the felt table? Is not the heard sound very different from its mechanical manifestations? And if despite this difference of appearance we are allowed to make an identification, postulating an object in the outer world (the physical table, the physical sound), then why should the observed difference between a thought and the impression of a brain process prevent us from making another identification, postulating this time an object in the inner (the *material* inner) world, viz., a brain process? It is, of course, quite possible that such a postulate will run into trouble and be refuted by independent tests (just as the earlier identification of comets with atmospherical phenomena was refuted by independent tests). The point is that the *prima facie* observed difference between thoughts and the appearance of brain processes does *not* constitute such trouble. It is also correct that a language based upon the assumption that the identification has already been carried out would differ significantly from ordinary English. But this fact can be used as an argument against the identification only after it has been shown that the new language is *inferior* to ordinary English. And such disproof should be based upon the fully developed materialistic idiom and not on the bits and pieces of materialese that are available to the philosophers of today. It took a considerable time for ordinary English to reach its present stage of complexity and sophistication. The materialistic philosopher must be given at least as much time. As a matter of fact, he will need more time, as he intends to develop a language that is coherent, fully testable, and that gives an account of the most familiar facts about human beings as well as of thousands more recondite facts that have been unearthed by the physiologist. I also admit that there exist people for whom even the reality of the external world constitutes a grave problem. My answer is that I do not address them, but I presuppose a minimum of reason in my readers: I assume they are realists. And assuming this I try to point out that their realism need not be restricted to processes outside their skin—unless, of course, one already presupposes what is to be established by the argument, that things inside the skin are very different from those outside. Considering all this, I conclude that the argument from observation which is used with such great self-assurance both by metaphysicians and by their more sophisticated "analytic" successors is invalid.

It is quite entertaining to speculate about the results of an identification of what is observed by introspection. Observation of microprocesses in the brain is a notoriously difficult affair. Only very rarely is it possible to investigate them in the living organism. Observation of dead tissue, on the other hand, is applied to a structure that may differ significantly from the living brain. To solve the problem arising from this apparent inaccessibility of processes in the living brain, we need only realize that the living brain is already connected with a most sensitive instrument—the living human organism. Observation of the reactions of this organism, introspection included, may therefore be a much more reliable source of information concerning the living brain than is any other "more direct" method. Using suitable hypotheses, one might even be able to say that introspection leads to a *direct observation* of an otherwise quite inaccessible, and very complex, process *in the brain.*[154]

XII. Mind-Body Problem: Acquaintance, the "Given"

Against what has been said in the last section, it might be, and has been, objected that in the case of thoughts, sensations, feelings, the distinction between what they *are* and what they *appear to be* does not apply. Mental processes are things with which we are *directly acquainted.* Unlike physical objects, whose structure must be unveiled by experimental research and about whose nature we can make only more or less plausible conjectures, they can be known completely and with certainty. Essence and appearance coincide here, and we are therefore entitled to take what they seem to be as a direct indication of what they are. This objection must now be investigated in some detail.

In order to deal with all the prejudices operating in the present case, let us approach the matter at a snail's pace. What are the reasons for defending a doctrine like the one we have just outlined? If the materialist is correct, then the doctrine is false. It is, then, possible to test statements of introspection by physiological examination of the brain and to reject them as being based upon an introspective mistake. Is such a possibility to be denied? The doctrine we are discussing at the present moment thinks it is. And the argument is somewhat as follows:

When I am in pain, then there is no doubt, no possibility of a mistake. This certainty is not a purely psychological affair; it is not due to the fact that I am too strongly convinced to be persuaded of the opposite. It is much more related to a logical certainty: there is no possibility whatever of criticizing my statement. I might not show any physiological symptoms, but I never meant to include them in my assertion. I might not even show pain behavior, but this is not part of the content of my statement either. Now if the difference between essence and appearance were applicable in the case of pain, then such certainty could not be obtained. It *can* be obtained, as has just been demonstrated, hence the difference does not apply and the postulation of a common object for mental processes and impressions of physiological processes cannot be carried out.

The first question that arises in connection with this argument concerns the *source* of this certainty of statements concerning mental processes. The answer is very simple: it is their lack of content that is the source of their certainty. Statements about physical objects possess a very rich content. They are vulnerable because of the existence of this content. Thus the statement "there is a table in front of me" leads to predictions concerning my tactual sensations; the behavior of other material objects (a glass of brandy put into a certain position will stay there and not fall to the ground; a ball thrown in a certain direction will be deflected), the behavior of other people (they will walk around the table), etc. Failure of any one of these predictions may force me to withdraw the statement. This is not so with statements concerning thoughts, sensations, feelings, or at least there is the impression that the same kind of vulnerability does not obtain here. The reason is that their content is much poorer. No prediction or retrodiction can be inferred from them, and the need to withdraw them can therefore not arise. (Of course lack of content is only a necessary condition of their empirical certainty; in order to have the character they possess, statements about mental events must also be such that in the appropriate circumstances their production can be achieved with complete ease; they must be observational statements. This characteristic they share with statements concerning physical objects.[155])

The second question is how statements about physical objects obtain their rich content and how it is that the content of mental statements as represented in the current argument is so poor. The answer that will be given by some philosophers is that such a difference follows from the "grammar" of the mental statements. We *mean* by pains, thoughts, feelings, processes that are accessible only to the individual who has them and that have nothing to do with the state of the body. The content of "pain" or of "thinking of Vienna" is low because "pain" and "thought" are mental terms. If the content of these terms were enriched, and thereby made similar to the content of "table," then the terms would cease to function in the peculiar way in which mental terms do, as a matter of fact, function, and "pain," for example, would then cease to mean what is meant by an ordinary individual who in the face of the absence of physiological symptoms, of behavioral expressions, of suppressed conflicts still maintains that he is in pain.

It is clear that this answer does not bring us very far. It amounts to saying that the content of mental terms is low because this is the case in the language of which they are part. Is there a rationale for using a language of this kind? Can it be shown that a language which contains a dichotomy between mental terms and material terms is better than any other language? And in considering this question, we should take into consideration that the relative poverty of mental notions is by no means a common property of all languages. Quite the contrary, it is well known that people have at all times objectivized mental notions in a manner very similar to the manner in which we today objectivize materialistic notions. They did this mostly (but not always—the witchcraft theory of the Azande constituting a most interesting materialistic exception[156]) in an objective *idealistic* fashion[157] and can therefore be easily criticized or smiled about by some progressive thinkers of today. It is clear that such criticism is irrelevant to the present discussion, which asks for the *reason* behind the special treatment of mental terms and which then criticizes the linguistic reply "this is so in ordinary English" by asserting that there are other ordinary languages without this peculiarity. Indeed a closer factual investigation reveals that there is hardly any interesting language that is built in accordance with the idea of acquaintance. This idea is nothing but a philosophical invention. What is the rationale for this invention?

The rationale clearly cannot be any *fact*. For this purpose consider the apparently factual argument that *there is knowledge by acquaintance,* that we *do* possess direct and full knowledge of our pains, of our thoughts, of our feelings, at least of those that are immediately present and not suppressed. It is clear that this argument is circular. If we do possess knowledge by acquaintance with respect to mental states of affairs, if there seems to be something "immediately given," then this is the *result* of the low content of the statements used for expressing this knowledge. Had we enriched the notions employed in these statements in a materialistic (or an objective-idealistic) fashion, *as we might well have done,*[158] then we would no longer be able to say that we know mental processes by acquaintance. Just as with material objects, we would then be obliged to distinguish between their *nature* and their *appearance,* and each judgment concerning a mental process would be open to revision by further physiological (or behavioral) inquiry. The reference to acquaintance cannot therefore justify our reluctance to use the knowledge we possess concerning mental events, their causes, their physiological concomitants (as their physiological content will be called *before* the materialistic move) for enriching the mental notions.

What has just been said deserves repetition. The argument that we attacked was as follows: there is the *fact* of knowledge by acquaintance. This fact refutes materialism, according to which there would be no such fact. The attack consisted in pointing out that although knowledge by acquaintance may be a fact, this fact is the result of certain peculiarities of the language spoken and therefore alterable. Materialism [and, for that matter, also an objective spiritualism like Hegel's (see his *Phänomen-ologie des Geistes*)] recognizes the fact and suggests that it be altered by including our knowledge of human beings (physical knowledge and physio-logical knowledge alike) in the mental notions. It therefore clearly cannot be refuted by a repetition of the fact. What must be shown is that the suggestion is undesirable, or that acquaintance is desirable.

We have here discovered a rather interesting feature of philosophical arguments. The argument presents what seems to be a fact of nature, viz., our ability to acquire relatively complete and secure knowledge of our own states of mind. We have tried to show that this alleged fact of nature is the result of the way in which any kind of knowledge (or opinion) con-cerning the mind has been incorporated and is being incorporated into the language used for describing facts: this knowledge, this opinion is not used for *enriching* the mental concepts, but for making predictions in terms of the still unchanged and poor concepts. Or, to use terms of technical philosophy, this knowledge is interpreted instrumentalistically, and not in a realistic fashion. The alleged fact referred to above is there-fore a projection into the world of certain peculiarities of our way of building up knowledge. Why do we (or philosophers using the language described) and perhaps even common people proceed in this fashion?

One proceeds in this fashion because one holds a certain philosoph-ical theory. According to this theory (which was mentioned in the intro-duction), the world consists of two domains, the domain of the outer, physical world, and the domain of the inner, mental world. The outer world can be experienced, but only indirectly. Our knowledge of it will therefore forever remain hypothetical. The inner, mental world can be directly experienced because we are in direct contact with it. We *are* this world. The knowledge gained in this fashion is therefore complete and absolutely certain. This, I think, is the philosophical theory behind the method described in the last section.[159]

Now I am not concerned here with the question of whether this theory is correct or not. It is quite possible that it is true (though there are many indications that it is not). What I *am* interested in here is the way in which the theory is presented.[160] It is not presented as a hypothesis that is open to criticism and that can be rationally discussed. In a certain sense it is not even *presented*. It is rather incorporated into the language in a fashion that makes it inaccessible to empirical criticism—whatever the observations, they are not used for enriching the mental concepts that will therefore always refer to entities knowable by acquaintance.

This procedure has two results. It hides the theory and thereby removes it from criticism, and it creates what looks like a very powerful fact in its favor. As the theory is hidden, the philosopher can even *start* with this fact and reason from it, thereby providing a kind of inductive argument for the theory. It is only when we examine what *independent* support there exists for the theory that we discover that the alleged fact is not a fact at all, but rather a reflection of the way in which empirical results are handled. We discover that "we were ignorant of our . . . activity and therefore regarded as an alien object what had been constructed by ourselves."[161]

This is an excellent example of a point that was mentioned in the introduction, viz., the circularity even of those philosophical arguments that rest upon what seems to be incontrovertible *empirical* evidence. The example is a warning. It warns us never to be too impressed by such evidence unless we have first investigated the source of its apparent strength. Such an investigation may reveal that its strength is a result of having silently adopted the very same principles the evidence is supposed to support, and it may thereby destroy the evidence. This is also the only place where a priori considerations rightfully enter the theory of knowledge. We shall now examine the matter in somewhat greater detail.

XIII. Methodological Considerations

There are some philosophers who agree that the *fact* of acquaintance cannot be used as an argument against the materialist (or against any other "internal realist"). Their reasons are not those given above, but rather the realization that none of the situations described in the ordinary idiom can be known by acquaintance. Realizing this, they look for arguments that remain valid in the face of adverse facts, and they therefore appeal to norms rather than to facts. They suggest the construction of an *ideal language* containing statements of the desired property. In this they are guided by the idea that knowledge must possess a solid, immutable foundation. The construction of such an ideal language has sometimes been represented as a task of immense difficulty and as worthy of a great mind. Altogether the adherents of certainty are in the habit of believing that certainty is difficult to attain and that it needs long and patient research to uncover principles, or even single statements, that are not endangered by the fallibility of human discourse. As I have tried to show elsewhere, the very opposite is the case.[162] Certainty is one of the cheapest commodities, and it can be obtained in a second once the problem has been set up in the proper fashion. Thus in the present case it is not necessary to demand that the elements of the ideal language one wants to construct be *already existing* statements of the ordinary idiom that possess the desired property (Russell's "canoid patch of colour" indicates that he conceived the task in this unnecessarily restricted fashion). Nor is it necessary to give an account of complex perceptions in terms of simple sensible elements (the investigations of the Gestalt· school of psychology suggest that such composition would indeed be an extremely difficult matter). Why should the attempt to find a safe observation language be impeded by such inessential restrictions? What we want is a series of observation statements leading to knowledge by acquaintance. Such statements can be obtained *immediately* by taking any observation statement and eliminating its predictive and retrodictive content as well as all consequences concerning public (material or nonmaterial) events occurring simultaneously. The resulting string of signs will be still observational, it will be uttered on the same objective occasion as was its predecessor, but it will be incorrigible, and the object described by it will be known "by acquaintance." This is how acquaintance can be achieved. Now let us investigate some consequences of this procedure.

Such an investigation is hardly ever carried out with due circumspection. What happens usually is this. One starts with a sentence that has a perfectly good meaning, such as "I am in pain." One reinterprets it as a statement concerning what can be known by acquaintance. One overlooks that such a reinterpretation drastically changes the original meaning of the sentence, and one retains in this fashion the illusion that one is still dealing with a meaningful statement. Blinded by this illusion, one cannot at all understand the objection of the opponent who takes the move toward the "given" seriously and who is incapable of getting any sense out of the result. What are these objections? Just investigate the matter in some detail. Being in pain I say "I am in pain," and, of course, I have some independent idea as to what pains are: they do not reside in tables and chairs; they can be eliminated by taking drugs; they concern only a single human being (hence being in pain, I shall not get alarmed about my dog); they are not contagious (hence being in pain, I shall not warn people to keep away from me). This idea is shared by everybody else, and it makes people capable of understanding what I intend to convey. But now I am not supposed to let this idea contribute to the meaning of the *new* statement, expressed by the same sentence, about the immediately given; I am supposed to free this meaning from all that has just been said; not even the idea that a dreamed pain and a pain really felt are different must now be retained. If all these elements are removed, then what do I mean by the new statement resulting from this semantic canvas cleaning? I may utter it on the occasion of pain (in the normal sense); I may also utter it in a dream with no pain present; I may use it metaphorically, connecting it with a thought (in the usual sense) concerning the number two; or I may have been taught (in the usual sense of the word) to utter it when I have pleasant feelings and therefore utter it on these occasions. Clearly, all these usages are now legitimate, and all of them describe the "immediately given pain." Is it not evident that using this new interpretation of the sentence I am not even in principle able to derive enlightenment from the fact that Herbert has just uttered it? Of course I can still treat it as a *symptom* of the occurrence of an event that in the ordinary speech would be expressed in the very same fashion, viz., by saying "I am in pain." But in this case I provide my own interpretation, which is very different from the interpretation we are discussing at the present moment. And we have seen that according to this interpretation the sentence cannot be taken to be the description of anything definite. The situation is not at all changed by Herbert's exclaiming, "but I at least know what my pains are!" First of all, if "pain" is used in the new sense, the sentence just uttered does not convey any definite information and might as well be replaced by a moan; and secondly, even Herbert cannot know what cannot be known in principle. Of course he is in pain (in the normal sense). However, this remark is not any more helpful than is the remark that a howling dog is in pain. For what we are discussing is not what is and is not going on in the world, but how what is going on is to be *described*. And the point of the above discussion is that the special sentences that are supposed to establish a connection between our knowledge and the world *mean nothing,* that they *cannot be understood* by anyone, and that they are therefore *completely inadequate* as "foundations of knowledge." Now if the given were a reality, then this would mean the end of rational, objective knowledge. Not even revelation could then teach us what admittedly cannot be known in principle. Language and conversation, if they existed, would become comparable to a cat serenade, all expression, nothing said, nothing understood. Fortunately enough, the given is but the reflection of some procedures that can be carried out *but that can also be avoided.* It can be eliminated by building up language in a more sensible fashion. Briefly, the procedure adopted is as follows.

We start with the very trivial remark that our talk should be informative and rich in content, and it should be possible to test every part of it. This trivial remark introduces an a priori element into the theory of knowledge.[163] However, as opposed to the a priori elements we have discussed in the introduction, this element is very simple, its consequences can be appreciated quite easily, and it is obvious at a single glance that the opposite procedure is most undesirable: if we want to talk at all, then we should say *something;* otherwise we might as well stay silent. Moreover, there is no reason why talk about minds should be less informative than talk about material objects. Knowledge by acquaintance, however, far from being an indisputable fact that refutes materialism, now turns out to be the consequence of a very undesirable procedure. This finishes our discussion of the last argument against the materialist.

XIV. Synonymy

It is a common feature of almost all objections against materialism that materialistic *synonyms* are demanded for mentalistic terms, and that materialism is damned if it cannot produce such synonyms. More recently, Feigl has weakened the demand for synonymy to *co-extensionality*. Both the stronger and the weaker demand require materialism to be able to mimic certain parts of the "grammar" of the mentalistic terms. Feigl's demand, for example, implies that the connotation of a materialistic analysis must not contain elements negated in the connotation of the corresponding mental terms, and that the connotation of the analyzed mental terms must not contain elements negated in the materialistic analysis. We have seen that these requirements are not restricted to the mind-body problem. They are supposed to apply wherever lower-level states of affairs are explained in terms of high-level theories. The explanans must then be able to produce either synonyms for the terms of the explanandum or at least phrases that are co-extensional with the terms of the explanandum.[164] And these requirements are especially emphasized in those cases where the lower-level theory uses *observational* notions: unless the higher-level theory can produce either synonyms for these notions, or at least terms that are co-extensional with them, it cannot be said to be an empirical theory, as it is then not able to establish a connection with experience. It is this last consideration that seems to have prompted Feigl and others to retain the demand for co-extensionality between new terms and old (observational) terms.

In Sec. IX we produced general arguments to the effect that both the demand for synonymy and the weaker demand for co-extensionality are unreasonable, and we also considered the disadvantages of correlation hypotheses. The specific argument with which we are concerned here has to be developed independently for natural languages and for scientific theories. Take the latter first. Why should more general theories, which are in most cases introduced in order to remedy the discovered weaknesses of less general theories, be capable of providing synonyms for the descriptive *terms* of their predecessors? Adequate representation of *facts* is a sufficient criterion for accepting any scientific theory. Is it assumed that such representation will be achieved only by a theory that mimics certain features of all its predecessors? Who guarantees that the *concepts* of these predecessors are faithful mirrors of reality despite the fact that the theories themselves are in need of improvement? Of course they are confirmed; but as is well known, theories exhibiting a very different con-

ceptual structure may be confirmed by the same evidence:[165] both Newtonian mechanics and general relativity are confirmed by the observations
concerning the path of Jupiter, yet it is impossible to find exact synonyms or even co-extensionals, in terms of general relativity, of all the
primitive descriptive terms of Newtonian mechanics. Moreover, we very
often discover in the course of scientific progress that what we thought
existed did not really exist. This shows that even the demand for co-
extensionality is too strong. And the remark that without some kind of
connection with earlier notions we do not really know what we are talking
about[166] assumes that the subject matter of a theory or of a point of view
is defined by the concepts of its predecessors rather than by the theory or
point of view itself. This gives too much power to ideas we have already
replaced with better ideas. Why should we need the classical notions in
order to explain relativity? And if we need them [which is in any case a
logical impossibility (see the example of Sec. IV)], then how was the subject matter of the classical theory defined? We are here involved in an infinite regress, unless we admit, as we must, because knowledge has a finite
history, that the first point of view that ever came into existence was also
capable of defining its own subject matter, which would then be needed
for defining any subsequent subject matter. But is one really to assume
that we must constantly return to the ideology of Cro-Magnon and
Aurignac in order to explain properly what we are talking about?

The remark that some connection of meaning must be established at
least with the *observational* terms of the previous theories (or of the ordinary language, in case we *start* theorizing) assumes that *terms are observational by virtue of their meanings,* and that a radical enough change of
meaning might turn them into unobservables. This theory, which we
might call the semantic theory of observation, introduces a priori elements
of a very undesirable kind into our knowledge. We shall therefore adopt a
different account of observation and assume observability to be a pragmatic concept: a statement will be regarded as observational because of
the *causal context* in which it is being uttered, and *not* because of what it
means. According to this theory, "this is red" is an observation sentence,
because a well-conditioned individual who is prompted in the appropriate
manner in front of an object that has certain physical properties will
respond without hesitation with "this is red"; and this response will
occur independently of the *interpretation* he may connect with the statement [he may interpret it as referring to a property of the *surface* of the
object, as property of the *space between* the object and the eye (as did
Plato), as a *relation* between the object and a coordinate system in which
he himself is at rest]. All we need in order to provide a theory with an
observational basis are statements satisfying this pragmatic property. We
do not need to connect these statements with the observational statements
of a different theory in order to give them meaning. Their meaning they

obtain from the theory to which they belong. Thus all we need in order to provide a neurophysiological theory of pains with empirical content are sentences that are uttered almost immediately upon the occasion of pain in the physiological sense. We do not need to interpret these sentences by referring to what we meant by "pain" in a more primitive stage of knowledge. We may interpret them as expressing certain fairly complicated central states. In no case, therefore, is it necessary for a newly suggested theory to be capable of providing either synonyms or co-extensionals for more ancient, but at the same time, perhaps, more familiar, terms. It is for this reason that I have emphasized on various occasions that the progress of knowledge may be by *replacement*, which leaves no stone unturned, rather than by subsumption, and that I have criticized as much too restrictive the current theories of explanation and reduction. A scientist or a philosopher must be allowed to start completely from scratch and to redefine completely his domain of investigation.

What has just been said applies with even greater force when the concepts to be imitated belong to some natural language. Scientific theories are testable; they are the result of systematic improvement in the light of repeated serious tests. They are therefore factually relevant and can be trusted either to be correct or to reveal their shortcomings on further investigation. Yet despite all this we have seen that they must not be regarded as unalterable yardsticks for any future theory, and that their descriptive concepts must not be regarded as perfect mirrors of reality to be copied by any future theory. Now if testable and highly confirmed scientific theories can and must be treated in this fashion, if no part of the language connected with them can be regarded as sacrosanct, then this can be even less so with idioms whose testability is in doubt and that may be without factual content. To base one's inquiry upon such idioms without having first investigated the correctness of the underlying ideology and the way in which this cosmology is expressed (testable, or in a manner that allows for adaptation to any possible situation) is not very different from accepting the advice of a soothsayer who is respected by the community in which one lives because of his high reputation. Moreover, it would be much more advisable not to try looking for treasures in the slums of some natural language that has first *grown* without regard to testability and has then been *modified* with the purpose of preserving some more recent point of view, but to build new theories in a planned fashion and to impart to them all properties one deems methodologically desirable. Finally, we must remind the reader that those who want to relate the concepts of a newly invented theory to ordinary concepts seem to overlook that there are many different "ordinary languages," and that no method has as yet been given of choosing between them. Clearly, philosophy cannot be made dependent upon such an arbitrary procedure.

Let us now evaluate, from the point of view of the remarks just made, the "Requirements and Desiderata for an Adequate Solution of the Mind-Body Problem" as they are found in Feigl's monumental essay *The Mental and the Physical*.[167]

Requirement 1: An "analytical study" is demanded of the meanings of the terms "mental" and "physical." *Comment:* What is or what should be the purpose of any inquiry concerning the mind-body problem? Its purpose is to advance our state of knowledge concerning human beings. How can this be achieved? By criticizing the existing ideas. Now we have learned that what is needed for the purpose of such criticism are *alternatives* that deviate as radically as possible from the accepted knowledge. Analysis is not required to bring about such deviation.[168] All we want is the guarantee that the alternatives allow for as few synonymies as possible. Whether or not they satisfy this condition can be found out *intuitively,* on the basis of the very same intuitions that guide an analysis, but without the laborious procedures involved in such analysis.

Requirement 2: An "analysis of the mind-body relation is to be sought" that takes due cognizance of the results of scientific investigation. *Comment:* Putting the matter this way means prejudging the issue by uncritically swallowing the familiar assumption that there is something corresponding to the mental terms. Is the possibility to be denied that we may discover that there are no "minds" in the sense in which this word occurs in the familiar mentalistic ideology, and that there is therefore no mind-body relation to be analyzed? Imitating requirement 2, one might put the problem of combustion and calcination in the form: to find a molecular account of the constituents of phlogiston which excludes the possibility that phlogiston might not exist. The rejoinder that specific "mental" events exist because they can be *observed* was dealt with in Sec. XI.

Requirement 3: The solution of the problem must render "an adequate account of the *efficacy* of mental states, events, and processes in the behavior of human organisms." *Comment:* Same as for requirement 2. Feigl asserts that "to maintain that planning, deliberation, preference, choice, volition, pleasure, pain, displeasure, love, hatred, attention, vigilance, enthusiasm, grief, indignation, expectations, remembrances, hopes, wishes, etc., are not among the causal factors which determine human behavior is to fly in the face of the commonest of evidence, or else to deviate in a strange and unjustifiable way from the ordinary use of language." Just this kind of "flying in the face of the commonest of evidence" and this "deviation . . . from the ordinary use of language" are required in the course of any detailed examination of the validity of such evidence and of such use. (See again Sec. IX.)

Requirement 4: "A most important requirement for the analysis of the mind-body problem is the recognition of the *synthetic,* or *empirical,* character of the statements regarding the correlation of psychological to neurophysiological states." *Comment:* This procedure implies the retention of the mental terms and co-extensionality between them and the physical terms. It has already been commented upon in Sec. IX and in the present section. Concerning the remark that an identification of mental states with brain states that goes beyond the assertion of extensional identity, but asserts also identity of nature, would eliminate the empirical content of the solution, it is to be said that a materialistic account must, of course, not be *ad hoc,* and that the new predictions made by it, if it is completed, are ample replacement for whatever empirical content might otherwise have been lost by such identification. Besides, if there are *no* mental events in the usual sense, then the identification hypothesis defended by Feigl either ceases to be meaningful or becomes false. In the first case, no contribution to the empirical content is made. In the second case, the negation of the statement will be a consequence of materialism (taken together with some obvious definitions),[169] and it will therefore contribute to its empirical content.

Requirement 5: (*a*) "The need for a criterion of scientific meaningfulness." *Comment:* This point being beyond the scope of the present paper, I can only assert dogmatically that I do not see the need for any such criterion. (*b*) "The recognition that epistemology, in order to provide an adequate reconstruction of the confirmation of knowledge claims, must employ the notion of immediate experience as a confirmation basis (the 'given' cannot be entirely a myth!)." *Comment:* As I tried to show in Sec. XII, the "given" is the result of a methodological procedure that tries to remove from attack a certain myth concerning human beings and at the same time provides sham facts in its support. This procedure is incompatible with realism [and therefore incompatible with requirement (*c*) below] and with the demand for testability. That it is incompatible with realism I have tried to show in my essay "Das Problem der Existenz Theoretischer Entitäten" [*Probleme der Wissenschaftstheorie* (Vienna, 1961)] and in my contribution to the *Essays in Honor of Karl Popper.* That it is incompatible with the demand for testability is obvious.[170] (*c*) "The indispensability of a *realistic,* as contrasted with operationalistic or phenomenalistic, interpretation of empirical knowledge." *Comment:* This is the only requirement I can accept. It is inconsistent with requirement 5(*b*).

Requirement 6: "The 'meat' of an adequate solution of the mind-body problem will consist in a specific analysis of the characteristics and the relations between the attributes of the mental (and especially the phenomenal) and the *physical.*" *Comment:* As against requirements 2, 3, and 4.

All these demands, and those discussed in the preceding sections, put far too great trust in the ideas that have been developed by our ancestors and conveyed to us by our nurses and philosophy teachers. The demand for synonymy, for example, takes it for granted that the theory which we choose as the basis for comparison is suitable for describing the world. But do we not invent theories in order to find means for description that are even more suitable, that are more closely adapted to the facts, and that will therefore behave differently from what is contained in the received opinion? And if this is our intention, is not then lack of synonymy (or of any other and weaker relation to previously spoken idioms) a sign of *success* rather than of failure (provided, of course, the new theory is empirically adequate, a matter that cannot be decided by semantic considerations)? Or are our savage forefathers (or our not-so-savage English teachers or physics teachers) to be the only ones who are allowed to build a language from scratch, and is all we can do imitate them as well as possible? These questions should be considered by those who want to judge new theories by the extent to which they mimic old beliefs and old languages rather than by the extent to which they mimic the *world,* and who expect that the limitations of these old languages, had they existed, would long ago have been discovered without any help from alternative points of view and without the attempt to invent, and to work out in detail, theories that turn upside down whatever seems to be obvious, highly confirmed, "natural," and therefore an almost direct expression of the truth.

XV. Experience

In the preceding sections we have examined various arguments dealing with what one might call the nature of mental processes. It turned out that all these arguments were based on the dogmatic retention of the one or the other particular hypothesis or set of hypotheses. Even the case of acquaintance could be explained as being due to the conservation of one particular theory of the mind and its relation to the rest of the world. Arguments from observation were not exempt from this general criticism. Quite the contrary, it emerged that observational facts which apparently support a mentalistic point of view are capable of reinterpretation and cannot therefore be taken finally to decide the issue. The result can be generalized (and has already been generalized in the introduction): observational findings are not at all final barriers for theories, although they are usually presented as such. Observational findings can be reinterpreted, and can perhaps even be made to *lend support* to a point of view that was originally inconsistent with them.[171] Now if this is the case, does it not follow that an objective and impartial judge of theories does not exist? If observation can be made to favor *any* theory, then what is the point of making observations, of carrying out time-consuming experiments, of spending millions of dollars on the purchase of new and better experimental equipment?

There are philosophers who believe that the possibility to reinterpret the results of observation does indeed imply the futility of all experimentation and the subjectivity of all knowledge, and who regard this as a *reductio ad absurdum* of the idea of the theory dependence of all observation. Positively they demand that observational statements retain at least part of their meaning in the face of theoretical advance, and they support this demand partly by reference to the above *reductio,* partly arguing that every observational statement contains an unalterable *factual core.* The theory held by these philosophers might be termed a *semantic theory of observation:* observational statements have a special meaning. This meaning must not be changed, or else the statements cease to be observational. Clearly, if this account is correct, we cannot be permitted to reinterpret our observations in any manner we like. Rather we must keep the chosen interpretation stable and must make *it* the measure of meaning for all theoretical terms.

The first problem arising in this connection is how the meaning of the observation statements is to be determined. Two procedures deserve mention. According to the first procedure, the meaning of an observation statement is determined by its "use." According to the second procedure, the meaning of an observation statement is determined by what is "given" (or "immediately given") before either the acceptance or the rejection of the statement. We shall call these procedures the *principle of pragmatic meaning* and the *principle of phenomenological meaning,* respectively. Neither principle leads to a satisfactory solution of the problem of the meaning of observational terms.

The principle of pragmatic meaning, taken by itself, cannot distinguish observational terms from other terms of a given theory. Observational statements are often derived from theoretical principles and related to them in many other ways. This is also part of their use. However, this part must be disregarded if the intention is to establish an invariant core of meaning for observational terms. Proceeding in this fashion means using a principle that goes *beyond* an analysis of usage.

The principle of phenomenological meaning takes over where the principle of pragmatic meaning seems to fail. It admits that the meaning of an observational statement is not completely determined by the way in which it is used by the speaker of a certain language. It points out that, apart from behaving in a certain manner, man also has feelings, sensations, and more complex experiences. It assumes that the core of the observational statements is determined by these experiences: in order to explain to a person what "red" means, one need only create circumstances in which red is experienced. The things experienced, or "immediately perceived," in these circumstances completely settle the question concerning the meaning of the word "red."

In order to get some insight into the implications of this principle, let us first take the phrase "immediately given" in its widest sense. The properties of the things that are "immediately given" in this wide sense can be "read off" the experiences without difficulty, i.e., the *acceptance* or *rejection* of any description of these things is uniquely determined by the observational situation. The question arises (and is answered in the affirmative by the principle of phenomenological meaning) whether this amounts to a determination of the *meaning* of the description accepted (or rejected).

Before proceeding to a more detailed examination, I would like to make it clear that the answer to this question must be negative. Meanings are the results of conventions. Whatever the facts, we may choose these conventions in various ways, and we may thereby confer different meanings upon a given expression. Even an appeal to the techniques of scientific research cannot alter the situation: first, because these techniques contain conventional elements; second, because they never lead to unique results, but leave various alternatives open; third, because at least for an empiricist these techniques assume a well-defined observation language and cannot therefore be used to *define* an observation language. The idea that a simple look can decide the interpretation of an observational expression is therefore not only unrealistic; it is impossible in principle.

To bring out this impossibility, let us examine the matter in somewhat greater detail. Let us consider the relation between an immediately given situation, or a *phenomenon P,* and (the acceptance of) a statement *S* whose meaning is assumed to be uniquely determined by that phenomenon. Let us call the relation between the statement and the phenomenon at the time when the meaning is being determined the relation of phenomenological adequacy, and express it by $R(P, S)$. The interpretation of *S* is then determined by reference to circumstances when $R(P, S)$ obtains.

Now we must keep in mind that the procedure of comparing phenomena with sentences is supposed to introduce a language *ab ovo.* There is no teacher available who already knows the language and who uses the phenomena in the procedure of teaching the meaning of its main descriptive terms. Nobody points out that *S* "fits" *P.* The phenomena must speak for themselves. This means that there must be, in addition to *S* and *P,* a further phenomenon *R,* indicating that *S* fits *P.* Moreover, it is not enough that there is just another element in the visual (or in the mental) field. The element *R* must be relevant, it must fit the whole situation in such a manner that it enables us to say that *S* fits *P.* We therefore need still a further phenomenon, guaranteeing relevance *R';* the same arguments apply to this phenomenon, hence we need still another phenomenon *R'';* and so on, *ad infinitum.* Result: the observer must perform infinitely many acts of observation before he can determine the meaning of a single observation statement. Also, if the *uttering* of the statement is supposed to be guided by phenomena, it will forever be impossible to utter an observation statement. We *do* observe; hence the idea that meanings are determined by phenomena, and the corresponding idea that the truth of observation statements is determined by attention to phenomena, must be rejected. It is no good repeating, as Feigl is in the habit of doing, "but I *experience P!*," for the question discussed is not what is *experienced* (this might be decided by a psychologist looking on from the outside), but whether what is experienced has been used correctly for the transference of meanings or is being described correctly. And we have shown that this question cannot be answered by appealing to the relation of phenomenological adequacy.

The idea that introspection provides an interpretation of certain sentences puts the cart before the horse. It is, of course, true that some sentences that stand in the relation of phenomenological adequacy to phenomena also possess an interpretation. However, this interpretation is not conferred upon them because they fit the phenomena, but is an essential presupposition of their fitting. This is easily seen when considering signs whose interpretation has been forgotten; they do not any more fit the phenomena that previously evoked their acceptance. It follows that the principle of phenomenological meaning would in most cases either lead to interpretations that are different from those considered by its champions or be inapplicable. And it would be inapplicable in exactly those cases in which it is supposed to provide us with an interpretation— i.e., in the cases of signs that have not yet been given any meaning.

The apparent success of the principle and the understanding allegedly gained with its help are due to a surreptitious use of the very same meanings the principle is supposed to provide. A red sensation is created. The observer watches carefully, says "red," and assumes that this informs him of the complete meaning of the word; he assumes that he now knows what redness really is, and that he has obtained his knowledge merely from the process of watching what is going on in the phenomenal field. Of course he can do this only because he knows how to use the word "red" in the first place. But reinterpreting this word in such a fashion that it seems to refer to something "given,"[172] forgetting the transition to the "given," and acting as if nothing at all had happened may indeed create the impression that a direct determination of meanings by a comparison with the given is possible.

But does introspection perhaps play a selective role? Is it perhaps possible that, given a phenomenon P and a class of *interpreted* sentences, the relation of phenomenological adequacy might allow us to select those sentences that correctly describe P (in the phenomenological sense)? I believe that introspection (inspection of phenomena) cannot play even this more modest role of a selector. I shall discuss three reasons. The first is the existence of "secondary interpretations":[173] I may have a strong inclination to call the vowel *e* "yellow." The important thing is that I feel this inclination only if yellow carries with itself the usual meaning. But according to this usual meaning, "yellow" is not applicable to sounds. A second reason is the existence of phenomenological situations whose adequate descriptions are self-contradictory. An example of such a situation has been described by Edgar Tranekjaer-Rasmussen.[174] The third, and most important, reason is, however, this: given a phenomenon, there are always many different statements (expressed by the same sentence) that will be found to fit the phenomenon. Assume the phenomenon to be the appearance of a table. In this case, both "there seems to be a table" and "there is a table" will be phenomenologically adequate. But so will be any statement that is obtained from "there is a table" by reducing its infinitely many consequences one by one. Each statement belonging to this manifold we shall call a *spectral statement;* the manifold itself, the *spectrum* associated with the phenomenon in question. The existence of the spectrum of a phenomenon makes it especially obvious that phenomena alone cannot even *select* interpretations, but that additional considerations are needed. A very common additional procedure whose separate character is hardly ever realized is to accept that spectral statement which can be interpreted as a description of the "given." It must be emphasized that this is indeed an *independent* step for which the phenomena do not provide the smallest hint. In Sec. XIII we gave our reasons why this step should not be taken. This being the case, we are now confronted with the problem of which of the infinitely many phenomenologically adequate statements should figure as our observation statement. Clearly, the principle of phenomenological meaning does not give us the slightest guidance in this matter. This is, of course, also evident from more general considerations: if we consider signs in isolation, then any interpretation that we confer upon them is a matter of convention. The same applies if we do not consider them in isolation, but as parts of a complicated linguistic machinery, since this machinery can always be enriched by additional rules.

To sum up: the meaning of an observational term and the phenomenon leading to its application are two entirely different things.[175] Phenomena cannot determine meanings. It may, of course, happen that strict adherence to one interpretation and the rejection of different accounts will lead to a situation where the relation between phenomena and propositions is indeed one-one. In such a situation, a distinction between phenomena and interpretations on the one side and phenomena and objective facts on the other side cannot readily be drawn, and the principle of phenomenological meaning will appear to be correct. However, appearances and reality are two different things. Alternative accounts *are* possible, and as soon as they are suggested, the proper interpretation will have to be chosen on the basis of some criteria, and it will not be possible to refer to phenomena alone as a source of meaning.

So far we have used the term "introspection" in the wide sense of "attendance to what is easily described." But our analysis also applies if we use a more sophisticated idea of what is immediately given—for example, if it is assumed that the given is not directly accessible, but must either be found by a special effort or will appear only under special conditions (perhaps when the reduction screen is used)—for the result of the special effort, or the things appearing under the special conditions, will again be phenomena, and we have already shown that phenomena cannot determine interpretations.

It is important to see this conclusion in its proper perspective. We are considering the idea that observation statements must possess an unchangeable core of meaning. We inquired how this core of meaning can be determined. We examined two answers to this inquiry, the principle of pragmatic meaning and the principle of phenomenological meaning. The result was in both cases that additional conventions are needed. This specific result agrees with the general idea that meanings are not facts, but results of conventions (either of explicit conventions or of the conventions accepted by the choice of a certain language as a medium of communication). The choice of an observation language is therefore up to us. How shall we proceed in order to arrive at a satisfactory decision in this matter?

One restriction has already been noted in Sec. XIII: the observation language must not be "about the given." A second restriction derives from the fact that observation statements follow from theories. This establishes a *connection* between the interpretation of observation statements and the interpretation of theoretical statements. It does not tell us, however, which interpretation is primary, or whether a question of precedence is at all involved. The idea of an observational core that we are examining at the present moment suggests that the interpretation of observation statements is primary, and that it is by reference to its principles that theoretical notions are explained. We must now investigate the consequences of such a procedure.

Before doing this, let me again repeat that whatever core is accepted must be based upon stipulations and cannot be determined by such simple procedures as looking, smelling, or attending to one's toothaches. It is my experience that people who have followed the abstract argument just outlined and who have perceived all the difficulties inherent in the simpleminded "look and see" approach return to their original position as soon as they are left alone (one might call this Hooke's law of mental elasticity). It shall therefore be repeated: you say you *know* what pains are, and you *know* what it means to say "I am now in pain," and you also say you know this from the occasions when you have experienced pain? Then you must remember what I have pointed out to you, that your appeal is to an impossible procedure that can be improved only with the help of additional stipulations. You admit this, adding now that, of course, you are not talking about some complex central process but about the "immediately given." But this way has been closed in Sec. XIII. It was argued there that talking about the given amounts to not talking at all. The given, therefore, is out. What you have to do is to make a choice between the infinitely many possible spectral statements that can be expressed by the sentence "I am in pain." Is there any possibility of *restricting* this choice and thereby justifying the idea of a core in an objective manner?

This brings us to the last red herring lying in our way. It has been assumed that speaking ordinary English amounts to having already made the choice. The observational notions of this idiom have a well-defined meaning. They are neither about the given, nor about central processes. They do indeed possess a core beyond which we must not go if we want to continue speaking this idiom. Now it is clear that our problem cannot be solved in this fashion. What we wanted was an objective criterion that uniquely determines the core. The reference to ordinary English does not provide such a criterion. It replaces the problem of choosing between different spectral statements by the problem of choosing between different languages incorporating different spectral statements. The choice is still open; it has only been formulated in a fashion that dulls our critical abilities and suggests that no further choice is necessary. The very same criticism applies to the attempt to determine the core by an analysis of the notion of observability. Again we have to point out that there are many different notions of observability, some actual, some not yet considered, and that it is up to us to take our choice. Wherever we turn—phenomena, idioms, behavior—we are always thrown back upon ourselves and made to realize that any restriction we choose to put upon our observational notions *is entirely our own work.* And this is not surprising. After all, it needs *thought* to determine meanings, and thought is present neither in our sensations, nor in our behavior, nor in our language.

There are, of course, good reasons why this stipulative element is so often overlooked. The realization that the core upon which the empiricists base the objectivity of their enterprise is the result of conventions takes all conviction out of their condemnation of a procedure that reinterprets observational statements in accordance with theoretical advance. This procedure was condemned because it seemed to imply that an objective and impartial judge of theories does not exist. What emerges now is that exactly the same accusation must be raised against the semantic theory.

It is now time to investigate the consequences of some particular definitions and stipulations of a core. This investigation will reveal that any attempt to postulate a limit to what can be incorporated into the observational notions leads to synthetic a priori principles and must therefore be rejected by an empiricist. A much more detailed examination is as follows:

Let S be a statement expressing the totality of conditions which must be satisfied in order that a term may be regarded as observational, and let C be the corresponding totality of properties in things that make them observational. It is not often the case that a detailed account is given of these conditions and properties. Yet whenever observability in principle (*direct* observability, that is) is denied to a certain term, the existence of such conditions and properties is implicitly assumed. Now consider a theory T which implies that $\overline{C}$. If this theory is correct, then the conditions of observability are not only not satisfied, *they have never been satisfied*, and empirical theories are therefore impossible. Every statement is, then, either a statement of pure mathematics (logic included) or a metaphysical statement.

Now I would like to emphasize at this point that such a consequence, taken by itself, does not seem to me to be as unacceptable as it will be to some empiricists. I am quite prepared to admit that research, even research that is *prima facie* empirical, may in the end lead to the result that observation is impossible and that an empirical science is a chimera. Nobody can say in advance that the world is open to knowledge by human beings. Such openness presupposes a very special and highly delicate correlation between the outer world and the world of consciousness. The assertion that the world is knowable through *experience* is even more restrictive and may therefore quite easily turn out to be false. An empirical science would, of course, be impossible under these circumstances. It is a prejudice to assume that this means the end of *all* objective knowledge. Empirical knowledge is only one of many forms of knowledge. Still, it is one thing to admit this possibility of limits to empiricism and quite a different thing to assume that such limits have actually been exhibited by the above argument. The argument is based upon some highly arbitrary definitions, and the only real content it might pretend to possess is its connection with ordinary English. This is not a strong enough basis to make us give up empiricism or admit that empiricism might possess some limits. Quite the contrary, the result will make us very doubtful both of the idea of a core and of the adequacy, for philosophical purposes, of "ordinary English."

There is one way that may be used in order to get out of this difficulty: theories which entail that $\overline{C}$ *are forbidden*. This procedure, which, as we pointed out, is an arbitrary stipulation, introduces the synthetic a priori and is therefore unacceptable for an empiricist. The detailed argument is as follows: C is a property that guarantees observability. Hence it cannot be a property that obtains under all circumstances; it must be a synthetic property and S must be a synthetic statement. For an empiricist, this means that the truth value of S is to be determined by empirical research.

The results of empirical research cannot be foreseen. S might turn out to be false. Such a possibility is excluded by the present move. This move is therefore not justifiable on the basis of experience. It puts *restrictions* upon the empirical method in order to guarantee the perennial truth of S and the perennial presence of C at least in some domain: S is now a synthetic statement whose truth can be established in a nonempirical fashion. That is, it is now a synthetic a priori statement.

There is a more sophisticated procedure that is widely accepted and that leads to exactly the same result, viz., to the exclusion of theories entailing that $\overline{C}$. However, this result is achieved by it in a more indirect manner, by a general method of interpretation that from its very nature could never produce such theories. I am, of course, referring to the double language model. According to this model, only those theories can be accepted as cognitively significant which obtain their interpretation from connecting a part of the corresponding formalism with some observation language. Now if the observation language used is built in accordance with conditions C, that is, if the meaning of its descriptive terms contains the notions of S as essential parts, then this procedure will, of course, never lead to a theory entailing $\overline{S}$. The criticism is the same as before. S is a synthetic statement. We know that it will be retained forever. The empirical method cannot provide such knowledge. S is therefore again made a synthetic a priori statement. Empiricism, at least in the form in which it was presented in the 'thirties ("all a priori statements are analytic") does not allow for the synthetic a priori. Hence the double language model is inconsistent with the principles of the very same empiricism it wants to represent in a formal manner. (That it is inconsistent with the method we have outlined in Sec. VIII is evident.[176])

We sum up our arguments against the idea of a core: first, it was pointed out that the interpretation of the observation language cannot be settled on the basis of observations, but must be settled on the basis of additional conventions. Then we discussed the specific conventions connected with the idea of a core. It was shown that no reason can be given for the choice of these conventions, and that their adoption has undesirable consequences (synthetic a priori). To this we may add the fact (which will not be discussed here) that the idea of a core and the conventions supporting it are inconsistent with actual scientific practice. These considerations create grave difficulties for the idea of a core.

It is somewhat puzzling to find that a point of view which is adhered to with almost religious fervor by so many reasonable people should be so weak. I think an explanation for this rather surprising phenomenon might be provided by looking at the origin of the idea of a core. Like so many ideas that later on became part and parcel of the empiricist tradition, this idea arose from a certain metaphysical point of view. Parts of this point of view have been sketched in the introduction and in Sec. XII. When empiricism entered its radical and antimetaphysical stage, the metaphysics was dropped and the idea of a core was left without support.

Exactly the same phenomenon can be observed when considering the idea that all our knowledge comes from the senses, or is based upon sensations. Originally this idea was connected with the idea of a real world, with properties that are independent of observation and that develop according to laws of their own. Human beings, so it was assumed, are part of this world. Thoughts, sensations, perceptions are (not necessarily material) parts of human beings. Empiricism, in this context, is a fairly special hypothesis concerning the relation between sensations (perceptions) and events that are external to the experiencing organism. The hypothesis asserts that perceptions and sensations, but not thoughts, are correlated with outer events. The demand for observation and experience that is the core of all empiricism can here be justified very easily: if our sensations and our perceptions are true indicators of what is going on in the world, and if thought will only accidentally hit upon the truth, then any theory about the world will have to be tested by a procedure involving sensations; it will have to be tested on the basis of experience. The additional assertion that only empirical theories have factual relevance can, however, not be derived: pure thought may accidentally hit upon the truth and is therefore factually relevant (we are now dealing with matters outside the domain of logic and mathematics). It is only when we want to find out whether it has actually succeeded in hitting upon the truth that the services of experience are required. Experience, therefore, does not function as an instance providing factual relevance. It is only an instance that allows us to discover whether a theory *that is already factually relevant* but not yet empirical, and that may therefore be called a metaphysical theory, is also true.

Now assume that the idea of a real external world is dropped as being "metaphysical." Such a procedure may be the result of arguments. Yet it leaves empiricism in a most unsatisfactory state. Knowledge is now assumed to consist of two elements: thought and sensation. Thought correlates and connects sensations, and it obtains whatever content it possesses from this correlating and connecting function. If this hypothesis is combined with the abovementioned hypothesis of the objective point of view, then the prominence given to sensations can, of course, be justified.[177] If the combination is not achieved, or is expressly rejected (as it is, for example, in the nontheological parts of the philosophy of Berkeley), then such a justification is no longer possible. One might believe that the *regularity* in the display of sensations can be used for giving them a preference over thought and for asserting that it is they that form the basis of our knowledge, whereas thought possesses only the auxiliary function of ordering them properly. But it is clear that without a certain regularity of thought itself, such ordering would not be possible. It is also overlooked that thought can bring about a more regular display of sensations and must therefore be more regular in appearance than are sensations, at least occasionally: the subjective version cannot by itself justify the preference that is given to sensations.

Nor is there sufficient reason, within this version, for an empiricist criterion of meaning. If all that is given are sensations and thoughts, and if both obey laws of their own, then the attempt to give meaning to sensations by interpreting them on the basis of thought (a procedure that is automatically carried out by the central nervous system and that therefore seems to be much more natural) is as acceptable as the usual procedure, which interprets thought by reference to sensations. The fact that despite its arbitrary character the subjective version is nevertheless regarded as an acceptable philosophy brings it uncomfortably close to irrational enterprises that are joined by people not because of their inherent excellence, but because of their popularity. This is, of course, also true of the idea of a core. It is time that this idea be abandoned and replaced by a more reasonable account of observation.

We must again emphasize that we are at complete liberty to present such an account. The idea of a core is without support. It may sound good; it may also be intuitively appealing. However, I think that any more detailed examination will show it to be a very unsatisfactory idea indeed. Thus nothing stands in the way of an alternative account of observation.

The account that I am now going to develop is by no means new. It was known in the 'thirties, when it was developed in considerable detail and with crystal clear arguments.[178] I shall call this account the *pragmatic theory of observation*. The theory admits that observational sentences assume a special position. However, it puts the distinctive property where it belongs, viz., into the domain of psychology: observational statements are distinguished from other statements not by their meaning, but by the circumstances of their production. There is no time here to give a more detailed account of these circumstances.[179] What should be clear is that these circumstances are open to observation and that we can therefore determine in a straightforward manner whether a certain movement of the human organism is correlated with an external event and can therefore be regarded as an indicator of this event. As opposed to many alternative accounts, the pragmatic theory of observation takes seriously the fact that human beings, apart from being called upon to invent theories and to think, are also used as measuring instruments. We must not assume that they possess a special ability in this latter case. Of course they will *interpret* certain physical and psychological processes that are going on within them. However, this does not make these processes less physical or psychological; for example, it does not turn them into processes that "speak," being able, as it were, to give meaning to sentences that have not yet received any interpretation. It is interesting to note that here certain *empirical* ideas, and especially the idea of a core, seem to be closely connected with the old *theological* notion that man has special properties not found anywhere else in nature. Only the empiricist applies this notion at the wrong place; he applies it to the *physical* and *psychological* parts of human beings. Not even scholastic philosophers have held such an extreme point of view. Science, however, has long taken it for granted that the human body, the human mind, and perhaps all of man can be explained on the basis of materialistic principles. The idea that phenomena can lead to meanings contradicts this point of view from the very beginning and opposes to scitific research the dictum of a philosophical doctrine, or rather the remnants of the effects of a philosophical doctrine that long ago was given up as "metaphysical."

The pragmatic theory of observation restores to science the right to examine human beings according to its own ideas. Moreover, it assumes that the *interpretation* of the observation sentences is determined by the accepted body of theory. This second assumption removes the arbitrary barriers and the a priori elements characteristic of the idea of the observational core. It encourages us to base our interpretations upon the best theory available and not to omit any feature of this theory. We now examine some consequences of this point of view.

The most important consequence of the transition to the pragmatic theory of observation is the reversal that takes place in the relation between theory and observation. The philosophies we have been discussing so far assumed that observation sentences are meaningful *per se*, that theories which have been separated from observation are not meaningful, and that such theories obtain their interpretation by being connected with some observation language that possesses a stable interpretation. According to the point of view I am advocating, the meaning of observation sentences is determined by the theories with which they are connected. Theories are meaningful independent of observations; observational statements are not meaningful unless they have been connected with theories. This consequence is incomprehensible on the basis of an empiricist criterion of meaning. Yet it is by no means as absurd as it sounds. A particular experience is a process like any other process in the world. It may be distinguished by the fact that it contributes causally to the production of a particular *sentence*. However, even in this respect, it is similar to what is going on inside an automatic device, a photoelectric cell, a measuring instrument whose indication may also be in the form of a *sentence* rather than in the form of the movement of a hand. It is clear that sentences thus produced, being physical processes, can receive meaning only by the fact that they are connected with a theory. It is therefore the *observation sentence* that is in need of interpretation and *not* the theory.

Now if this is the case, then the role of observation in the choice of theories must be drastically reconsidered. It is usually assumed that observation and experience play a theoretical role by producing an observation sentence that by virtue of its meaning (which is assumed to be determined by the nature of the observation) may *judge* theories. This assumption works well with theories of a low degree of generality whose principles do not touch the principles on which the ontology of the chosen observation language is based. It works well if the theories are compared with respect to a background theory of greater generality that provides a stable meaning for observation sentences. However, this background theory, like any other theory, is itself in need of criticism. As has been pointed out, such criticism must use alternative theories. Alternatives will be the more efficient the more radically they differ from the point of view to be investigated. It is bound to happen, then, at some stage, that the alternatives do not share a single statement with the theory they criticize. The idea of observation that we are defending here implies that they will not share a single observation statement either. To express it more radically, each theory will possess its own experience, and there will be no overlap between these experiences. Clearly, a crucial experiment is now impossible. It is impossible not because the *experimental device* would be too complex or expensive, but because there is no universally accepted *statement* capable of expressing whatever emerges from observation. *But there is still human experience as an actually existing process,* and it still causes the observer to carry out certain actions, for example, to utter sentences of a certain kind. Not every interpretation of the sentences uttered will be such that the theory furnishing the interpretation predicts it in the form in which it has emerged from the observational situation. Such a combined use of theory and action leads to a selection even in those cases where a common observation language does not exist. This means that our acceptance of very general points of view (as opposed to our acceptance of specific assumptions) is a *practical action*. It is an action based upon the identification of two types of behavior, viz., (1) the "behavior" of certain selected sentences that are predicted by the theory (this behavior may be pictured as a robot built in accordance with the theory and the necessary initial conditions, but which has no sense organs), and (2) the behavior of the very same sentences as uttered by a human observer who does not know the theory. A general cosmological theory, then, does two things. It provides a way of physically mimicking actually occurring physical processes; this one might call the *pragmatic* aspect of the theory. And it provides a way of seeing these processes as parts of a coherent whole; this might be called the *semantic* aspect. The processes mimicked by the selected sentences are, of course, the psychological processes going on in an observer who utters the sentences in the circumstances correlated with the initial conditions: the theory—an acceptable theory, that is—has an inbuilt

syntactical machinery that *imitates* (but does not *describe*) certain features
of our experience. This is the *only* way in which experience judges a general cosmological point of view. Such a point of view is not removed because its observation *statements* say that there must be certain experiences
that then do not occur. [The procedure is excluded by the pragmatic account of observation; according to this account, observation statements are
about the features described by the theory; they are not about experience
(except, of course, if the theory itself happens to be about experience, as is
the case with a psychological theory).] It *is* removed if it produces observation *sentences* when observers produce the *negation* of these sentences.
It is therefore still judged by the predictions it makes. However, it is not
judged by the truth or the falsehood of the prediction statements—this
takes place only after the general background has been settled—but by the
way in which the prediction sentences are ordered by it and by the agreement or disagreement of this *physical* order with the *natural* order of observation sentences as uttered by human observers, and therefore, in the
last resort, with the natural order of sensations. Such a judgment is, of
course, heavily anthropocentric. It makes an accidental pattern, the behavioral pattern of normally brought up human beings who have not too
many hallucinations, a measure of the way in which the universe is to be
seen. However, this anthropocentric element is completely harmless. It has
nothing to do with the *truth* of the theory. Moreover, the pragmatic aspect of a theory is not a sufficient condition of its adequacy. In order to be
able to expand our field of action, the theory must guide us into new domains. It must also make us critical of our actions so that we may find out
which actions are based on strong causal antecedents and which are not.
Only the latter ones will be valuable indicators of external events. In this
way, a theory rebuilds both our actions and our expectations without becoming completely circular. Of course the absence of circularity is based
upon an empirical hypothesis, to wit, that our sensation patterns exhibit
a certain stability and independence of theorizing. If every cosmology
could cause us to have its own pattern of sensations, then our choice of
theories could be based only upon formal considerations (simplicity) or
upon metaphysical considerations. It is possible to imagine a world in
which such a procedure would indeed have to be chosen. It has not yet
been established that our own world is not of this kind.

XVI. Conclusion

Our examination of empiricism has revealed that this doctrine (1) is
incomplete and (2) contains undesirable assumptions. The incompleteness
can be removed by regarding empiricism as a *cosmological hypothesis*
concerning the relation between man and the universe; this hypothesis
is in no way different from other scientific hypotheses, although proof of

its incorrectness would force us to reconsider the activity of scientific research in a very fundamental way. It is assumed by the hypothesis that there exists a real objective world that contains human observers, and that sensations, but not thoughts, are highly correlated with events in this world. Empiricism understood in this manner is radically different from the *philosophical* empiricism whose vague and dogmatic pronouncements are usually the basis of a discussion of the problem of knowledge, and parts of which (such as, for example, the idea of the "given") have been criticized above. The criticism has been made that the basis of the sciences cannot itself be a scientific theory. This objection can be safely removed by pointing out that the sciences are doing very well in the art of basing research concerning certain theories upon other theories. A version of empiricism that has the character of a scientific theory is therefore possible. Being more critical, being capable of transformation by research, being capable of providing arguments for many fundamental assumptions, such a version would also seem to be preferable. The undesirable assumptions of empiricism, on the other hand, which have been our main concern here, can be removed by the use of a theoretical pluralism.

This introduces the topic of the meaning of observation statements. According to the cosmological hypothesis mentioned above, sensations and perceptions are *indicators,* and they are in this respect on a par with the indications of physical instruments. Like these indications, they are in need of interpretation, which must be chosen in such a way that the resulting statement (1) says something and (2) is testable. The conditions just stated exclude an interpretation in terms of the "given." Also (3) the interpretation chosen must make impossible the perennial retention of factual statements. This condition eliminates the idea of an observational core. Finally (4), the interpretation must be such that it correctly reports the situation of which the sensation or the perception is supposed to be a good indicator. This means it must be provided by the theory used for describing the outer world. All these demands are satisfied by the pragmatic theory of observation, which is thereby shown to be a natural consequence of the cosmological version and of some very trite "methodological" remarks.

Considering now the principles according to which a decision between two different accounts of the external world can be achieved, we start by pointing out that in the case of low-level theories, we may proceed in the usual manner, with the help of *crucial experiments.* In this particular case, the background theories will provide a common interpretation for the observation statements of *both* theories and thereby make crucial experiments possible. It is different with high-level theories, for example, with theories concerning the nature of the basic elements of the universe. Such theories may not share a single observational statement. Three procedures may be used in this case. The first consists in the invention of a still more general theory describing a common background that defines test statements acceptable to *both* theories. (The possibility of introducing new crucial tests in this way is a further reason why the invention of more general theories should *not* be made dependent on the refutation of less general theories.) The second procedure is based upon an internal examination of the two theories. The one theory might establish a more direct connection to observation, and the interpretation of observational results might also be more direct. The third procedure was discussed in the last section and consists in taking the pragmatic theory of observation seriously. In this case, we accept the theory whose observation sentences most successfully mimic our own behavior. All these procedures must, of course, be examined in greater detail.

I cannot conclude this essay without giving at least a hint as to how its results are related to more general questions of philosophy. The fact that I have frequently discussed scientific theories must not be misunderstood. I have discussed them not because I want to restrict myself to the philosophy of science, but because I regard scientific theories as excellent examples of actual knowledge. Philosophy has always had the tendency to proceed from a stable system and to judge thought by the criteria implicit in such a system. It is clear that such criteria will fail when applied in the attempt to improve matters in a fundamental way. This is why a consideration of the sciences is important, for despite the great amount of conservatism that is still contained in this enterprise, we have here criticism and progress through revolutions that leave no stone unturned and no principle unchanged. These changes are not so completely arbitrary and unreasonable as are the changes of ideology following the death of a tyrant or some other "great man." They are argued and brought about according to very simple methodological rules. Collecting these rules, eliminating from them the remaining traces of dogmatism and blind ideology would seem to be the starting point of a type of knowledge that is open to improvement, and therefore humanized. It would also seem to have important applications in domains outside epistemology.

As was pointed out in Sec. III, some ideas that play a decisive role in the empiricist tradition are of a quite different origin and may be found in quite different philosophies. They are also active in fields outside philosophy. Is not a tyranny the natural correlate of the idea of absolute knowledge, and is not indoctrination the method of teaching most appropriate to it? Conversely, is not the idea of fallibility and the correlated demand for a multiplicity of ideas, and the hope that truth will arise from the civilized clash of such ideas the essence of all democracy? The connections reach also into more recondite fields. The history of artistic ideals, to take only one example, is very similar to the (much more recent) history of empiricism. It starts with a comprehensive doctrine that provides *reasons* for many procedures; it terminates with a few scattered remains between which theoretical aestheticians then strenuously try to establish some kind of coherence. Thus the monistic idea of a perfect work of art whose parts all collaborate in harmony makes sense at a time when the arts have still a ritualistic function and when the single end requires perfectly balanced means. The idea was retained after the "emancipation" of the arts, and it is now supported either subjectively (by reference to the inner unity of being of the artist—which leads to a great deal of arbitrariness) or not at all. All this shows that theoretical monism is the reflection, within the domain of theoretical knowledge, of a *much more general point of view*, which at a time when abstract distinctions were not yet known and when *metabasis* was not yet a crime left its traces in almost all human activities. Conversely, we may guess that the theoretical pluralism that we propagate in epistemology may lend itself to generalization and may then lead to an outlook in the arts and in religion, as well as to a new, comprehensive ideology that assembles the scattered remains of a long-forgotten tribal ideology and unites them in a truly humanistic system of belief. Partial successes have already been achieved. In politics, the idea of democracy is established in at least some minds, and the battle of tongues that goes with it in at least some countries. In the arts, the idea of alternatives has been introduced with excellent arguments by Bertolt Brecht.[180] It is up to us to develop this idea further and to turn it from a general philosophical principle into an active ingredient of all aspects of our life.

Notes

For support of research, the author is indebted to the National Science Foundation and to the Minnesota Center for the Philosophy of Science.

1. J. F. W. Herschel writes:

An immense impulse was now given to science and it seemed as if the genius of mankind, long pent up, had at length rushed eagerly upon Nature, and commenced, with one accord, the great work of turning up her hitherto unbroken soil, and exposing her treasures so long concealed. A general sense now prevailed of the poverty and insufficiency of existing knowledge in *matters of fact;* and, as information flowed fast in, an era of excitement and wonder commenced to which the annals of mankind had furnished nothing similar. It seemed, too, as if Nature herself seconded the impulse; and while she supplied new and extraordinary aids to those senses which were henceforth to be exercised in her investigation—while the telescope and the microscope laid open *the infinite* in both directions—as if to call attention to her wonders, and signalise the epoch, she displayed the rarest, the most splendid and mysterious, of all astronomical phenomena, the appearance and subsequent total extinction of a new and brilliant fixed star twice within the lifetime of Galileo himself. [J. F. W. Herschel, *The Cabinet of Natural Philosophy* (Philadelphia, 1831), Secs. 106-107.]

2. For a more detailed discussion of this feature, see Sec. XV below.

3. In what follows, the term "theory" will be used in a wide sense, including ordinary beliefs (e.g., the belief in the existence of material objects), myths (e.g., the myth of eternal recurrence), religious beliefs, etc. In short, any sufficiently general point of view concerning matter of fact will be termed a "theory."

4. A splendid example of the existence of different point of view hiding behind the appeal to observations is provided by the early history of the Royal Society in London. It is commonly agreed that "the foundation of the Royal Society was one of the earliest practical fruits of the philosophical labors of Francis Bacon" [Sir Archibald Geikie, *The Records of the Royal Society*, 3rd ed. (London, 1912). See also Bishop Sprat, *History of the Royal Society* (London, 1677), which by its very chauvinism makes clear the general attitude of those interested in the welfare of the society. For a condensed account, see L. T. More, *Isaac Newton, A Biography* (New York: Dover Publications, Inc., 1962), Chap. 13]. Charles II granted a royal coat of arms to the Society with the motto *Nullius in verba*, emphasizing the factual and observational bent of its members. Hooke and Newton again and again voiced their opinion concerning the importance of observation. Empiricism seemed to be the doctrine uniting all the members of the Society. Yet a closer look reveals many different attitudes concerning the use of observations, ranging from a crude philosophy that would demand the collection and preservation (possibly in a museum) of any fact, however abstruse, to the most sophisticated empiricism of Newton, whose "phenomena" were not observable states of affairs pure and simple (in a more modern technical sense they are even unobservable) but *laws* represented mathematically and made plausible by reference to appearances. For details, see Sec. II.

5. The fact that almost any philosophical doctrine may find realization either in a *cosmology*, that is, in a theory of the universe that is capable of sensual representation, and/or in a *theory of man*, which may also be sensually realized in a corresponding society—this fact makes it very clear that the procedure leading to the adoption of a philosophical position cannot be *proof* (proof shows that no other position could possibly be realized), but must be a *decision* on the basis of preferences. We must admit that philosophical positions correspond to views of the universe that, if worked out in detail, may be filled with perceptual content and may therefore be said to be blueprints for possible, *and realizable*, universes. These positions admit of a *genuine choice*.

Philosophers have habitually judged the situation in a very different manner. For them, only *one* of the many existing positions was true and, therefore, possible. This attitude, of course, considerably restricts the domain of responsible choice. The point made in the text and slightly elaborated here would seem to imply that the problem of responsible choice enters even the most abstract philosophical matters and that ethics is, therefore, the basis of everything else. For details, see my *Knowledge Without Foundations* (Oberlin: Oberlin College, 1961).

6. A similar point has been made by Arne Naess in a criticism of my "Explanation, Reduction, and Empiricism," in *Scientific Explanation, Space and Time*, Herbert Feigl and Grover Maxwell, eds., Minnesota Studies in the Philosophy of Science, Vol. III (Minneapolis: University of Minnesota Press, 1962).

7. Not even this very modest statement is correct. The Hippocratean corpus, to mention only one earlier example, is full of explicit exhortations to observe in detail and to refrain from hypotheses [for quotations, see George Sarton, *History of Science*, Vol. I (Cambridge: Harvard University Press, 1959) pp. 356 ff.]. The example of Aristotle is too well known to be mentioned here.

8. (a) One of the mechanisms bringing about such coherence is the partial de-

pendence of perception upon belief. What we receive from the outer world (and from the so-called "inner" world) are certain *clues,* which most of the time are pretty vague and indefinite. Perception is the result of the reaction of the total organism to these clues. In this reaction, the knowledge acquired, the beliefs held, the emotional condition of the receiver, his fears and his expectations, play a most important role. It is these that are (partly) responsible for the formation of well-defined wholes out of indefinite patterns of stimuli.

Just consider the appearance of a lake on a bright summer day. There are small areas of brightness where the sunlight is reflected from the wave crests, and these areas are separated by a darker background. At some places, a well-defined shadow seems to fall upon this ever-changing pattern of extreme brightness and darkness. These are the clues. Yet what we see is something very different. We see a continuous, uninterrupted surface and a boat traveling along *on* this surface. The objects seen, the lake and the boat, are fairly independent of the *details* of the arriving pattern, which means that the tendency to perceive a well-defined objective situation may make the observer see things that are not really there. [For details, see, for example, M. D. Vernon, *The Psychology of Perception* (Baltimore: Penguin Books, Inc., 1962).] Initially unrelated impressions are combined into wholes, and the belief bringing about the combination will lead to perception of *objects* even in cases where many of the constituent impressions are missing. This is how a belief may give a well-defined outline to what is perceived only vaguely, indistinctly, by a person lacking it.

The very same process is responsible for the existence of *genuine observational reports* concerning devils and gods. [For gods, see E. R. Dodds, *The Greeks and the Irrational* (Berkeley: University of California Press, 1951), pp. 14 ff. For an explanation of why these observational reports are rarely taken seriously, and why their existence is so frequently denied by historians, see the remarks at the beginning of fn. 9, below.] We are all aware of thoughts, impulses, feelings that run counter to our conscious intentions. Usually we disregard them, for they do not occur in a very coherent fashion. They may appear and disappear without any apparent reason. It is quite different with a person believing in the existence of demons. He would perceive a meaningful pattern in such occurrences; they would appear to him as the result of the attempts of some demon to corrupt him. Considering the astounding plasticity of the human mind, this belief *could even bring about a more regular display of such alien occurrences;* or, to use psychoanalytic terminology, it could bring about a consistent intrusion of the id into the domain of the ego. Expectation, fantasy, fear, and mental illnesses flowing from them (hearing voices, having the feeling that one's behavior is brought about by an alien agency, split personality, etc.) would do the rest. [For observational reports, see Karl Jaspers, *Allgemeine Psychopathologie* (Berlin: Springer Verlag, 1959), pp. 75-123.] Demons would have become directly observable. *And this has actually happened.* Just consider the case of a frustrated woman who has become mentally unbalanced, who hears voices, who cannot distinguish dream from reality, who has nightmares, and who, on the basis of such phenomena, has deluded herself into believing that she had intercourse with the devil. (There was a time when dream occurrences were regarded as indicative of the *real* state of one's soul!) She may develop—and women in such a situation actually *have* developed—a phantom pregnancy. Can there be any more direct evidence in favor of the existence of demonic influence? [For an excellent and much more detailed discussion of the psychological phenomena we have just outlined, see Jules Michelet, *Satanism and Witchcraft* (New York: Wehman Bros., 1939).]

The theory of the dependence of perception upon belief is by no means as fanciful as it may appear to a radical empiricist. That primitive people, whose life is governed by a powerful myth, live in an observational world very different from our own is shown by their art. It has been assumed for some time, no doubt under the influence of empiricism, that the "primitive" character of these productions is due to a lack of skill: these people live in the same perceptual world as we do, but they are unable to produce adequate copies of it. This assumption has been refuted. One element of the refutation consists in showing that the primitive artist may on occasion exhibit a quite considerable skill, but that he refuses to use this skill for the creation of what *we* are inclined to call a "realistic" picture.

Another point is that realism of representation is an impossible doctrine. It assumes that there is only one correct way of translating occurrences in the three-dimensional real world into situations portrayed in an altogether different medium. The world is as it is. The picture is not the world. What, then, does the realist demand? He demands that the conventions *to which he is accustomed* (and which are only a meager selection from a much wider domain of conventions) be adopted. That is, he makes *himself* the measure of the reality of things—the very opposite of what the realistic *doctrine* would allow. [For details, see Ernst Gombrich, *Art and Illusion* (New York: Pantheon Books, Inc., 1960), esp. Chap. 8. See also the very interesting first chapter of Bruno Snell, *The Discovery of the Mind* (Cambridge: Harvard University Press, 1953).]

The role of traditions and of the conventions deriving from them, including those that look "primitive" to us, form a most interesting part of the theory of the arts as well as of the general theory of perception. We find that conventions are used not only in the absence of the object, but exert their influence *even when a direct account of the visible object is attempted.* Gombrich (*op. cit.*, pp. 81 f.) tells the following story:

> . . . The fate of exotic creatures in the illustrated books of the last few centuries is as instructive as it is amusing. When Dürer published his famous woodcut of a rhinoceros, he had to rely on secondhand evidence which he filled in from his own imagination, coloured, no doubt, by what he had learned of the most famous exotic beast, the dragon with the armoured body. Yet it has been shown that this half invented creature served as a model for all renderings of the rhinoceros, even in natural history books, up to the eighteenth century. When, in 1790, James Bruce published a drawing of the beast in his *Travels to Discover the Source of the Nile,* he proudly showed that he was aware of the fact: "The animal represented in this drawing is a native of Tcherkin, near Ras el Feel . . . and this is the first drawing of the rhinoceros with a double horn that has ever yet been presented to the public. The first figure of the Asiatic rhinoceros, the species having but one horn, was painted by Albrecht Dürer, from the life. . . . It was wonderfully ill-executed in all its parts, and was the origin of all the monstrous forms under which that animal has been painted ever since. . . . Several modern philosophers have made amends for this in our days; Mr. Parsons, Mr. Edwards, and the Count de Buffon have given good figures of it from life; they have indeed some faults, owing chiefly to preconceived prejudices and inattention. . . . This . . . is the first that has been published with two horns, it is designed from the life, and is an African."

If proof were needed that the difference between the medieval draftsman and his eighteenth-century descendant is only one of degree, it could be found here. For the illustration, presented with such flourishes of trumpets, is surely not free from "preconceived prejudices" and the all-pervading memory of Dürer's woodcut. We do not know exactly what species of rhi-

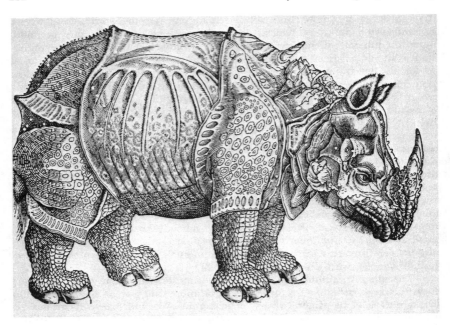

A. Dürer, *Rhinoceros* (woodcut). (*Photograph from The Bettmann Archive.*)

W. Daniell, *Rhinoceros in Its Native Wilds* (engraving). (*Photograph from The Bettmann Archive.*)

noceros the artist saw at Ras el Feel. . . . But I am told that none of the species known to zoologists corresponds to the engraving claimed to be *al vif!*

What we have just discussed was not unknown to philosophers. Thus in his essay "Wahrheit und Lüge im aussermoralischen Sinn" (*Werke,* Schlechta, ed., Vol. III, pp. 239 f.), Nietzsche makes the following observation:

> In the last resort it is only the rigid and regular web of concepts which makes it clear to a person awake that he is as a matter of fact awake. . . . The waking day of a mystically excited people as are the older Greeks is indeed more similar to a dream than is the day of a sober mind; and this is due to the always present miracle as assumed by the myth. When every tree can at any time talk as a nymph, or when a god can rob virgins under the cover of a bull, when the goddess Athena is suddenly seen in person, driving side by side with Peisistratos in a splendid carriage through the market places of Athens—and this the honest Athenian did in fact believe—then like in a dream everything is possible at any moment, and the whole of nature surrounds man as if it were only a masquerade of the gods.

(b) Another element contributing to the *observational* character of a myth is that considering the extent to which human welfare is supposed to be affected by demons (colliding with a chair you will at the very most break your leg; colliding with an emissary of Satan you may end up in eternal hellfire), I am prepared to venture the hypothesis that at the time in question—when the sense of reality was frequently disturbed, when dreams were considered reality and phantasies and evil thoughts put on one level with causal agencies of actual events in the outer world, when the incidence of hallucinations, phantasies, illusions seemed to be much more frequent than it is today, when the fate of one's

African rhinoceros, the "real thing." (*Photograph from The American Museum of Natural History.*)

soul was the supreme motive of almost all actions and this not only for a select group but for everybody alike (think of the crusades a few centuries earlier!)—I venture the hypothesis that in this time of partly accepted and partly enforced conformity, the "demonic," or "spiritual," usage of mental terms was *much more solidly established,* much more "successful," than is the language of mental events—or even the language of material objects—of today. And I support this hypothesis by pointing out that the firmness, the solidity, the regularity of usages are a function of the firmness of the beliefs held as well as of the needs that these beliefs satisfy: what is regarded linguistically as the "firm ground of language" and is thus opposed to all speculation is usually that part of one's language that is closest to the most basic ideology of the time and expresses it most adequately. *That part will then also be regarded as being of great practical importance.* Being intimately connected with the practical life it will also be used for the expression of observational results, and it will be the only point of view that is used in this fashion. It is clear that it will receive tremendous empirical support; almost any practical action is relevant to the belief, and almost any result will therefore confirm it.

(c) A third reason contributing to the observational accuracy of a myth may be seen in the fact that people do not outstep its boundaries. An Indian who believes that the gods will support him in his hunting prepares well in order to be worthy of divine support. He tries to maintain a certain level of physical fitness; he sleeps long before the day of hunting; he starts with a prayer designed to calm his soul—and it is only *after* all these tasks have been completed that he starts on his enterprise. [For these details, see Paul Radin, *Primitive Man as a Philosopher* (New York: Dover Publications, Inc., 1957).] An empiricist should not really criticize such a procedure—especially when he adopts the *ceteris paribus* clause and insists that a theory is always valid within a certain domain of experimentation and must not be applied outside. [See Werner Heisenberg, "Der Begriff der abgeschlossenen Theorie in der modernen Naturwissenschaft," *Dialectica,* Vol. II (1948), 331-336.]

9. It goes without saying that the identification of myth and complete nonsense is only another consequence of the empiricist attitude that identifies "good" knowledge on the one side with what is obtained from observation, and, on the other side, with what is contained in the most recent textbooks. Myths, such as the myth concerning the existence of a hierarchy of evil spirits, are not contained in the most recent textbooks of physics, chemistry, psychology, etc. Hence they cannot be observational. It seems that a historian has the duty to establish such an assertion by *factual historical research* and not by deductions from physics textbooks and a doubtful philosophical doctrine. However, quite apart from being ill-founded, the doctrine that a myth is entirely removed from reality is also most implausible. Can one really believe that a point of view that guides the whole life of a community, that influences its most trivial decisions, that regulates its actions, forms its dreams, is used in the interpretation of what is done and seen—can one really believe that such a doctrine will be without observational support? That the point of view of a hard-working people threatened by destruction will be more realistic than the cosmology of a carefree group of abstract thinkers has been argued by Hegel in his observations on the relation between master and slave. [For a detailed discussion, see G. Lukács, *Der Junge Hegel* (Zurich: Europa Verlag, 1948), Chap. 3.] Moreover, the thesis of the essential primitiveness of a myth is also very dangerous. It weakens resistance against a *clever* myth that is presented, say, in mathematical language and proceeds in a pseudo-empirical fashion. It has been the fate of many empiricists that after a

lifelong battle against interesting physical theories and uninteresting common-place mistakes they became the victims of a well-presented myth. Pascual Jordan's interest in occult phenomena and W. Pauli's in Jungian psychology are most conspicuous examples.

As opposed to such superficial and prejudiced accounts, it must be empha-sized that a myth is far from being a chaotic and irrational heap of ideas that sometimes agree and sometimes do not agree with the facts, without there being the slightest feeling of discomfort in the latter case. [For detailed arguments con-cerning the inner coherence and consistency of mythological thought, see Ernst Cassirer's splendid introduction to Vol. II of his *Philosophie der symbolischen Formen* (Berlin: Bruno Cassirer Verlag, 1925). Cassirer's positive theory does not seem to me to be acceptable, however. For a more acceptable account, see Henri Frankfort, *et al.*, *The Intellectual Adventure of Ancient Man* (Chicago: Uni-versity of Chicago Press, 1946), as well as the more detailed investigations of Breasted, Evans-Pritchard, and other anthropologists. On the whole, one can say that the attitude of the empiricists has been a great hindrance to proper and un-prejudiced investigation of the structure of mythological thought. Philosophy of science is not the only field that has been confused by that school.] Quite the contrary, a myth at its best is a most carefully worked out and well-balanced point of view (better balanced, sometimes, than are some of the slipshod methods now in use in certain parts of the quantum theory) whose parts are related to each other in a manner that allows for the explanation of almost any conceivable situ-ation—conceivable, that is, for the person who has accepted it and is firmly com-mitted to its pattern of explanation.

The empirical character of a good many mythological stories is revealed, among other things, by the fact that a proper interpretation turns them into astronomical theories.

> We can see . . . how many myths, fantastic and arbitrary in semblance . . . may provide a terminology of image motives, a kind of code which is (just now) beginning to be broken. It was meant to allow those who knew (*a*) to determine unequivocally the position of given planets in respect to the earth, and to one another; (*b*) to present what (empirical) knowledge was of the fabric of the world in the form of tales about how the world began.

[Giorgio de Santillana, *The Origins of Scientific Thought* (Chicago: University of Chicago Press, 1961), p. 14. See also his criticism of some contemporary inter-pretations of mythical thinking in "On Forgotten Sources in the History of Sci-ence," in A. C. Crombie, ed., *Scientific Change* (London: William Heinemann, Ltd., 1963), pp. 812-828.] This being the case, confirming evidence for the myth would, of course, turn up every day.

The same is true of myths concerning terrestrial phenomena. Thus every witch trial, every confession of presence at a witches' sabbath, every otherwise un-explained occurrence of pestilence, floods, disease of cattle, impotence, night-mares, unfaithfulness of wives (who, of course, immediately interpreted, and therefore *experienced*, whatever struggle, fear, temptation was going on within them as the intrusion of some partly hostile, partly friendly demon), malforma-tion at birth, schizophrenia—all these things would give direct proof of the exist-ence of incubi, succubi, and other evil spirits, so that finally denying them would seem to be as nonsensical, arbitrary, whimsical as the denial of the existence of tables and chairs appears to some "philosophers" of today. It is far from true, as has been suggested by empiricist historians, that the evidence upon which the whole structure was based either did not exist, or was faked, or was the result of fear and intimidation. Witch trials were no more prejudicial to the accused than were the trials of the French Revolution, the trial of Dreyfus, or the Scopes trial. The evidence available at the time in question was as real as is the contemporary evidence in favor of complementarity, although it must be admitted that being more direct and being connected with some very strong and troubling experiences, as well as with the threat of eternal damnation, it was much more impressive. The *existence* of this evidence, which, I must repeat, was spectacular and overwhelming, constitutes a grave problem for anyone who supports his predilection for empiricism by his predilection for the sciences and who thinks that it is the emphasis upon *observations* which precipitated scientific progress.

10. The same is true if we compare the so-called "ordinary language" of some contemporary philosophers and the language of a society that is governed by a myth. Philosophers have argued that the ordinary language correctly reflects the facts as it deals with "topics which constantly concern most people in some practical way" [G. J. Warnock, *English Philosophy in 1900* (London: Oxford University Press, 1956), pp. 150 f.]; and Warnock goes on to mention "perception, the ascription of responsibilities, the assessment of human character" as being topics of this kind. Now these topics may be of interest to the gentlemen living free and unconfined in the various colleges of Oxford and entertaining themselves now with a little gossip, now with the discussion of some esoteric "puzzle"; but they are hardly essential for a community of people surrounded by wild animals, threatened by bad weather, by illness, exposure, hunger, and intimidated by all sorts of strange occurrences. Topics of this kind form, indeed, a superstructure that is only loosely connected with reality.

11. Bacon, *Novum Organum,* 95.

12. We are, of course, at all times discussing systems that are supposed to be factually relevant. The domain of *pure mathematics* is outside the scope of the present paper.

13. For details, see below and Sec. XV.

14. Reading Agassi, *Towards an Historiography of Science* [*History and Theory* (The Hague: Mouton & Co., 1963), Beiheft 2], has provided me with the following splendid quotation from Voltaire [*Newton and Leibniz* (Glasgow, 1764), p. 59]: "Strange! We know not how the earth produces blades of grass; how a woman conceives a child; and yet we pretend to know how ideas are produced!"

15. For a somewhat different, though related, account, see Feibleman, "Philosophical Empiricism from the Scientific Standpoint," *Dialectica,* Vol. XVI (1962), 5 ff.

16. This simple truth is overlooked by many impatient defenders of the orthodox philosophy of the quantum theory who proudly compare the sophistication of, say, the wave mechanics with the "generalities" of Bohm and Vigier. For details, compare Sec. VIII of "Problems of Microphysics," in Robert G. Colodny, ed., *Frontiers of Science and Philosophy* (Pittsburgh: University of Pittsburgh Press, 1962).

17. It is nowadays quite frequently assumed that "if one considers the history of a special branch of science, one gets the impression that nonscientific elements . . . relatively frequently occur in the earlier stages of development, but that they gradually retrogress in later stages and even tend to disappear in such advanced stages which become ripe for more or less thorough formalization" [H. Groenewold, "Non-Scientific Elements in the Development of Science," *Synthèse* (1957), p. 305]. Our considerations in the text would seem to show that if this is indeed the *actual* development, then it can result only in a well formalized, precisely expressed, and completely petrified metaphysics.

18. The way in which this point of view is asserted is, of course, quite immaterial. Even the fact that its categories might be present in our *perceptions* does not render its adoption less arbitrary. It only shifts the point of arbitrariness from thought to the senses.

19. Two theories will be called incommensurable when the meanings of their main descriptive terms depend on mutually inconsistent principles. Examples are the impetus theory and Newton's celestial mechanics, Newton's celestial mechanics and special relativity, and geometrical optics and wave optics. For details concerning the first example, see my "Explanation, Reduction, and Empiricism," Sec. V, in *Scientific Explanation, Space, and Time, op. cit.*

20. Considering this feature of the argument, I cannot understand why Feigl should find "neo-Wittgensteinian" traces in my reasoning ["The Power of Positivistic Thinking," presidential address to the meeting of the Western Division of the American Philosophical Association, *Proceedings of the American Philosophical Association* (1963)]. "Wittgensteinian" analyses are characterized by the fact that they are restricted to a single "language game"; they are *monistic*. Alternatives are brought in, not in order to arrive at a better theory through a criticism of the existing one, but rather *in order to get better insight into the existing theory.* And getting better insight into the existing theory (the existing "language game") means revealing its hidden strength, i.e., its capability to deal with the problems that have arisen and to remove the impression that a revision might be needed. A Wittgensteinian uses alternatives with a *dogmatic,* or a *conservative,* purpose. Knowing where the argument is going to lead, he does not really take them seriously, although he may admit that they possess an important therapeutic function. Our own point of view is radically different. Alternatives *are* taken seriously. They are all candidates for a better theory in the future. Their function is *critical* and *progressive,* that is, they are used with the purpose of finding the *weaknesses* of the customary set of ideas. The discussion of such weaknesses does not have an exclusively dramatic purpose, viz., to let the accepted point of view shine forth the brighter once the philosophical fog has been removed. A greater difference can hardly be imagined. Moreover, I feel I must object to the tendency, implicit in the above characterization of Feigl's, to forego historical accuracy in favor of hero worship. The arguments demanding intersubjectivity of observation statements (and corresponding high theoretical content) are not at all original with Wittgenstein; they were formulated before him, and in a much clearer and simpler way. History, and not fashion, should decide who has achieved what and who is therefore worth listening to.

21. For details, see my "Explanation, Reduction, and Empiricism," in *Scientific Explanation, Space, and Time, op. cit.,* pp. 34-40.

22. An example is Galileo's argument in his *Dialogues Concerning the Two Chief World Systems* showing that the moon cannot have a perfectly spherical surface. For the methodological principles involved in the situation, see Sec. VI below: the history of the Scientific Revolution is an excellent example of the necessity of a theoretical pluralism for the purpose of progress.

23. As has been pointed out (fn. 9), it is difficult to find correct factual descriptions of the nature of mythical thinking. It is equally difficult to find an even halfway satisfactory account of the Scientific Revolution, and ecpecially of the procedures of Galileo. In the case of a myth, the argument goes like this: a myth is different from what is accepted science today, which has been obtained on the basis of observation and straight thinking. Hence a myth is nonobservational and crooked thinking. In the case of Galileo, the argument has the following form: Galileo contributed to classical mechanics, which we still use. What is

still used by science today has been obtained on the basis of observation and straight thinking; hence Galileo too must have obtained his results by observation and straight thinking. It is this a priori argument (the statement that all good ideas are obtained by observation constituting the a priori element) rather than empirical historical research that leads to assertions like "Galileo refuted the Aristotelian dogmas respecting motion, by direct appeal to the evidence of sense, and by experiments of the most convincing kind" (Herschel, *op. cit.*, p. 85). This "Galileo myth" according to which Galileo busily rushed around making experiments and "climbed the leaning tower of Pisa with one one-hundred pound cannon ball under one arm and a one-hundred pound cannon ball under the other" [ironical remark in Herbert Butterfield, *The Origins of Modern Science* (London: G. Bell & Sons, Ltd., 1957), p. 81] has been refuted by more recent historical research. "In general," writes E. J. Dijksterhuis in his *Mechanization of the World Picture,* C. Dikshoorn, trans. (London: Oxford University Press, 1961), p. 338, "one has always to take stories about experiments by Galileo as well as by his opponents, with some reserve. As a rule they were performed only mentally, or they are merely described as possibilities." More especially, there is evidence that proves the "complete baselessness of the belief tenaciously maintained by the supporters of the Galileo myth, namely, that he discovered the law of squares by performing with falling bodies a number of measurements of distance and time, and noting in these values the constant ratio between the distance and the square of time. This conception is further flagrantly contradicted by all that is known about the degree of importance that Galileo attached to scientific experiment" (*ibid.*, p. 340).

However, the situation is even worse. There were weighty physical arguments against the motion of the earth that Galileo defended. These physical arguments were an immediate consequence of the then popular Aristotelian physics that was confirmed by a great variety of commonly known observational results. [For details, see my "Realism and Instrumentalism," in *The Rational Approach, Essays in Honor of Karl Popper* (New York: The Free Press of Glencoe, Inc., 1963), esp. fn. 4, where the difficulties of the Aristotelian theory at the time of the discussion of the Copernican system are put in the proper light.] Was it not natural, in these circumstances, to assume that a realistic interpretation of the Copernican theory was out of the question and that the merit of this theory consisted merely in having found a coordinate system in which the problem of the planets assumed an especially simple form? This move could not be countered by a purely philosophical criticism and by the demand that every theory be interpreted in a realistic fashion. A *philosophical* criticism of the instrumentalistic interpretation of the Copernican theory quite obviously could not remove the inconsistency between a Copernican universe and the Aristotelian dynamics. Nothing less than a new theory of motion would do, a theory of motion, moreover, that could not be as strictly empirical and common sensical as the Aristotelian theory and could therefore count on strong opposition. This was clearly realized by Galileo: "Against the principles of the conventional cosmology, which were always brought out against him, he needed an equally solid set of principles, indeed, more solid—because he did not appeal to ordinary experience and common sense as his opponents did" [Giorgio de Santillana, *The Crime of Galileo* (Chicago: Chicago University Press, 1955), p. 31]. See also the pages following the quotation, as well as my article on the philosophy of nature in *Fischer Lexicon, Band Philosophie* (Frankfurt-am-Main: S. Fischer Verlag, 1958).

Galileo never succeeded in fully developing these principles, and he was therefore always open to the objection that his philosophy rested upon a nonexistent theory of motion. The assertion that Galileo succeeded in deriving the new physics from his observations is therefore incorrect for more reasons than one.

Galileo's trust in the adequacy of the telescope is another, and perhaps an even more striking, example of the gap that exists between historical accounts and the actual facts. When Galileo insisted that his telescope revealed what was going on in the heavens, he could support his belief neither by experience nor by reference to theory. There was no optical theory that could provide a correct account of the phenomena seen when looking through a telescope. The first attempt at such a theory was made by Kepler, who prepared geometrical optics. But geometrical optics deals only with some idealized properties of the radiation in a telescope; it does *not* deal with the visual impressions received by the observer who puts his eye to the ocular. It is therefore woefully inadequate when regarded as a theory of what is seen (as was realized already by Berkeley in his *New Theory of Vision*). A theory of telescopic *vision* has been developed only very recently and is still far from complete. *Theories,* therefore, could not support Galileo's trust in the telescope. The *facts,* on the other hand, were plainly against it. Everybody knew of the strange phenomena that occurred when lenses were not held at the proper distance, and even when the proper distance was preserved, lenses could not always be trusted. Moreover, it was very difficult to repeat Galileo's results, and he had to insist that his own telescopes be used. Is the use of a telescope under these circumstances not very similar to the use of a medium who has sometimes been successful, who has also often failed, and the principles of whose success are unknown? For details and further literature, see V. Ronchi, "Complexities, Advances, and Misconceptions in the Development of the Science of Vision: What Is Being Discovered?" in Crombie, *op. cit.,* pp. 542-561.

24. It is customary to present the Aristotelian physics and astronomy as a system of thought that was inconsistent with a great variety of experimental results but was still retained for ideological reasons. Copernicus and Galileo, so it is said, took these observational results seriously and therefore gave up Aristotelianism (and the corresponding Ptolemaic system of the universe) and devised a new system that took all facts into account. The Aristotelians stuck to their theory despite many unpleasant facts. Galileo, however, was more honest; he respected facts and therefore tried to manufacture a theory that was in better agreement with them.

There is hardly an ounce of truth contained in this favorite fairy tale of empiricism. First of all, there exists not a single theory that is not in some difficulty or other and is yet not retained in the hope that the difficulty may at some time turn out to be soluble in its terms. Hans Reichenbach [*The Rise of Scientific Philosophy* (Berkeley: University of California Press, 1951), pp. 101 f.] makes much of Newton's refusal to publish his theory while it was still inconsistent with the facts. "Rather than set any theory, however beautiful, before the facts, [Newton] put the manuscript of his theory into his drawer." He continues by saying that "the story of Newton is one of the most striking illustrations of the method of modern science." Now quite apart from the fact that the historical reason given by Reichenbach for the delay in the publication of the *Principia* does not seem to be the correct one (see, for example, More, *op. cit.,* Chaps. 9 and 11), the attitude praised would seem to be most imprudent. It encourages the theoretician to give up in the face of difficulties and propagates a dogmatic reliance upon certain observational data (which, in the case of astronomy, are the result of rather laborious calculations) and upon a certain specific derivation, or derivation sketch, of these data from the theory. The later history of Newton's theory shows the advantages connected with a healthy dogmatism that is not based upon the alleged existence of a proof of absolute correctness, but rather upon the hope that that theory will in the end prevail.

There did exist phenomena (such as the great inequality of Jupiter and Saturn) that seemed to be inaccessible to treatment on the basis of Newton's theory until Laplace finally gave a satisfactory solution. The more recent history of the kinetic theory is another case in point. The difficulties that this theory had been shown to possess, including proofs of internal inconsistency, had led to the belief that it was finished. "It is not at all easy for us today," writes Smoluchowski in a most instructive paper ["Gültigkeitsgrenzen des zweiten Haupsatzes der Wärmetheorie," in *Oeuvres*, Vol. II (Krakow, 1927), pp. 361 ff.] "to recall the mood that was prevalent toward the end of the last century. At that time the scientific leaders in Germany and in France were convinced—with very few exceptions—that the atomistic kinetic theory of matter had ceased to be of any importance. Considering the great successes of thermodynamics, the . . . second law . . . had been raised to the rank of an exactly valid, absolute dogma that was true without exception. And as the kinetic theory had run into certain difficulties in the attempt to arrive at a satisfactory interpretation, especially as regards the irreversible processes, one had condemned it together with the atomistic hypothesis. . . . All this has changed today." Why? Because some physicists were not afraid of the difficulties; because they did not put their manuscript into the drawer, but continued to work on the theory until they were finally able to show its *supremacy*, and not only its *adequacy*.

It sometimes needs much unreasonable faith to be able to bring out the best in a theory. The advice implicit in Reichenbach is that we should be more defeatist, or in any case much less persistent. But it is impossible advice. There does not exist a single theory that is not confronted by prima facie refuting instances. And therefore there does not exist a single theory that would not immediately have to end up in the drawer of the one who suggested it. Of course one cannot simply ignore the prima facie refuting instances (although it is sometimes wise; after all, one cannot do everything at the same time). One must attempt to explain why there is an impression of inconsistency with experimental results. This was done by the Aristotelians—which brings us back to the topic of this footnote: the relation between the Aristotelian point of view and the attitude of the defenders of the new physics.

It is not at all correct that the Aristotelians stuck to their point of view despite the facts, and that the defenders of the new physics took the lesson of the facts seriously. Both parties were confronted with serious difficulties, and the question as to which difficulties were greater is a nice one indeed. (For details, see again my essay "Realism and Instrumentalism," in *The Rational Approach, op. cit.*) Common sense, broad observation was quite definitely on the side of the Aristotelians. The troubles of the calender could have been improved by additional epicycles (as a matter of fact, almost any quasiperiodic motion could have been accounted for in this manner; the reason is that the method of epicycles is nothing but a geometrical application of Fourier's more general theorem). Thus it is quite clear that observations could never have *forced* anyone to adopt a new physics, a physics, moreover, that to start with was inconsistent with many fundamental observations.

25. Places of publication are as follows: *Scientific Explanation, Space and Time,* Minnesota Studies in The Philosophy of Science, Vol. III (Minneapolis: University of Minnesota Press, 1962); *Frontiers of Science and Philosophy,* Pittsburgh Studies in the Philosophy of Science, Vol. I (University of Pittsburgh Press, 1962); *Problems of Philosophy, Essays in Honor of Herbert Feigl* (Min-

neapolis: University of Minnesota Press, 1964); *Proceedings of the Aristotelian Society*, New Series, Vol. LVIII (1958); *ibid.*, Supp. Vol. XXXII (1958).

26. This is asserted by Agassi in a paper he let me read prior to its publication. There seems to be a great similarity between Agassi's point of view and that of this essay.

27. In his book *The Physicist's Conception of Nature*, Arnold J. Pomerans, trans. [London: Hutchinson & Co. (Publishers), Ltd., 1958], pp. 105, Werner Heisenberg states:

> Questions and answers, observations and determinations, are no longer directed at a general, metaphysical and theological understanding, but are delimited with modesty. . . . *This modesty was largely lost during the nineteenth century.* Physical knowledge was considered to make assertions about nature as a whole. Physicists wished to turn philosophers. . . . Today physics is undergoing a basic change, the most characteristic trait of which is a return to its original self-limitation.

Starting with the last part of this composite quotation, I must register my surprise at hearing that today physicists want to refrain from generalizing their physics into philosophy. There has been hardly any age in the history of the sciences when so many physicists posed as philosophers. Not only is it asserted that ancient philosophical ideas, such as the idea of determinism or of the reality of the external world, are outmoded, but we also witness the propagation of a new *Allerweltsprinzip*, the idea of complementarity that is supposed to be applicable to all domains and is regarded as the legitimate successor to classical philosophy. Heisenberg himself is not at all averse to applying this principle to theology:

> It has certainly been the pride of natural science since the beginning of rationalism to describe and to understand nature without the concept of God, and we do not want to give up any of the achievements of this period. But in modern atomic physics we have learned [how?] how cautious we should be in omitting essential concepts just because they lead to inconsistencies [*Syllabus of the Gifford Lectures* (1956), esp. p. 16].

There is, of course, a "basic change," and there is also the belief that this change has brought physics closer to experience (see the end of this section). Suffice it to say that the belief is not justified, and that there are good reasons to assume that the quantum theory is much more loosely connected with the domain of facts than was any previous theory. [For this assertion, see my *Problems of Microphysics, op. cit.*, Sec. X-XI.]

The assertion, however, that the scientific metaphysics started in the nineteenth century is simply incorrect. The identification of empiricism with the philosophy of nature is much older. It is present in Newton and is definitely established at the time of Laplace [see Alexandre Koyré, *From the Closed World to the Infinite Universe* (Baltimore: Johns Hopkins University Press, 1957), esp. Chaps. 2, 7, 9, and 11; for a brief summary, see my article "Naturphilosophie," in Fischer, *Encyclopädie der Philosophie* (Frankfurt-am-Main: S. Fischer Verlag, 1958)]. It is far from correct to say that Galileo was a collector of details and averse to the consideration and defense of general points of view. The opposite seems to be much closer to the truth. Convinced of the essential truth of the Copernican system and of the cosmology connected with it, he used every means, fair or foul, to bully others into accepting it. Bertolt Brecht's Galileo seems to be much closer to the historical figure than is the desiccated mythological portrait presented by most historians of science. (See also fn. 23, above.)

28. Agassi, *op. cit.*, p. 3.

29. ". . . It is a traditional policy of inductivist historians of science to pretend that in science there was only one revolution, the Renaissance revolution against prejudice and superstition which started the smooth development predicted by Bacon" (*Ibid.*, p. 25). See also T. S. Kuhn, *The Structure of Scientific Revolutions* (Chicago: University of Chicago Press, 1962), esp. Chaps. 1 and 11.

30. I am sure many thinkers will agree with Samuel Sambursky, who says [*The Physical World of the Greeks*, Merton Dagut, trans. (New York: Collier Books, 1962), p. 111]: "Experiment plays a far greater part in modern physics, so much so that it is now the final arbiter of every theory." This is correct insofar as modern science experiments in order to discover the limitations of the theories it uses. [This idea is already present in Newton, although for him experiment is also the starting point of investigation—see fn. 36 concerning his rules of experimentation.] Yet experiment is given a much more important role by Aristotle in the formulation of theories. His theories build upon an experience that is taken as an unanalyzed whole, and they are about bodies that are essentially observable [see fn. 23]. The laws of these theories repeat what is indicated by experience; they do not deal with unobservable processes. Aristotle's theory is therefore much closer to experience than are the succeeding ideas of "modern science." It is only recently that we could witness a return to the Aristotelian manner of theory building. See also the final paragraphs of Sec. I above.

31. "It is clear that not all the contrarieties constitute 'forms' and 'originative sources' of body but only those which correspond to touch" (*De Generatione et Corruptione*, 329b 9 ff.) The emphasis upon touch is irrelevant in this connection. It is the result of a specific theory of sensation that was quite common among the later Presocratics and was also held by Plato. What *is* important is the quite explicit dependence existing here between physics and philosophy, notably the theory of knowledge; physics deals with perceptible bodies *per definitionem* [see also Plato, *Timaios*, 31b3 ff.]; only perceptible qualities are admissible among the constituents of the elements [for details concerning this most interesting theory, see F. Solmsen, *Aristotle's System of the Physical World* (Ithaca: Cornell University Press, 1960), esp. Chap. 17].

It is worthwhile pointing to the great similarity that exists between this principle of the Aristotelian physics and the contemporary principle of meaning invariance [see Secs. III and VIII below] which demands that the meaning of observational terms remain unchanged in the course of scientific progress and be the measure of meaning for all theoretical terms. It should also be remembered that the state of quantum mechancial systems is characterized by "observables," which in their turn are defined with reference to possibilities of observation. [Not all properties are characterized in this fashion, however; the mass, the charge, isospin, strangeness, and parity play the role of parameters and are in this respect similar to the primary qualities of the Newtonian physics. The only difference is that now the "secondary qualities," i.e., position, momentum, angular momentum, etc., cannot any longer be regarded as primary qualities vaguely described but must be regarded as completely defined by the corresponding observables.]

32. Dijksterhuis, *op. cit.*, p. 30.

33. Considering the abstract nature of the theory of potentiality and actuality, which was an essential part of the Aristotelian doctrine of motion, this assertion must, of course, be taken with a grain of salt.

The reappearance of the concepts of actuality and potentiality in quantum theory are, on the other hand, a further indication of the conceptual similarities existing between this theory and the Aristotelian physics. See David Bohm, *Quantum Theory* (Englewood Cliffs, N.J.: Prentice-Hall, Inc., 1951), pp. 132, 138, 175, 331, 333, 385, 451, 609, 620, and 625, as well as Werner Heisenberg, "The Development of the Interpretation of the Quantum Theory," in *Niels Bohr and the Development of Physics*, W. Pauli, et al., eds. (London: Pergamon Press, Ltd., 1955), pp. 13 ff.

34. Koyré writes:

In the Newtonian world and in Newtonian science, . . . the conditions of knowledge *do not* determine the conditions of being [as they did in Aristotle—see fns. 31-32]; quite the contrary, it is the structure of reality that determines which of our faculties of knowledge can possibly (or cannot) make it accessible to us. Or, to use an old Platonic formula: in the Newtonian world and in Newtonian science, it is not man but God who is the measure of things. [Alexandre Koyré, "Influence of Philosophical Trends on the Formulation of Scientific Theories," in *The Validation of Scientific Theories*, P. G. Frank, ed. (Boston: Beacon Press, Inc., 1954), pp. 192-201, esp. p. 199.]

Newton's (or rather Descartes') law of inertia is the most obvious manifestation of the "transcendent" character of Newton's theory [the term "transcendent" is due to W. C. Kneale, *Probability and Induction* (London: Oxford University Press, 1949), esp. pp. 92-110]. It describes a behavior that is never encountered in the real world and is therefore unobservable in principle. (See also Koyré, *op. cit.*, pp. 196 f. and Dijksterhuis, *op. cit.*, pp. 30 f.) It is the contention of the present essay that a large part of contemporary physics is again formulated in a fashion that makes the structure of nature dependent on possibilities of observation. (For details, see *Problems of Microphysics, op. cit.*, Secs. X-XI). There are three major differences, however. First, this dependence is *represented* (though not actually introduced) as the result of factual research rather than of methodological procedure as it was in the case of Aristotle. ". . . We have accepted this state of affairs," writes Heisenberg [*The Physicists' Conception of Nature, op. cit.*, p. 25], "because it describes our experiences adequately." Second, what counts as experience is now something much more complex than was the everyday experience of Aristotle. It is constituted by states of affairs describable in terms of classical physics. [See also the remarks in the final paragraph of the present section.] Third, this return to an observer-centered physics lacks the clarity and detachment of the Aristotelian procedure and contains a good deal of sentimentality ["modern man confronts only himself—he no longer has partners or opponents" (*ibid.*, pp. 22 f.)].

Another point that most distinctly indicates the distance existing between the basic entities of the world picture of classical mechanics and experience is the reappearance of the problem of secondary qualities that had first troubled Democritus [Dem. Fr. 125; Galen, *De Medicina Empirica;* see the discussion in G. S. Kirk and J. E. Raven, *The Presocratic Philosophers* (London: Cambridge University Press, 1960), p. 424] but that seems to have disappeared in the Aristotelian physics. (The *minima naturalia* of Aristotle possess the same qualities as the elements that they constitute, and the latter are characterized in a purely observational fashion; see Dijksterhuis, *op. cit.*, pp. 205 ff.) It is this problem that is the motor of British empiricism from Locke to Hume. Locke provides a partly satisfactory solution [see Chaps. 7, 8, and 21 of his *Essay Concerning Human Understanding*, but compare also his much shorter *Elements of Natural Philosophy*, especially Chaps. 11 and 12, which contains a dogmatic account of what is argued in the *Essay*]. According to this solution, secondary qualities are distinct from primary qualities not intrinsically, but only by their mode of presentation; they are the way in which very complex primary qualities affect human observers, who give a *total response* containing no indications as to the detailed structure of the primary situation that brought it about. Furthermore, they are *characterized* by reference to this total effect rather than to their own intrinsic properties. In this theory, the *nature* of the object (which is entirely primary) is still independent of the reactions of the observer; a causal relation is assumed to exist. However, what is *observed* is only in some cases attributed to the object itself in the form in which it is observed (primary qualities). Secondary qualities are based upon impressions too indistinct to allow an inference as to the actual structure of the object.

It is well known how Berkeley attacked this distinction by attacking the idea of an observer-independent reality, and how he reintroduced the principle that the conditions of being are exclusively determined by the conditions of knowledge. As I have tried to point out in *Problems of Microphysics, op. cit.*, Sec. X, the quantum theory is the almost successful attempt to carry out Berkeley's program, classical states of affairs replacing the sensations upon which Berkeley wanted to base his own physics. It is very interesting to see that a purely observational physics is possible only on the basis of classical situations. No doubt this is due to the fact that there are no simple laws connecting sensations that have been pre-empted of any component of thought, whereas there do seem to exist laws connecting classical situations. [Berkeley, who realized this, had to use a material object language in order to state the laws connecting sensations (the laws are such that they lead to the appearance of material objects); he was *not* able to separate his philosophy from physics or perhaps to suggest *new* physical laws.]

35. See the explicit requirements of the inductive style that were put forth by Brouncker, Boyle, and Hooke and incorporated in the regulations of the Royal Society. For this point, see fn. 60 of Agassi, *op. cit.*

36. Rule IV [Book III of the *Principia*, Floriam Cajori, ed. (Berkeley: University of California Press, 1934), p. 400]: "In experimental philosophy we are to look upon propositions inferred by general induction from phenomena as accurately or very nearly true, notwithstanding any contrary hypothesis that may be imagined, till such time as other phenomena occur, by which they may either be made more accurate, or liable to exceptions." [For a criticism of this rule, which is much more liberal than many other rules that were to follow it and which constitutes the essence of what we have called a "radical empiricism," see below, Sec. VII.] There follows an enumeration of the *phenomena* (pp. 401-405) and of the *propositions* derivable from these phenomena by mathematical deduction. The concluding *Scholium Generale* then asserts (p. 574):

> . . . What is not deduced from phenomena is to be called a hypothesis; and hypotheses, whether metaphysical or physical, whether on occult qualities or mechanical, have no place in experimental philosophy. In this philosophy, particular propositions are inferred from the phenomena and afterwards rendered general by induction . . . to us it is enough that gravity does really exist, and acts according to the laws which we have explained, and abundantly serves to account for all the motions of the celestial bodies, and of our sea.

All this gives the impression that a rigorous derivation of important physical laws from facts of observation is possible, and has been actually achieved. On this feature of the derivation, see Albert Einstein, "On the Method of Theoretical

Physics," in *Ideas and Opinions,* Carl Seelig, ed. (London: Alvin Redman, Ltd., 1956), p. 273; also J. M. Keynes, "Newton the Man," in *Essays and Sketches in Biography* (New York: The Noonday Press, 1956), esp. p. 283. On the notion of a phenomenon, see below.

37. It is interesting to see how the attitude toward his theory has influenced the evaluation of Newton as a human being. Newton simply could do no wrong. Criticism and revelation of personal features incompatible with what a great scientist is supposed to be were suppressed. His opponents, and even those who, though not opposed to his ideas, did yet occasionally arouse his wrath, were painted pitch black. Bailey, in his *Life of Flamsteed,* did try to correct this picture in one particular case by publishing Flamsteed's diary as well as part of his correspondence with Newton. More especially, he tried to refute, and in the opinion of scholars succeeded in refuting, the then very common assumption that Flamsteed had willfully and jealously withheld data necessary to Newton, and in this way prevented an earlier publication of the *Principia.* Brewster, who fully participated in the "idolatry which had grown up about [Newton's] character" (More, *op. cit.,* p. 410) censures this undertaking most severely. "It was reserved for two English astronomers [Flamsteed and Bailey], the one a contemporary, the other a disciple, to misrepresent and calumniate their illustrious countryman" (*Life of Newton,* Vol. I, p. xi). Empirical research into the life of this allegedly greatest empiricist simply was not permitted. [For details, see the presentation of the many incidents resulting from this uncritical attitude in More, *op. cit.;* also the first chapter of Margaret 'Espinesse, *Robert Hooke* (Berkeley: University of California Press, 1956).] Evidence was suppressed, and only very slowly was it possible to come to a more just appreciation of the person and the achievements of this most extraordinary man. See also Keynes's decisive essay referred to in fn. 34.

38. See Bavink, *Ergebnisse und Probleme der Naturwissenschaften* (Zürich, 1954), p. 34.

39. M. Born, *Natural Philosophy of Cause and Chance* (London: Oxford University Press, 1949), pp. 129-133. For the evaluation of these results, see *ibid.,* p. 13.

40. *Encyclopädie der philosophischen Wissenschaften,* Lasson, ed. 1920, pp. 235 ff.: "As is well known the laws of absolutely free motion have been discovered by Kepler. . . . Since then it has become customary to say that Newton had been the first one to find proof for these laws. There has never been a more unjust transference of fame from one author to another. Concerning these matters I have to say [that] mathematicians are agreed that Newton's formulae can be derived from Kepler's laws. . . ." [As is indicated in the text above, they are indeed "agreed" on that point.] As opposed to many empiricists who "usually avoid any reference to Newton's perturbation theory" (Agassi, *op. cit.,* p. 2), Hegel boldly faces the issue: "It is recognised that the *real content* of Newton's theory—whose recognition has by the way made superfluous *and has even refuted* [my italics] much of what had been regarded as his essential principles and as a contribution to his fame—what he has added to the content of Kepler's laws, is contained in the principle of *perturbation*" (Lasson, *op. cit.,* p. 237). However, with this remark, which would seem to refute the above polemics, the matter is dropped. And we hear again that "the Newtonian formulae . . . show the distortion and inversion due to reflective thinking stopping midway in its activity (p. 241)."

It is customary to use Hegel as an example designed to frighten naughty empiricist children who have the tendency to depart from the straight path of virtue. The above passage would seem to show that Hegel and this straight path have much in common. Indeed, the latter so much emphasizes the aspect of application that for it the distinction between Newton's basic theory and the theory of perturbation would be as fundamental as it seems to be for Hegel.

41. Pierre Duhem, *Aim and Structure of Physical Theory*, P. P. Wiener, trans. (Princeton: Princeton University Press, 1954), Chap. 6, Sec. IV. See also K. R. Popper, "The Aim of Science," *Ratio*, Vol. I (1957).

42. See his Herbert Spencer Lecture, "On the Method of Theoretical Physics," delivered at Oxford in June, 1933, and reprinted in *Ideas and Opinions, op. cit.*, pp. 270-276.

43. According to Einstein:

. . . The fictitious [noninductive] character of fundamental principles is perfectly evident from the fact that we can point to two essentially different principles both of which correspond to experience to a large extent [the reference is to Newton's theory of gravitation and to the general theory of relativity]. This proves at the same time that every attempt at a logical deduction of the basic concepts and postulates of mechanics from elementary experiences is doomed to failure.

(*Ibid.*, pp. 274 f.) See also his "Physics and Reality," *op. cit.*, pp. 300, 307, as well as Popper, "The Aim of Science," *op. cit.*

44. In the mathematical argument that is usually presented, the premises are Kepler's laws and the conclusion is the formula kMm/r^2. The mathematics is, of course, faultless. Not so the interpretation of the premises and the conclusion. If we interpret the premises as expressing Kepler's laws, then we cannot base an inductive argument on them because they are *false*. If, on the other hand, we interpret these formulae as describing the behavior of a single mass point in the close neighborhood of a much larger mass, then their truth cannot be denied. However, in this case the formulae describe an imaginary process, and not what is going on in the real solar system. The conclusion will then also describe such an imaginary process. The step from actual observational astronomy to a real physical law governing the planets is still not made. See also my comments in *Current Issues in the Philosophy of Science*, Herbert Feigl and Grover Maxwell, eds. (New York: Holt, Rinehart & Winston, Inc., 1961), p. 39.

45. The method described was used by many scientists in order to give their theories the appearance of being not hypothetical, but based upon experience. This deceived a good many historians and philosophers and made them believe, on the basis of "historical evidence" that a radically empirical attitude was indeed the distinctive property of the new physics. This shows quite clearly that even history cannot rest content with a mere *recording* of the facts "as they have actually happened," but must proceed to an *analysis* on the basis of a well-defined point of view.

46. More, *op. cit.*, p. 500.

47. C. R. Weld, *History of the Royal Society*, Vol. I (London: J. W. Parker, 1848), p. 219.

48. Quoted in *ibid.*, p. 113.

49. *Ibid.*, p. 93.

50. This credulity is, of course, a quite natural consequence of the demand to approach matters *without prejudice*, which means without a principle of selection and criticism.

51. *Ibid.*, pp. 107 f.

52. Quoted from Isaac Newton, *Papers and Letters on Natural Philosophy,* I. B. Cohen, ed. (Cambridge: Harvard University Press, 1958), p. 111.

53. Proposition recorded in the journal book of the Society; quoted from More, *op. cit.,* p. 93.

54. See 'Espinasse, *op. cit.,* Chap. 2.

55. Quoted from Newton, *Papers and Letters, op. cit.,* p. 114. The uneasiness of some of Newton's own contemporaries should be kept in mind when dealing with Goethe's *Farbenlehre.* This is a book which most clearly reveals the split existing between the scientific practice of the classical physicists and their house philosophy [which Goethe adopts as a basis for criticism, but "whose false maxim not to admit anything theoretical" he rejects as being too narrowminded; see *Farbenlehre,* Gunther Ipsen, ed. (Leipzig, 1927), p. 615]. I shall not try to defend Goethe's *positive* theory, which is definitely inferior to Newton's. Nor shall I defend all of his arguments, some of which are based upon a serious misunderstanding of the experimental procedure. But it is necessary to point out (1) that his criticsm of Newton's theory is completely in line with that version of empiricism which was made the theoretical (though not the practical) measuring stick of all solid knowledge, and (2) that his own theory shares certain very important features with the phenomenological theories of the later nineteenth century. For example, his intention to treat the optics of turbid media as an *Urphaenomen* coincides with the refusal to admit a deeper analysis of the structure of heated substances.

That Goethe's theory is empirically adequate, and that it may also be internally consistent is implied in H. von Helmholtz's evaluation of it [*Popular Scientific Lectures,* First Series (New York: David McKay Co., Inc., 1898), pp. 29-51]. What is attacked is his demand "that nature must reveal her secrets out of her own free will" (p. 45) as well as certain implausible elements of the theory. The demand is regarded as "essentially wrong in principle" (p. 45). The fact that Goethe defends it is ascribed to his being a poet. "Goethe, . . . as a genuine poet, conceives what he finds in the phenomenon the direct expression of the idea" (p. 40). The step "into the region of abstract conception," which for the realist Helmholtz is essential to the scientific method, "scares the poet away" (p. 45). But the very same step was criticized by Newton's own empiricist contemporaries; it was criticized later on by Berkeley, as well as by his followers in the nineteenth century; it was criticized by the pedestrian opponents of the kinetic theory of matter who could hardly be accused of possessing a poetic turn of mind. Helmholtz's criticism of a narrow observationalism, his assertion to the effect that "a natural phenomenon is not considered in physical science to be fully explained until you have traced it back to the ultimate forces which are concerned in its production and its maintenance" (p. 45), his conviction that science attacks "the belief in the direct truth of our sensation" (p. 47)—all this is unexceptionable. What cannot be accepted is his opposition to Goethe's theory as being too "poetical" when he must have known that there existed a strong observationalist trend *within science itself.* He must have known that Goethe's insistence "that Newton's facts might be explained on his own hypothesis, and that therefore Newton's hypothesis was not fully proved" (p. 43) coincided with the objections of Hooke and of others. There is even a definite resemblance between certain parts of Goethe's position and some alternatives that were discussed at Newton's own time in order to exhibit the hypothetical character of his point of view. Goethe's practical turn of mind, his alleged fear of the abstract, his being a poet, all these facts are therefore quite beside the point.

Tyndall's essay [I am here quoting from *Fragments of Science* (New York:

Appleton-Century-Crofts, Inc., 1900). Unfortunately I have been unable to use the English edition and my quotations are therefore retranslated from the German; reference is made to the translation by Anna von Helmholtz and Estelle du Bois-Raymond (Braunschweig: Friedrich Vieweg und Sohn, 1895)] is somewhat less radical, but no less interesting. There is an insinuation concerning Goethe's poetic turn of mind (pp. 86 f.), but the point is not worked to death as it is by Helmholtz. It is admitted that many of Goethe's thoughts "are not familiar to the exact scientist. In order to understand them an amount of loving care is needed that is not required by the exact sciences" (p. 57). This is a very interesting admission. It is admitted that the exact scientist disregards many aspects that can be observed, that is, it is admitted that he is not a radical empiricist. Even more interesting is Tyndall's accusation that Goethe "turned around the proper sequence of thought, and tried to make the result of a theory its basis" (p. 63). Tyndall's own noninductivist turn of mind is here revealed very clearly. He accepts the kinetic theory of matter; he admits that "completely unphysical speculation may stimulate experiment and thereby become a means for the discovery of important physical results" (fn. 53 of his book on the theory of heat). But why turn on Goethe, when many of his fellow scientists proceeded in the same fashion, or at least pretended that they proceeded in this fashion. Moreover, his criticism of Goethe's treatment of turbid media (pp. 66 f.) applies with exactly the same force to those physicists who rejected the speculations of the kinetic theory and insisted that thermodynamics retain the phenomenological mode of description, and for whom the blue of the sky, for example, had to be regarded as being incapable of further analysis.

To sum up: the attitude of almost all physicists toward Goethe's theory of colors makes explicit some "inner contradictions" of classical physics. The *philosophy* of this period demands that theories be kept close to observational results (for details, see Agassi, *op. cit.*). This suggests, and is said to suggest, the existence of an experience that is independent of theoretical advance and is the neutral and unprejudiced measure of the adequacy of theories. The practical procedure is very different. A theory is adopted, usually because it seems to be attractive for some reason or other. Experience is *reinterpreted* in order to be "close" to this theory, and the theory is then defended by empirical argument, by pointing out how closely it fits the facts. Two wishes are here satisfied at one stroke—the wish to retain the theory and the wish to retain empiricism. Empiricism thus turns out to be a most elastic doctrine, a doctrine that, like Barth's theology, can be adapted to any situation and that is yet given the air of harsh exclusiveness in order to make conformity with it appear worthwhile and satisfaction of its demands a virtue. Goethe has looked through many features of this complex procedure. He writes: "for how should it be possible to hope for progress in the sciences if what is inferred, guessed, or merely believed to be the case can be put over on us as a fact?" (*Farbenlehre, op. cit.*, p. 393.)

56. Recommendations for improving the teaching of mathematics, contained in a letter to Mr. Hawes, treasurer of St. Christ's Hospital in London. Quoted from More, *op. cit.*, pp. 402 f. In connection with this advice, one cannot help remembering the absurd stories that the Royal Society sometimes received from "unprejudiced" observers abroad and that were duly recorded (see the above example, text to fn. 51). One is also reminded of Kant's sometimes quite vivid descriptions of foreign countries that he had never seen and of his evaluation of the alleged knowledge of people who had traveled far and were therefore supposed to have seen a lot: "Knowledge of the world demands more than just seeing the

world! One must know what to look for in the foreign countries!" [Immanuel Kant, *Physische Geographie,* Paul Gedan, ed. (Leipzig, 1905), p. ix].

57. *Op. cit.,* p. 614. It is important to repeat that Goethe himself was critical of the "false maxim of the Society not to deal with anything theoretical" (p. 615) and that he perceived that the future development, that is, the uncritical acceptance, *as a fact,* of Newton's results depended quite essentially on this maxim. "Hooke contradicts immediately . . . , he maintains that his theory of vibrations would do just as well, and perhaps even better. In this connection he promises to adduce new phenomena, and other important things. However, he does not think of developing Newton's experiments; *he also regards the phenomena described as facts, which leads to much silent advantage for Newton*" (p. 615). It seems to me that Goethe has spotted exactly one important feature of classical empiricism, viz., the fact that any theory, however abstract, can be defended by reference to experience, provided the theory is allowed to pervade the very results of observation. He also notices that this procedure is used in an underhand manner and that the close resemblance between theory and observation is not admitted.

58. Isaac Newton, *Papers and Letters, op. cit.,* p. 48.

59. Linus, *op. cit.,* p. 151.

60. Kuhn, *op. cit.,* p. 34. Ernst Mach [*The Principles of Physical Optics* (New York: Dover Publications, Inc., n.d.), p. 87] and others have criticized Linus for an "inexact" repetition of the experiment. However that may be, the "semicircular lines," if they appear at all, appear only under very special conditions, and it is, of course, a matter of conjecture to assume that these very special conditions reveal things as they really are.

61. Or even as a *positivist* [Brewster, *Memoir* (Edinborough, 1855), Vol. II, p. 532].

62. The idea of complementarity is very similar in this respect. See the comments at the end of Sec. I above and Sec. XI of my *Problems of Microphysics, op. cit.*

63. See the quotation at the end of fn. 55.

64. See E. A. Burtt, *The Metaphysical Foundations of Modern Physical Science* (London, 1931), p. 218.

65. This is, of course, true for all statements concerning dispositions. See K. R. Popper, *Logic of Scientific Discovery* (New York: Basic Books, Inc., 1959), Sec. XIV.

66. See fn. 65.

67. It would be most interesting to relate here the discussions in connection with the second law of thermodynamics and the reasons that were given against the kinetic theory. That the attempt to refer everything back to experience was one of the reasons why the fundamental principles of mechanics were still so unclear has been asserted by L. Boltzmann: "The lack of clarity in the principles of mechanics seems to be connected with the fact that one did not at once start with hypothetical pictures framed by our minds, but tried to start from experience. The transition to hypotheses was then more or less covered up and it was even attempted artificially to construct some kind of proof to the effect that the whole edifice was free from hypotheses. This is one of the main reasons for the lack of clarity" [*Vorlesungen über die Principe der Mechanik,* Vol. I (Leipzig: Johann Ambrosius Barth, 1897), p. 2]. For more detailed analyses, see *Populäre Schriften* (Leipzig: Johann Ambrosius Barth, 1905). It is very much to be regretted that

there does not yet seem to exist any comprehensive account of Boltzmann's most interesting theory of knowledge.

68. Many of his arguments had been anticipated by Berkeley. See K. R. Popper, "Berkeley as a Precursor of Mach," *British Journal for the Philosophy of Science,* Vol. IV (1953), reprinted in *Conjectures and Refutations* (New York: Basic Books, Inc., 1963), pp. 166 ff.; see also J. Myhill, "Berkeley's 'De Motu,' an Anticipation of Mach," in *George Berkeley,* University of California Publications in Philosophy, Vol. XXIX (Berkeley: University of California Press, 1957). Berkeley fully realized that Newton's alleged "deduction" from phenomena cannot be demonstrative, "for all deduction of this kind depends on a supposition that the Author of Nature always operates uniformly and in a constant observance of those rules *we* take for principles, which we cannot evidently know" (*Principles of Human Knowledge,* Sec. CVII).

69. See his *Mechanik.*

70. See fn. 29.

71. *Ideas and Opinions, op. cit.,* p. 291. In his first paper, Einstein quite explicitly recognized this feature of physical theory. Rather than start with an enumeration of the *facts,* he presented a few *general principles* (constancy of light velocity, principle of relativity). He was criticized for proceeding in this fashion by many experimentalists who regarded his ideas as mere speculations and unworthy of the title of a physical theory. Especially in England, it took a long time before the theory was finally accepted (Rutherford was opposed to it; the final acceptance was due among other things to Eddington's elegant presentation, which appealed to the mathematicians as well as to the empirical success of the general theory). Even philosophers confronted the solid facts of classical mechanics with the "formalisms" of relativity. [An example is Nicolai Hartmann's influential *Naturphilosophie* (Berlin: Walter de Gruyter, 1948). For a presentation and criticism of Hartmann's views on relativity, see my essay "Nicolai Hartmann's Naturphilosophie," *Ratio,* Vol. V (1963), 81-94. For Einstein's own procedure, see the very illuminating autobiographical remarks in *Albert Einstein, Philosopher-Scientist* (Evanston: Library of Living Philosophers, Inc., 1949), p. 52.] More recently there have been assertions to the effect that the Michelson-Morely experiment was unknown to Einstein when he developed his theory. Thus Michael Polanyi reports [*Personal Knowledge* (Chicago: University of Chicago Press, 1958), pp. 10 f.] that Einstein had authorized him in 1954 to publish the statement that "the Michelson-Morely experiment had a negligible effect on the discovery of relativity." [For a discussion of this historical problem one should also read Adolf Grünbaum, "The Genesis of the Special Theory of Relativity," in *Current Issues in the Philosophy of Science, op. cit.,* pp. 13 ff.]

The recent publication of R. S. Shankland's conversations with Einstein ["Conversations with Albert Einstein," *American Journal of Physics,* Vol. XXXI (1963), 47 ff.] seems to add support to Professor Polanyi's statement. I quote without comment: "When I asked him how he had learned of the Michelson-Morley experiment, he told me that he had become aware of it through the writings of H. A. Lorentz, but *only after 1905* had it come to his attention! 'Otherwise,' [Einstein] said 'I would have mentioned it in my paper' " (p. 48; see also fn. 3 on that page). "He continued to say the experimental results which had influenced him most were the observations on stellar aberration and Fizeau's measurements on the speed of light in moving water. 'They were enough,' he said" (p. 48). "I asked Professor Einstein where he had first heard of Michelson and his experiment. He replied, 'This is not so easy. I am not sure when I first heard of the Michelson experiment. I was not conscious that it had influenced me directly during the seven years that relativity had been my life. I guess I just took it for granted that it was true.' However, Einstein said that in the years 1905-1909, he thought a great deal about Michelson's result, in his discussions with Lorentz and others and in his thinking about special relativity. He then realised (so he told me) that he had also been conscious of Michelson's result before 1905 partly through his reading of the papers of Lorentz and more because he had simply assumed this result of Michelson to be true" (p. 55).

Considering all this, it is not at all fair to make Einstein responsible for the new role radical empiricism assumed in the twentieth century. [Such assertions have been made by P. W. Bridgman, *Logic of Modern Physics* (New York: The Macmillan Company, 1927). See esp. Chap. 1.] Niels Bohr remarks: "We have recently experienced such a revision in the rise of the theory of relativity which, by a profound analysis of the problem of observation, was destined to reveal the subjective character of all the concepts of classical physics" [*Atomic Theory and the Description of Nature* (London: Cambridge University Press, 1934), p. 97]. Bohr also emphasizes the "profound inner similarity between the problems met with in the theory of relativity and those which are encountered in the quantum theory" (p. 5).

It is true that the new theory made spatio-temporal properties accessible to empirical investigation and eliminated entities that had been found to be unobservable. However, this was possible only after the fundamental role of electromagnetic processes, and especially of the propagation of light in the vacuum had been set down as a *new hypothesis*. [See Einstein's assertion to the effect that his own theory "turns (the difficulty of the aether hypothesis) into a principle," i.e., makes the far-reaching assumption that the aether cannot be observed *because it does not exist*. "Zur Elektrodynamik bewegter Körper," *Annalen der Physik,* Vol. XVII (1905), reprinted in Lorentz, *et al.*, *Das Relativitätsprinzip* (Leipzig: Johann Ambrosius Barth, 1923).] This new hypothesis provided an interesting explanation of the existing experimental results, but it also went far beyond them by assuming the nonexistence of processes incompatible with the principle of relativity. Moreover, this hypothesis was set down quite explicitly as a new attempt and *not* as a deduction from facts. Nor was it presented as a result of a "careful analysis of the notions of space and time," as the matter is usually described. As Professor Grünbaum has shown most convincingly ["Operationalism and Relativity," in *The Validation of Scientific Theories, op. cit.,* pp. 84-94; see also Chap. 12 of Popper, *Open Society and Its Enemies,* 4th ed. (Princeton: Princeton University Press, 1962), p. 20], (the high testability of) the new theory was not the result of a consistent application of the demand to remove unobservables, or of the even more radical principle that what does not show itself in observation does not exist; it was rather a *side effect* of the new assumptions used, which are much more compact than the bundle of ideas surrounding the aether. Relativity, and especially the general theory, is the crowning achievement of the classical, or non-Aristotelian, trend in physical reasoning, which insists in practice on a strict separation between the conditions of knowing and the conditions of being. Einstein is the first physicist who makes this character of classical physics *explicit* by deviating both from the inductivistic procedure and the inductivistic *style*.

72. *Atomic Theory and the Description of Nature,* p. 66.

73. The abbreviations used in the text are: *A* for *Atomic Theory, op. cit.; E* for Bohr's essay in P. A. Schilpp, ed., *Albert Einstein, Philosopher-Scientist, op. cit.; D* for Bohr's article in *Dialectica* (1948), 312 ff.; and *R* for his article in *Physical Review,* Vol. XLVIII (1936).

74. Bohr wants to construct a language whose basic framework will not have

to be rebuilt, but which has "holes," as it were, for the filling in of future knowledge. Concerning this "hole theory of language," see fn. 151 below.

75. For the "monsters," see Stephan Körner, ed., for The Colston Research Society, *Observation and Interpretation* (London: Butterworth & Co., Ltd., 1957), p. 184.

76. This need not at all be incorrect. Reporting on his investigation of the Portsmouth Papers, Geoffrey Keynes concludes that "as one broods over these queer collections, it seems easier to understand—with an understanding which is not, I hope, distorted in the other direction—this strange spirit, who was tempted by the devil to believe, at the time when within these walls he was solving so much, that he could reach *all* the secrets of God and Nature by the pure power of mind—Copernicus and Faustus in one" ["Newton the Man," read by Keynes at the Newton Tercentenary Celebration at Trinity College, Cambridge, on July 17, 1946, and therefore not revised by the author, who had written it some years earlier, and published in *Essays and Sketches in Biography, op cit.*, p. 290. See also Chaps. 2, 7, 9, 11 of Koyré, *From the Closed World to the Infinite Universe, op. cit.*]

77. My discussion of the tenets of the logical empiricists is prompted by the fact that their theory of explanation most clearly reflects the implicit attitude of most scientists. However, this applies only to the general outlines of their doctrine, not to the details, which are of no interest whatever in the evaluation of the sciences. Moreover, a refutation of the general outlines makes a more detailed discussion superfluous.

78. *Logic of Scientific Discovery, op cit.*, Sec. XII. The decisive feature of Popper's theory, a feature that was not made clear by earlier writers on the subject of explanation, is the emphasis he puts upon the initial conditions and the implied possibility of two kinds of laws, viz., (1) laws concerning the temporal sequence of events and (2) laws concerning the space of initial conditions. In the case of the quantum theory, the laws of the second kind provide very important information about the nature of the elementary particles, and it is to them and not to the laws of motion that reference is made in the discussions concerning the interpretation of the uncertainty relations. On this point see also E. L. Hill, "Quantum Theory and the Relativity Theory," *Current Issues in the Philosophy of Science, op. cit.*, pp. 429 ff., as well as my comments, pp. 441 ff.

79. Hempel, "Studies in the Logic of Explanation," as reprinted in Herbert Feigl and May Brodbeck, *Readings in the Philosophy of Science* (New York: Appleton-Century-Crofts, Inc., 1953), p. 321.

80. Ernest Nagel, "The Meaning of Reduction in the Natural Sciences" as reprinted in A. C. Danto and Sidney Morgenbesser, *Philosophy of Science* (New York: Meridian Books, Inc., 1960), p. 301.

81. An objection to this formulation is that theories which are consistent with a given explanandum may still contradict each other. This is quite correct, but it does not invalidate my argument, for as soon as a single theory is regarded as sufficient for explaining all that is known (and represented by the other theories in question), it will have to be consistent with all these other theories.

82. In an earlier publication [*Proceedings of the Aristotelian Society*, New Series, Vol. LVIII (1958), 147], I called a very common description of the results of meaning invariance the *stability thesis*.

83. "I have been accused of the habit of changing my opinions in philosophy," writes Bertrand Russell [Preface to L. E. Denonn, ed., *Bertrand Russell's Dictionary of Mind, Matter, and Morals* (New York: Philosophical Library, Inc.,

1952)]. "I am not myself in any degree ashamed of having changed my opinions. What physicist who was already active in 1900 would dream of boasting that his opinions had not changed during the last half century? In science men change their opinions when new knowledge becomes available; but philosophy in the minds of many is assimilated rather to theology than to science. . . ." I have nothing to add to this statement, except perhaps that Russell seems to have been somewhat optimistic in his evaluation of the sciences.

84. This demand for a unified point of view and for institutions teaching this point of view and taking care of objectors has been one of the greatest obstacles for a proper appreciation of the democratic way of life. Especially nations unaccustomed to this way of life have regarded parliamentary discussion not as the thing itself but rather as a preparation for a future totalitarian unity, and they have become discouraged and disgusted when such unity did not appear. The additude of the majority of the Germans to the Weimar Republic is a case in point. See Golo Mann, *Deutsche Geschichte* (Frankfurt-am-Main: S. Fischer Verlag, 1950), pp. 718 ff.

85. It seems to me that a well-documented history of the evaluation of the Pre-Socratics would constitute a most valuable prolegomenon to a history of epistemology and metaphysics that is not any longer restricted in scope by the use of such categories as empiricism and rationalism, spiritualism and materialism, monism and pluralism, and the like. By drawing attention to details, or by regarding the question of the sources of knowledge as the most fundamental philosophical problem and the classification of answers to this problem as the most basic classification of philosophies, the traditional histories usually neglect the very decisive difference between critical and dogmatic philosophizing. Moreover, they are too much absorbed in minor quarrels to be capable of perceiving this difference. The Pre-Socratics knew of no "sources," or at least the question did not arise in their case. A philosopher's reaction to them will therefore bring out the really decisive ingredients of his thought. And considering the clarity and simplicity of Pre-Socratic thought [Joel and Heidegger notwithstanding], this reaction will have to possess a similar clarity and simplicity in order to seem adequate at all. It is for these reasons that the Pre-Socratics may be a much better mirror of later developments than any other philosophical school.

86. "Back to the Pre-Socratics," *Proceedings of the Aristotelian Society,* New Series, Vol. LIV (1959), reprinted in *Conjectures and Refutations, op. cit.* See also my *Knowledge Without Foundations, op. cit.*

87. St. Augustine, as quoted in Marshall Clagett, *Greek Science in Antiquity* (London: Abelard-Schuman, Limited, Publishers, 1957), p. 132.

88. Colin McLaurin, *An Account of Sir Isaac Newton's Philosophical Discoveries* (London: Buchanan's Head, 1750), p. 28.

89. *Wärmelehre* (Leipzig: Johann Ambrosius Barth, 1897), p. 364.

90. *Zwei Aufsätze* (Leipzig, 1912).

91. For a discussion of these objections, see ter Haar's article in *Reviews of Modern Physics,* Vol. XXVII (1957), 289-338, as well as Paul and T. A. Ehrenfest, *The Conceptual Foundations of the Statistical Approach in Mechanics,* M. J. Moravcsik, trans. (Ithaca: Cornell University Press, 1959). These objections, taken together with a common aversion against speculation, almost succeeded in extinguishing the kinetic theory. For details, see M. von Smoluchowski, "Die Gültigkeitsgrenzen des zweiten Hauptsätzes der Wärmetheorie," reprinted in *Oeuvres,* Vol. II (1927), pp. 361 ff.

92. *Natural Philosophy of Cause and Chance* (London: Oxford University Press, 1948), p. 109. For a criticism of this passage, see my *Problems of Microphysics, op cit.,* p. 238.

93. L. Rosenfeld, "Misunderstandings About the Foundations of the Quantum Theory," in *Observation and Interpretation, op. cit.,* p. 42.

94. See the discussions in *ibid.*

95. For details, see *Problems of Microphysics,* as well as my essay on measurement in *Observation and Interpretation.* See also the following brief and very intuitive application of the consistency condition to the theory of measurement: In their book on the nonrelativistic quantum theory [*Quantum Mechanics,* J. B. Sykes and J. S. Bell, trans. (Reading, Mass.: Addison-Wesley Publishing Company, Inc., 1958), p. 22], L. D. Landau and E. M. Lifshits point out that "the classical nature of the apparatus means that . . . the reading of the apparatus . . . has some definite value. This enables us to say that the state of the system apparatus electron after the measurement will in actual fact be described, not by the entire sum [$\Sigma A_n(q)\phi_n(\delta)$ where q is the coordinate of the electron, δ the apparatus coordinate] but by only the one term which corresponds to the 'reading' g_n of the apparatus, $A_n(q)\phi_n(\delta)$."

96. Heisenberg, *Physics and Philosophy* (New York: Harper & Row, Publishers, Inc., 1958), p. 44 (my italics).

97. The arguments that Heisenberg uses in this connection are presented and criticized in Sec. VIII and in fn. 144 of *Problems of Microphysics, op. cit.*

98. The factual basis of this evaluation and of the distinctions on which it is based has been studied with quite interesting results by T. S. Kuhn, *The Structure of Scientific Revolutions, op. cit.* To this should now be added Agassi's provocative *Towards an Historiography of Science, op. cit.*

99. See fn. 29.

100. As an example of such an evaluation, let us consider briefly the point of view developed in Nicolai Hartmann's *Philosophie der Natur* (Berlin: Walter de Gruyter, 1948). There are many similarities between the point of view of Hartmann and that of contemporary philosophers of science. Hartmann wants to base his investigations upon what he calls "phenomena," or "hard facts." He rejects a speculative philosophy of nature and repeats again and again that it is up to the *theory* to prove its mettle: "to fight against phenomena is sheer folly" and "phenomena are not in need of proof" (p. 361). And just like his Anglo-American colleagues, he has his own idea of what is and what is not a fact. In this way he can both sell his own philosophy and pride himself on paying due attention to "facts." Now what concerns us here is Hartmann's attitude toward periods of crisis. Such periods of "confusion," of "disjointedness" (p. 19) are not suitable for philosophical consideration. The material is much too heterogeneous; there are no uniting principles, and one must wait until "a kind of unity of the cosmological point of view has re-established itself" (p. 19), i.e., until all the different theories have been eliminated with the exception of one: theoretical monism is a necessary presupposition of a philosophy of nature in Hartmann's sense. Now if it should turn out that theoretical monism is incompatible with swift scientific advance, then this would show that Hartmann's philosophy can live only provided science is dead. (This character it shares with many contemporary "philosophies of science.") For some details, see fn. 105 and text. For the conservative function of a reinterpretation of "facts," see the last section. For details concerning Hartmann's philosophy of nature, see my review in *Ratio,* Vol. V (1963).

101. *Aim and Structure of Physical Theory, op. cit.,* Chaps. 9 and 10. See also Popper, "The Aim of Science," *op. cit.*

102. According to Ernest Nagel [*The Structure of Science* (New York: Harcourt, Brace & World, Inc., 1961), p. 338], reduction is "the explanation of a theory or a set of experimental laws established in one area of inquiry by a theory usually, though not invariably, formulated for some other domains." This implies that the conditions for explanation and the conditions for reduction coincide for Nagel.

103. Nagel, "The Meaning of Reduction," *op. cit.,* p. 302.

104. See Sec. 4.7 of Michael Scriven, "Explanation, Prediction, and Laws," in *Scientific Explanation, Space, and Time, op cit.* Similar objections were raised by Kraft and Rynin.

105. For details concerning this example, see Sec. VI of "An Attempt at a Realistic Interpretation of Experience," *op. cit.*

106. It must be admitted, however, that Einstein's original interpretation of the special theory of relativity is hardly ever used by contemporary physicists. For them the theory of relativity consists of two elements: (1) the Lorentz transformations and (2) mass-energy equivalence. The Lorentz transformations are interpreted purely formally and are used to make a selection among possible equations. Such an interpretation does not allow one to distinguish between the original point of view of Lorentz and the entirely different point of view of Einstein. According to it, Einstein achieved a very minor *formal* advance (this is the basis of Whittaker's attempt to debunk Einstein). It is also very similar to what application of the double language model will yield. Still, an undesirable philosophical procedure is not improved by the support it gets from an undesirable procedure in physics. Moreover, a philosophical method of interpretation should be elastic enough to be capable of presenting different theories (such as the theories of Einstein and of Lorentz), and it should not be built in a manner to prejudge the issue between them. (The above comment on the contemporary attitude toward relativity was made by Professor E. L. Hill in discussions at the Minnesota Center for the Philosophy of Science. I would like to use this occasion to thank the members of the Center, Professors Feigl, Hill, Maxwell, and Popper, for the many interesting and stimulating discussions I have been privileged to have with them.)

107. See, for example, Hermann Weyl, *Raum, Zeit, Materie* (Berlin: Springer Verlag, 1919), Sec. XX.

108. See fn. 71.

109. For details, see Paul and T. A. Ehrenfest, *op. cit.* See also Sec. VI below.

110. For Mach's objections, see Sec. III above. See also the references in fn. 91.

111. See Planck's report in his *Scientific Autobiography and Other Papers,* Frank Gaynor, trans. (New York: Philosophical Library, Inc., 1949), esp. pp. 20 f.

112. See fn. 106 and text.

113. This has been made very clear both by K. R. Popper, *Logic of Scientific Discovery, op. cit.,* Sec. X and by V. Kraft, *Erkenntnislehre* (Vienna, 1962), Chap. 1. For an evaluation of this part of Kraft's book, see my review in *British Journal for the Philosophy of Science,* Vol. XIII (1963), 319 f.

114. It is now generally admitted that the early logical positivism was much too radical and that many of its theses were untenable. Yet there is one feature of it that has not been sufficiently appreciated: this philosophy was not in-

timidated by the sciences. True, the sciences were given a most important position in philosophical reasoning, and it was demanded that due attention be paid to their results as well as to the methods used in arriving at these results. But these results were not regarded as sacrosanct and as exempt from criticism. Quite the contrary, one of the main tenets of the early positivists, and especially of Mach, was that the sciences themselves contained metaphysical components *and were therefore in need of revision.* Whether correct or not, this belief was responsible for some very interesting developments: both the quantum theory and the theory of relativity owe a lot to Mach's insistence upon the necessity of a *reform* of physics. And considering the need for alternatives in theorizing that will be established presently, such criticism would seem to be necessary for the advancement of the sciences.

Today this bold and critical attitude has almost completely disappeared. Instead of trying to be *reformers* of the sciences, contemporary philosophers are content to *imitate* them and to adapt their own ideas to the latest discoveries of the historians, or else to the latest fashion of the scientific enterprise. Thus we read in Hans Reichenbach's *Philosophic Foundations of the Quantum Theory* (Berkeley: University of California Press, 1946), p. vii: "All that is intended in this book is clarification of concepts; nowhere in the presentation, therefore, is any contribution towards the solution of physical problems to be expected." Carnap's idea that the task of philosophy consists in *explicating* the notions of science, in making them more precise rather than in changing them in any fundamental way, lies in the same direction. The contemporary movements of vulgar linguistics have contributed considerably to this attitude of wait and see. The result is a most regrettable decrease in the number of the rational critics of the sciences. Also, unwanted support is given in this way to the Hegelian thesis (which is now implicitly held by many historians and philosophers of science) that what exists has a "logic" of its own and is for that reason reasonable and exempt from criticism. Expressed in more pedestrian terms, this thesis amounts to supporting a method, or a hypothesis, on the ground that it is used by many, regarded with awe by even more, and has proved its mettle financially (Nobel prizes). It is nothing but conformism covered up in high-sounding language. Is it not clear that a successful defense of such conformism in the Middle Ages would have forever prevented the emergence of modern science with its so very different "logic"? Modern science is, after all, the result of a conscious criticism not only of the *theses,* but also of the *methods* of Aristotelian physics (see the introduction). For a thinker who demands that a subject be judged "according to its own standards," such a criticism is quite impossible; he will be strongly inclined to reject any interference and to "leave everything as it is" (Wittgenstein). It is somewhat puzzling to find that such demands are nowadays advertised under the title of a "philosophy" of science.

Against such conformism, it is of paramount importance to insist upon the normative character of the scientific method. Adopting this point of view, one cannot regard the arguments in Sec. IV as ultimately decisive. True, they reflect some of the properties of classical physics. But what must be established now is that these properties are *desirable* properties. What is needed is an evaluation of the consistency condition and the condition of meaning invariance on the basis of demands that can be discussed *independently of our actual practice* and that enable us to *evaluate*—that is, either to *praise* or to *condemn*—this practice.

115. The indefinite character of all observations has been made very clear by Pierre Duhem, *op. cit.,* Chaps. 4 and 6. For an alternative way of dealing with this indefiniteness, which to my knowledge has not been suggested before, see

Stephan Körner, *Conceptual Thinking* (New York: Dover Publications, Inc., 1959). Comparing Körner's theory with the two theories discussed in the present essay (the semantic theory of observation and the pragmatic theory of observation; for details, see Sec. XV), we might characterize the former as a theory that distinguishes observation statements from theoretical statements not by their *content* (as does the semantic theory), not by their *pragmatic context* (as does the pragmatic theory), but by their *logical form*. It would be of great interest to have available a more detailed development of this theory.

116. Among the philosophical methods that have been used for the purpose of preserving an old and familiar theory, we have mentioned these: transcendental deduction, phenomenological analysis, analysis of usage, analysis of observational results. All these methods work only in a system of thought that satisfies the consistency condition and does not contain pairs of alternatives. It follows that this condition not only *preserves* what is old and familiar, but also enables the dogmatic defender of such points of view to demonstrate why they are *worth preserving*. This is an excellent reason why philosophers who are interested in the development and improvement of our knowledge and in the elimination of irrelevant argument have started criticizing the consistency condition and have tried to invent a method for the construction of knowledge that not only does not contain it any more, but that *demands that it be violated*. Such a method, quite apart from being necessary for the discovery of past faults and future truths, cuts the ground from underneath all dogmatic philosophizing. It not only discourages *acceptance* of metaphysical arguments of the dogmatic kind; it also eliminates an essential presupposition of the *possibility* of such arguments and thereby altogether destroys dogmatic thinking.

Some philosophers went further than was really necessary and suggested that the principle of noncontradiction also be abandoned. Now they were correct in saying that the development of knowledge *negates* its own earlier stages. Of course, knowledge *itself* does not automatically lead to such self-negation, and it never would if consistency and meaning invariance were retained. The appearance of self-negation is the result of a *method* that is consciously applied by thinking beings and that demands the use of mutually inconsistent alternatives, of the actual *invention* of alternatives (which is a contingent event—dependent on intelligence, relaxation, and good luck—and not something that is *bound* to happen), and of *transitions* between these alternatives brought about by the discovery of inadequacies. It may also be admitted that the earlier stages are not wholly eliminated, but *preserved*. Still, they are not preserved in exactly the same form in which they were previously held, but are *reinterpreted* and thereby changed. A violation of the law of noncontradiction would be necessary only if this preservation were in accordance with meaning invariance—that is, if the negation would leave unchanged certain very essential features of the view negated—or if the competing alternatives were all held at the same time. The former case, which seems to be the one occurring within some dialectical philosophies, indicates that the apparent radicalism of these philosophies, their denial of noncontradiction, is the result of a conservatism in the remainder of their doctrine: the need for a special "dialectical logic" disappears as soon as "preservation" is not interpreted as preservation of *meanings*. For this point, see also K. R. Popper, "What Is Dialectic?" *Mind*, Vol. XLIX (1940), reprinted in *Conjectures and Refutations, op. cit.,* pp. 321 ff.

117. More detailed evidence for the existence of this attitude and for the way in which it influences the development of the sciences may be found in Kuhn, *The Structure of Scientific Revolutions, op. cit.* The attitude is extremely

common in the quantum theory. "Let us enjoy the successful theories we possess and let us not waste our time with contemplating what *would* happen if *other* theories were used"—this seems to be the motto of almost all contemporary physicists (See Heisenberg, *op. cit.*, pp. 56, 114, as well as the discussion of these passages in fn. 114 of *Problems of Microphysics, op. cit.*) and philosophers [see N. R. Hanson, "Five Cautions for the Copenhagen Critics," *Philosophy of Science*, Vol. XXVI (1959), 325 ff.; see also the discussion of Hanson's attitude in my *Problems of Microphysics, op. cit.*]. For Newton's methodological apology, see fn. 34.

Newton admits that, given a certain state of knowledge, it is always possible to invent a multitude of *hypotheses,* and he even makes the remark that such invention is "no difficult matter" (reply to Huyghens, quoted from *Papers and Letters, op. cit.*, p. 144). But he emphasizes that a distinction must be drawn between hypotheses and *theories,* which latter merely state in general and abstract terms (see the reply to Hooke, *ibid.*, p. 119) the results of experimental investigation. Such theories contain no arbitrary element; they are obtained not merely "from a confutation of contrary suppositions . . . but from experiment, concluding positively, and directly" (letter of July 8, 1672 to the *Philosophical Transactions, ibid.*, p. 93). They can be criticized only by reference to experiment, that is, either by repetition of the experiments that originally led to their enunciation or by the performance of new relevant experiments. If the experiments are valid, then "by proving the theory they must render all objections invalid" (*ibid.*, p. 94). Theories, therefore, possess a certainty and a uniqueness that are absent from hypotheses. "The best and safest method of philosophizing," however, "seems to be first to inquire diligently into the properties of things, and establish those properties by experiment, and then to proceed more slowly to hypotheses for the explanation of them. For hypotheses should be subservient only in explaining the properties of things, but not assumed in determining them. . . . For if the possibility of hypotheses is to be the test of the truth and reality of things, I see not how certainty can be obtained in any science; since numerous hypotheses may be devised, which shall seem to overcome new difficulties. Hence, it has been here thought necessary to lay aside all hypotheses . . . " (Reply to Pardies' second letter, *ibid.*, p. 106).

There are many who would admit today that theories are never uniquely and directly determined by experimental results (for an exception, see fn. 93). Even an apparently direct description of such results can be asserted as correct only to a certain degree of precision (unless, of course, one infuses precision into the experimental results themselves, thereby elevating them to the level of observationally intuitive laws). The objections of Hooke and others who asserted the possibility of alternative accounts were irrelevant not because alternative accounts were not possible, but because they chose to discuss *irrelevant* alternatives. (This was seen by Goethe; see fn. 57.) But it is still believed that Newton gave good advice when rejecting hypotheses as a measure of truth and pointing to experimental results as the only way of ascertaining the value of a general point of view. Two comments are in order: first, that we do, of course, also criticize experimental results, and that in this criticism, hypotheses (trivial hypotheses, maybe, but hypotheses nonetheless) play a decisive role; second, that some important experimental results can be produced only with the help of alternative hypotheses, which must therefore be developed even if the current point of view should seem to be a direct expression of fact. (The latter situation may be realized if an abstract presentation of experimental results is given. But such an abstract presentation has lost all force that is usually connected with experience.)

Experience is supposed to be the measure of hypotheses. Now if its results are abstractly formulated, then what we base our hypotheses on are the principles of such a formulation, *which are but other hypotheses;* we do not base them upon a neutral source of knowledge, "experience." Taking all this into account, one must conclude that the "hypothesophobia" of many empiricists is due to their having been thoroughly taken in by a certain hypothesis. [The term "hypothesophobia" is taken from Eduard von Hartmann's *Die Weltanschauung der modernen Physik* (Bad Sachsa, 1909), p. 222: "Physics can never reach certainty which is denied to all factual sciences and can be found only in the purely formal sciences. . . . Hypothesophobia is just as much a children's illness of physics as is the belief in the absolute certainty of its doctrine." In 1909 this was bold language indeed. In the meantime we have passed through two revolutions that changed the very foundations of physical thinking. Almost every hypothesis under the sun has been tried out. "Overthrowing physical laws was a favorite pastime of students of physics and was encouraged in the seminars," as Hugo Dingler wrote in despair (*Die Methode der Physik* [Leipzig, 1936], Introduction). Still it did not take too long for the eager young spirits to settle on a new doctrine, which is now again regarded as the solid and unshakable foundation of all future knowledge.]

118. It seems that on this point the attitude of many physicists coincides with the attitude of linguistic philosophers who also reject speculation inconsistent with the principles of their favorite language and who also defend this position by pointing to the empirical success of this language, insinuating that this success eliminates all alternatives, or at least makes them uninteresting.

119. Having witnessed these effects under a great variety of conditions, I am much more reluctant to regard them as mere curiosities than is the scientific community of today. For details, see my edition of Ehrenhaft's lectures of 1947, *Einzelne Magnetische Nord- und Südpole und deren Auswirkung in den Naturwissenschaften* (Vienna, 1948).

120. See Reinhold Fürth, *Zeitschrift für Physik,* Vol. LXXXI (1933), 143-162.

121. For these investigations, see Albert Einstein, *Investigations on the Theory of the Brownian Movement,* A. D. Cowper, trans. (New York: Dover Publications, Inc., 1956), which contains all the relevant papers by Einstein and an exhaustive bibliography by Fürth. For the experimental work, see J. Perrin, *Die Atome* (Leipzig, 1920).

It should be remembered that at present the relation between the kinetic and phenomenological theories is far from clear. What is used is a *mixture* containing parts of the first theory and parts of the second theory (together with still further assumptions), and the components of this mixture are changed from one particular case to the next. Consistency is by no means assured. Nagel's presentation [*op. cit.*] is in this respect quite inadequate.

122. Indirect refutations of the kind described occur much more frequently than is usually admitted. It does not seem to be an exaggeration to assume that all *decisive* refutations make use of alternatives. In any case, a refutation that is based upon a successful alternative is much *stronger* than is a refutation resulting from the direct comparison of theory and "facts." (And, of course, in matters of knowledge, one should adopt the principle that a stronger method is to be preferred to a weaker one. We want *good* theories and not a "spectrum," extending from interesting and clear ideas to ideas that are downright vague and boring.) The reason why a refutation through alternatives is stronger is easily seen. The direct case is "open," in the sense that a different explanation of the

apparent failure of the theory (of the inconsistency between the theory and certain singular statements) might seem to be possible. The presence of an alternative makes this attitude much more difficult, if not impossible, for we possess now not only the *appearance* of failure (viz., the inconsistency) but also an explanation, on the basis of a successful theory, of why failure *actually occurred*. We may, of course, still doubt the explanation, but our reason for doing so will have to be stronger than are the reasons we might give for not believing a direct refutation. They will have to contain an account of why an otherwise successful theory (we have to assume that the success of the refuting theory is comparable to the success of the theory whose validity is under discussion) should break down in one particular case. And this means that the original theory must be developed and expanded in order to be able to give an answer to this question.

A second reason for the importance of indirect refutations is this: as has been explained (fn. 24), there does not exist a single interesting theory that is not in some kind of trouble. Singular statements, and even experimental laws that *prima facie* refute the theory, can always be found fairly easily. We have also pointed out that this situation requires a healthy balance between dogmatism and skepticism. It is often better to wait and hope for the best than to throw up one's hands in despair and declare that the theory has been refuted. After all, the inconsistency might also have been due to faulty calculation, or else to incorrect observational results. This being the case, troublesome facts, taken by themselves, are almost never sufficient to eliminate the theory. What is needed is an alternative that "elevates the difficulty into a principle" (see fn. 71 above), fares well, both in the domain where the correctness of the original theory is without doubt and in new domains, and which, moreover, possesses some intrinsic advantages, such as greater simplicity, greater generality, etc. This particular way of *accounting for* a difficulty is needed in addition to the *existence* of the difficulty if a straightforward refutation is to be obtained.

T. S. Kuhn's *The Structure of Scientific Revolutions, op. cit.*, contains much interesting material on this point. He describes well the increasing feeling of frustration connected with the appearance of more and more difficulties and with the need for more and more auxiliary hypotheses. These are the preliminaries of a revolution. However, it does not emerge too clearly that *without a new theory*, the dabbling and patching up might go on forever and that people might perhaps become accustomed to the mess, or even invent a philosophical point of view containing argument why a mess is better than a coherent *new* theory.

123. The idea that "we are to look upon propositions inferred by general induction from phenomena as accurate or very nearly true, *notwithstanding any contrary hypothesis that may be imagined till such time as other phenomena occur*," which occurs in this form in Newton (see fn. 34 above), is almost universally accepted. The arguments in the text above show that it is inconsistent with the demand to increase the empirical content of whatever knowledge we possess. They also show, by the way, that a point of view that has been refuted *should not be given up* (it contributes to the empirical content of the successful theory) but should be developed so that it can, perhaps, cope with the troublesome situation.

124. As has been pointed out in Sec. II, the idea of complementarity is the result of the attempt to adapt classical concepts to the new experimental situation created by the discovery of the quantum of action. It is therefore to be expected that it will fit the facts available without the help of alternatives. The

invention of alternatives is now the more necessary, for it is almost certain that refuting facts will not be discovered in a direct fashion.

125. The quantum theory can be adapted to a great many difficulties. It is an open theory in the sense that apparent inadequacies can be accounted for in an *ad hoc* manner, by adding suitable operators or elements in the Hamiltonian, rather than by recasting the whole structure. A refutation of its basic formalism (i.e., of the formalism of quantization and of noncommuting operators in a Hilbert space or a reasonable extension of it) would therefore demand proof to the effect that there is no conceivable adjustment of the Hamiltonian or of the operators used which makes the theory conform to a given fact. It is clear that such a general statement can be provided only by an *alternative theory*, which, of course, would have to be detailed enough to allow for independent and crucial tests. This is how the attempts to develop theories in terms of so-called "hidden parameters" can be, and have been, justified methodologically.

126. Free from *fundamental* difficulties, that is. This must be emphasized, as there is no single physical theory of some consequence that is not beset by some difficulties. See fns. 24 and 122.

127. Rosenfeld, *op. cit.,* p. 44.

128. There exist at the present moment many different approaches within the theory of elementary particles, and only some of them are still in accordance with the idea of complementarity in its original form. This idea therefore seems to be losing its hold on the mind of some physicists. Still it is so widespread, and appeal to it is so popular, that a discussion is necessary.

129. See fn. 8. Some thinkers have expressed the view that discussion which is governed by various alternatives rather than by a single "paradigm" is "often directed as much to the members of other schools as it [is] to nature" (Kuhn, *op. cit.,* p. 13). Our considerations would seem to show that the situation is quite different. Far from being exclusively about "nature," the statements of a comprehensive point of view that is developed without any checking from alternatives may in the end be *about nothing at all* (it is correct under all imaginable circumstances) and merely *express* the metaphysics of the community of those who regard it as the only true reflection of nature.

130. For a detailed description of a once very influential myth, see H. C. Lea, *Materials Toward a History of Witchcraft,* A. C. Howland, ed., 3 vols. (New York: Thomas Yoseloff, Inc., Publisher, 1957), as well as *Malleus Malleficarum,* Montague Summers, trans. (London, 1928). For the absence of the relevant empirical evidence from many historical treatises, see fn. 9 above.

131. See fn. 8.

132. Analysis of usage, to take only one example, quite clearly presupposes the existence of certain regularities concerning usage. The more people differ in their fundamental ideas, the more difficult it will be to uncover such regularities. Hence analysis of usage will work best in a closed society that is firmly held together by a powerful myth, such as was the philosophy in the Oxford of about fifteen years ago. See also fn. 116.

133. Schizophrenics very often hold beliefs that are as rigid, as all-pervasive, and as unconnected with reality as are the best dogmatic philosophies. Only such beliefs come to them naturally, whereas a professor may spend his whole life in the attempt to find arguments that create a similarly closed state of mind (unless, of course, he is a linguistic philosopher with not too rich a vocabulary and not too inventive a mind). It would be most interesting to examine in detail

these structural similarities between dogmatic philosophies on the one side, and the thought structures erected by schizophrenics on the other.

134. The use of a single theory also creates what might be called the *illusion of positive evidence*. Assuming we consider alternatives some of which are refuted by the observations available, we realize that the remaining theories have only *one* thing in their favor: they have not yet been refuted. It is clear that no inference concerning their trustworthiness can be drawn from this premise. After all, there may be some very obvious refuting instance that we have overlooked. Now if we are dealing only with a single theory and consider no alternatives, this *negative* character of the survival of the single theory (it is one of those theories that have not yet failed) will easily be misunderstood as something *positive,* for example as *support* of the theory by the observations.

135. Using terms we used earlier (Sec. III), we might say that *normal science* is liable to become almost indistinguishable from a myth. It is only in a *crisis* that science approaches the ideal of objective knowledge and remains in touch with reality.

.136. The advice to pay proper attention to the facts, that is, to *repeat* observations and not to *replace* them, is rarely free from religious overtones. Instances of a less dogmatic attitude are met with strong moral disapproval. It would be quite interesting to investigate in greater detail the liturgical elements inherent in empiricism. [For a similar assertion concerning materialism, see R. G. Collingwood, *The Idea of Nature* (New York: Oxford University Press, 1960), p. 104.]

137. Wittgenstein has been one of the most eloquent defenders of this principle in recent philosophical thinking. However, he applies it in such a fashion that *progress* (and not only change) from one theory to a better one seems to become quite impossible. For this point, see my review of N. R. Hanson, *Patterns of Discovery* in *Philosophical Review,* Vol. LXIX (1960), 247-252.

138. For details and further references, see Sec. VI of "Explanation, Reduction, and Empiricism," *op. cit.*

139. We assume here that a dynamic, rather than a kinematic, characterization of motion has been adopted. For details, see *ibid.*

140. Direct observation, therefore, cannot at all teach us "what we see." *Theories* might teach us that what we thought was there did, in fact, not exist. An elaboration of this point leads to the refutation of what has been called the "paradigm case argument."

141. For vivid examples, see Karl Jaspers, *Allgemeine Psychopathologie* (Berlin: Springer Verlag, 1959), pp. 75-123.

142. Dialectical philosophers have always emphasized the need not to think in a "mechanical" way, that is, in a framework whose concepts are precisely defined *and kept stable in any argument,* and they have pointed out that arguments precipitating progress usually terminate in concepts that are very different indeed from the concepts in which the question was originally formulated. They have also paid due attention to the fact that the development of our knowledge presupposes the existence of at least two alternative systems of thought, of a thesis and an antithesis. The only objections that might be raised against them are (1) that they went too far and were prepared to eliminate the principle of noncontradiction (for details, see fn. 116), and (2) that they regarded thought as a *necessary development* that occurs independently of the human will. It may well be that the metaphysics behind this idea is correct and that our subjective thought is indeed the mirror image of the development of objective ideas. How-

ever, the occasional effectiveness of conservative measures, the difficulty to over-come familiar modes of thought, suggests that the only reason for the develop-ment of knowledge is the absence of more efficient measures aimed at suppressing novelties and of tyrants (political or otherwise) prepared to enforce such meas-ures.

It is interesting to note that the ordinary language (of intelligent people, that is) does not conform to the rigid pattern that some philosophers claim to have found in it. "Dialectics," says Hegel, I think correctly [Goethe, *Gespräche mit Eckermann*, Houben, ed. (Leipzig, 1948), p. 531], "is fundamentally nothing but the spirit of contradiction that lives in every human being . . . and that is a gift of considerable value if the purpose is to distinguish what is true from what is false."

143. Bacon, *Novum Organum*, Introduction.

144. Isaac Newton, *Papers and Letters, op. cit.,* p. 106.

145. "Der Gegensatz zwischen der mechanischen und der phänomenolo-gischen Physik," *Wärmelehre* (Leipzig: Johann Ambrosius Barth, 1896), pp. 362 ff.

In his *Logical Foundations of Probability* (Chicago: University of Chicago Press, 1950), Carnap quotes a statement by Einstein to the effect that "there is no logical way leading to these . . . laws, but only the intuition based upon sympathetic understanding of experience" [*Mein Weltbild* (1934), p. 168], and he expresses his "complete . . . agree[ment]" with this statement. "However," he continues, "I think we must be careful not to draw too far-reaching negative consequences from this fact. I do not believe that this fact excludes the possibility of a system of inductive logic with exact rules, or the possibility of an inductive machine with a . . . more limited aim" (p. 193). Now an inductive machine of the kind envisaged by Carnap would judge a theory that has already been in-vented on the basis of observational results. The details of the invention would not enter the judgment, nor would the metaphysical ideas that could be used either to support or to criticize the theory and that might perhaps even be used to replace the accepted interpretation of all the observations with a different interpretation, which means that metaphysics is given a psychological function only.

That the evaluation of a given theory on the basis of evidence is not a mat-ter involving this theory, the evidence, and nothing else, has been argued by Ernest Nagel [*Principles of the Theory of Probability, International Encyclo-paedia of Unified Science,* Vol. I, No. 6 (Chicago: University of Chicago Press, 1939), pp. 70 ff.]:

> . . . The practical decision [concerning the extent to which a theory is confirmed by evidence] is in part a function of the contemporary scientific situation. The estimation of the evidence for one theory is usually conducted in terms of the bearing of that evidence upon alternative theories for the same subject matter. When there are several competing theories, a decision between them may be postponed indefinitely, if the evidence supports them all with approximately the same precision. Furthermore, the general line of research pursued at a given time may also determine how the decision for a theory will turn out. For example, at a time when a conception of discon-tinuous matter is the common background for physical research, a theory for a special domain of research formulated in accordance with the dominant leading idea may require little direct evidence for it; on the other hand, a theory based on a continuous notion of matter for that domain may receive little consideration even if direct empirical evidence supports it as well as, or even better than, it does the alternative theory.

This description of *actual procedure* does take into account the fact that general ideas, "metaphysical ideas," as we might call them, such as the idea that the world is filled with continuous stuff, do affect the evaluation of the evidence relevant to a particular theory. Yet it is not at all clear whether the influence is psychological or occurs according to certain methodological rules. Moreover, Nagel's account leaves unanswered the question whether such an influence is at all *desirable* or is not simply the reflection of some *un*desirable metaphysical prejudices on the part of the scientists. Hence as long as his account is unsupported by methodological argument, it is not much better than Johnson's argument against Berkeley. See also Carnap's reply to Nagel, *op. cit.*, pp. 220 ff.

146. See fn. 114.

147. Rosenfeld, *op. cit.*, p. 42.

148. Johann Fischl, *Christliche Weltanschauung und die Probleme der Zeit* (Graz: Steirische Verlagsanstalt, 1946), p. 50. The fact that Professor Fischl and modern "mentalists" who have passed the purifying fire of logical positivism can still make use of the very same argument shows again the great elasticity of what is called "appeal to the facts."

149. Consider from this point of view the following criticism of Hegel's philosophy of nature: "He tried to anticipate by philosophy something which in fact could only be a future development of the natural sciences" [R. G. Collingwood, *op. cit.*, p. 132; see also Collingwood's criticism of materialism (*ibid.*, pp. 104 ff.)]. The obvious reply is that without "philosophical anticipation" there is no "future development" of the sciences.

150. It might be objected (and it has as a matter of fact been objected by my friends in the Minnesota Center for the Philosophy of Science) that such a procedure still presupposes the ability to discover the presence, or the absence, of synonymy, and thereby establishes the need for linguistic analysis that I have denied on various occasions. This may be so. But the same ability is needed by those who either argue in favor of synonymy or set out to discover the grammatical rules according to which some expression is being used. Moreover, the defenders of synonymy depend much more on their ability of semantical detection than we who favor *changes* of meanings. *They* must use conceptual systems containing terms that are *in all respects* identical with certain terms already in use. All *we* need is to make sure that the new terms will be in *some* respects different from the accepted concepts. Absurdity is a good guide in these matters. It is, of course, better to consider conceptual systems *all* of whose features deviate from the accepted points of view. But failure to achieve this in a single step does not entail failure of our epistemological program. Failure to achieve synonymy in any step entails failure of the program of the opponent.

151. Considering what has been said above, it would seem to be good tactics not to present one's theory in the form of identity hypotheses, but to develop it independently, and completely from scratch. There exist various arguments against such a procedure. [For what follows, see also my note in *Journal of Philosophy*, Vol. LX (1963).] It is pointed out, for example, that questions ("what is the relation between mental processes and material processes?") should be *answered*, that they should not be *avoided*. This is a most laudable demand, but let us examine where it leads. Let us assume that a necessary condition for the acceptance of a new theory is that it be capable of answering certain questions. Let us also assume that the postulate of meaning invariance forbids us to reinterpret the descriptive terms of the questions. We then obtain an a priori argument for retaining essential parts of the point of view behind the language in which the

question is formulated. And these parts are retained, not because they have been examined carefully and have been found to be satisfactory, but because of the peculiar way in which the new theory is being introduced. Hence if we object to a priori reasoning, we must either give up meaning invariance or stop asking certain questions, or both. Considering the magical effect words seem to have on all philosophers ["surely, you don't mean to say that this *pain* (this PAIN!!!) you experience now is a material process!"], it would seem to be good tactics to omit the mental terminology altogether, to develop the theory completely from scratch, to show what it can do, and not make its success lend apparent support to an ancient ideology by allowing it to be expressed in the terms of this ideology. If the theory is successful, then all we have to do is to let it speak for itself and wait until the opponents have died, which is in any case the only way of convincing *them*. (How could one possibly have convinced the adherents of witchcraft that they were the victims of an illusion? Descartes, whose theory implies the *denial* of the possibility of witchcraft ". . . prudently avoids special denial and contents himself with ignoring [evil spirits] . . . as unworthy of discussion" [Lecky, *History of the Rise of Rationalism in Europe,* Vol. I; see also Lea, *op. cit.,* p. 1358]. Acceptance of Descartes' theory was one of the reasons why the belief in witchcraft finally disappeared in continental Europe.)

Another objection is that without the use of mental terms the new theory would be without empirical content. Moreover, if "thought" is redefined so as to make "thoughts are material processes" analytic, then the thesis of materialism ceases to be of interest. The first part of this objection assumes that the observation language of a new theory must be about the observable entities of some previous theory. This is obviously impossible when the new theory *negates* the existence of such entities (see also the discussion of bridge laws in Sec. IV above). In this case, the new theory must provide *its own observation language.* Considering that it is always taken for granted that the previous theory could provide its own observation language, this should not be too difficult. The second point ("thoughts are material processes" must be empirical) is already taken care of: if "thought" is supposed to *retain* its archaic meaning, then the *negation* of this statement will follow from materialism and thereby contribute to *its* empirical content ("thoughts are not material processes" being empirical if and only if "thoughts are material processes" is empirical).

A third objection is that without use of already existing terminology, the new theory cannot be made understandable. The first question that arises in connection with this objection is how the "already existing terminology" was made understandable in the first place. Did it need another, and even more ancient, terminology? And if this was not the case, then why should not a new theory also be taught from scratch? Moreover, the common language in which some philosophers want to explain everything is taught to very small children. Is it assumed that a materialist will be incapable of doing what a small child does quite well? Or are the ideas of mental philosophy perhaps inborn? One sees that all the objections mentioned do not really touch the topic under discussion, but only reflect the prejudice of those who have learned *one* language and who, like true Greeks, regard as barbarians all those speaking a different language.

There is one special case in which the discussion of identity hypotheses taken together with the demand for meaning invariance would *not* endanger progress. This case would be realized by a language whose "mentalistic" terms are "open" in the sense that (1) their content admits of further specification and (2) a particular specification creates "materialistic" terms. In such a language, becoming a materialist does not necessitate throwing away past talk and past belief; it amounts only to making such talk *more specific*. This, if I understand him correctly, is the way in which J. J. C. Smart approaches the mind-body problem ["Sensations and Brain Processes," *Philosophical Review*, Vol. LXVIII (1959), 141-156].

The theory of language underlying this particular procedure assumes that an ordinary idiom, or some other language, while capable of specification, is yet correct as regards the *wide outlines* it draws at a certain time. It does not commit itself too much; it does not say anything about domains where information is not yet available; it contains "holes," as it were, that may be filled in by future research. I shall call this theory of language the "hole theory."

The hole theory is accepted by many philosophers. The so-called "paradigm case argument" quite essentially depends upon its being correct. Thus the existence of material objects is inferred from the fact that material object words are being used in the ordinary language [for details concerning such an argument, see Norman Malcolm, "Moore and Ordinary Language," in *The Philosophy of G. E. Moore*, P. A. Schilpp, ed. (Evanston: Library of Living Philosophers, Inc., 1942)]. The procedures leading to such use consist in the exhibition of ordinary situations and corresponding verbal instruction. The idea that existence can be derived from the success of these rather pedestrian methods assumes that the concept of a material object is, as it were, a repetition of the situations employed in the process of teaching (tables are things on which one can sit, under which one can hide, which do not move by themselves but need effort in order to be moved, and so on) and that *nothing hypothetical* is contained in this concept. This, it seems to me, is a very poor account even of the concepts of the ordinary language: the philosophers adopting the hole theory underestimate and misrepresent the idiom on which they base some of their most decisive arguments.

In this, of course, they follow the radical empiricists. They assume, without having the slightest right to do so, that the ordinary language is rather similar to an artificial empirical language that says only as much as has been found to be the case so far, *and not a tiny little bit more*. It is interesting to note, by the way, that Leibniz regarded the *German* (but *not* the Latin) of his time as a language of exactly this kind [see his "Unvorgreifliche Gedancken, betreffend die Ausübung und Verbesserung der Deutschen Sprache," in *Wissenschaftliche Beihefte zur Zeitschrift des allgemeinen Deutschen Sprachvereines* (Berlin, 1907), pp. 292 ff.]. Bohr's intention to create a *physical language* containing holes has been mentioned above.

It goes without saying that the hole theory is quite unrealistic. As regards the success of Bohr's own procedure, see Secs. X and XI of my *Problems of Microphysics, op. cit.* As regards the ordinary language, I think two examples will suffice, one showing that there are no "holes" in the ordinary language, at least not at the places where they are supposed to occur, the second showing that the frame is far from being established once and for all.

In a notorious passage, Susan Stebbing criticizes "the nonsensical denial of [the] solidity of" material objects that some authors inferred from the kinetic theory of matter and the electron theory [*Philosophy and the Physicists* (London: Methuen & Co., Ltd., 1937), p. 53. The argument used in this passage is mentioned with approval by J. O. Urmson in "Some Questions Concerning Validity," in *Essays in Conceptual Analysis*, A. G. N. Flew, ed. (London: St. Martin's Press, 1956), pp. 120 ff. Miss Stebbing, it reads here (pp. 121 f.), "shows conclusively that the novelty of scientific theories does not consist, as has been unfortunately suggested, in showing the inappropriateness of ordinary descriptive language." Except for such "trivial" uses (p. 122) Urmson is, however, critical of the argument]. She defends her charge of nonsense by pointing out that the kinetic theory does not make tables collapse when we sit upon them, that it does not make them swallow up objects that have been put on top of them. Using this kind of argument, she insinuates that "solidity" as used in the ordinary language *covers just these everyday occurrences* and nothing else. This is far from true. A look into the *Oxford English Dictionary* (where "solid" is defined as meaning, among other things, "free from empty spaces, cavities, interstices, etc.; having the interior *completely filled up and in* . . . of material substance; of a dense and massive consistency; composed of particles which are firmly and continuously coherent") teaches that the notion of solidity does not only make assertions about macro-occurrences; it also contains a micro-account of the objects to which it is being applied. That is, it not only asserts that they resist penetration, but it also asserts, or implies, that they resist penetration *because of their being full of compact stuff.*

As regards the second point, I wish only to remind the reader that observer independence has for a long time been regarded as an essential property of material objects, and that this feature has been refuted by the theory of relativity.

However, quite apart from being unrealistic, the hole theory is also very undesirable. It tends to decrease the content of the languages spoken, and thereby makes these languages vague, noncommittal, and unfit for the purpose of criticism. Taking all this into account, we must conclude that Professor Smart's attempt at least to clarify the mind-body problem, despite its initial plausibility, is bound to fail, as is any attempt that shows too great concern for the existing systems of thought.

152. "A Plea for Excuses," quoted from *Philosophical Papers,* J. O. Urmson and C. J. Warnock, eds. (London: Oxford University Press, 1961), p. 130.

153. Quite apart from what has been said in the text, the argument also proceeds from much too simplistic an interpretation of the naturalistic thesis. It is assumed that all a naturalist can offer are *processes,* or *objects,* and that it is with *them* that he must identify sensations and thoughts. Now it has to be admitted that the structure (or, to use a more recent term, the "grain") of materialistic processes is very different from the structure of sensations and thoughts. But there are also *properties* of such processes, and relations between them and their structure that need not at all differ from that of thoughts and sensations. Take the case of a fluid. It consists of numerous particles, each of them carrying out a very complicated motion (I shall neglect the quantum theory and discuss classical materialism only). Yet the *density* of the fluid does not at all possess this complicated structure. It is a statistical property that ceases to be well defined in domains where the "grain" of the fluid becomes noticeable. Or take the color (physical) of a solid object. Again this property is defined in such a fashion that it cannot be meaningfully applied to a trio of particles belonging to the object. Of neither property would it make sense to say that it contains an element moving on a curved path; nor would it make sense to assert of them that they contain parts. There are other properties, such as the energy of the system, its entropy, its electrical capacity, that have no location except the location indicated, in a vague sense, by the system as a whole. Or consider the potential energy of a system relative to another one—this property is not situated anywhere in space. Still, when considering it we are not at all leaving the materialistic point of view. The case of haziness can be handled in a similar fashion: assuming sensations to be structural properties of material processes rather than these processes themselves, we must remember that for every property there are conditions under which it ceases to be well defined. Thus, for example, the scratchability of a solid object (Mohs-

scale) ceases to be well defined when the object starts melting. Similarly, we may assume that sensations are properties of central processes that occasionally cease to be well defined and thereby escape this part of the argument.

I should like to emphasize that I am not proposing the above as the *correct* physiological account of sensations. Whether such an account exists and how it would look in detail I do not know. My point is that it is a *possible* account, that the argument is not applicable to it, that it has not yet been excluded by independent considerations, and that its production at this general level of discussion must therefore be regarded as a refutation of the argument. Thus the argument is too crude, even if its purpose were only to show that a materialistic *analysis* of mental *terms* (rather than a materialistic *explanation* of mental *processes*) is impossible.

154. This may be regarded as a partial justification of Lord Russell's dictum that we always see our own brain. Concerning realism, see my paper "Das Problem der Existenz Theoretischer Entitäten," in *Probleme der Wissenschaftstheorie, Festschrift für Viktor Kraft* (Vienna: Springer Verlag, 1961), pp. 35-72.

155. The phrase "knowledge by acquaintance" is used with at least two meanings. It means (1) the same as *observational* knowledge. In this sense we know the properties of tables and chairs by acquaintance. And it means (2) observational knowledge that is complete and *incorrigible*. It is acquaintance in the second sense that I am discussing in the present section.

156. Evans-Pritchard, *Oracles and Witchcraft Among the Azande* (London: Oxford University Press, 1937).

157. For examples, see J. H. Breasted, *Development of Religion and Thought in Ancient Egypt* (New York: Harper & Row, Publishers, Inc., 1959); see also Frankfort, *et al., op. cit.,* and E. R. Dodds, *The Greeks and the Irrational* (Berkeley: University of California Press, 1951).

158. Almost every philosopher will nowadays protest against such a procedure. I think that all the main objections have been dealt with in the text (and will be dealt with below). The only objection left would be based upon the principle that one should not continue using a certain *word* when the *concepts* have been changed in a radical way; having shown that a materialistic pain and an "ordinary" pain would be two very different things indeed, the defender of the established usage (or of what is regarded as the established usage by him) might advise the materialist not to use the word "pain," which for him rightfully belongs to ordinary English and not to physiology. Now quite apart from the fact that this would mean being very squeamish indeed, and unbearably "proper" in linguistic matters, the desired procedure *cannot be carried out.* The reason is that changes of meaning occur too frequently and cannot be localized in time. Every interesting discussion, that is, every discussion which leads to an advance of knowledge, terminates in a situation where some decisive change of meaning has taken place. Yet it is not possible, or it is only very rarely possible, to say *when* exactly the change took place. Moreover, a distinction must be drawn between the *psychological circumstances* of the production of a sentence and the *meaning* of the statement that is connected with the sentence. A new theory of pains will not change the pains; nor will it change the causal connection existing between the occurrence of pains and the production of "I am in pain," except perhaps very slightly. It *will* change the meaning of "I am in pain." Now it seems to me that observational terms should be correlated with causal antecedents and *not* with meanings. The causal connection between the production of a "mental" sentence and its "mental" antecedent is very strong. It has been taught in our youth. It is the basis of all observation concerning the mind. To sever this con-

nection is a much more laborious affair than is a change of connections with meaning. The latter change all the time anyway. It is therefore much more sensible to establish a one-one correlation between observational terms and their causal antecedents than between such terms and the always variable *meanings*. This procedure has great advantages, and it can do no harm. An astronomer who wishes to determine the rough shape of the energy output (dependence on frequency) of a star by looking at it will hardly be seduced into thinking that the word "red" that he uses for announcing his results refers to *sensations*. Linguistic sensitivity may be of some value. But it should not be used to turn intelligent people into nervous wrecks.

159. This theory need not be explicitly formulated, nor need the people using the method be aware of its existence. Yet its influence need not on that account be negligible. Quite the contrary, just as in the case of social commands and prohibitions, "the most effective . . . control is that which is not noticed; which is not overt or formal; which is, as it were, 'closer to us than breathing, nearer than hands or feet' " [H. E. Barnes and Howard Becker, *Social Thought from Lore to Science,* 3rd ed., Vol. I (New York: Dover Publications, Inc., 1961), p. 11]. Influential, but not explicitly formulated, theories are in their effect very similar to Kant's categories. See also fn. 160.

160. As has been pointed out, such presentation need not be the result of an explicit decision that has been consciously formulated and then used for building the language in this specific way. However, the procedure does not therefore become more acceptable. Whether hidden or not, the *result* is open to criticism.

161. Kuno Fischer, *Immanuel Kant und seine Lehre* (Heidelberg, 1889), p. 10. As shown in a most excellent manner by Fischer, it was Kant who first recognized that certain alleged facts are not properties of nature but of our mental organization, or of the method used by us in building up knowledge. He also assumed that this mental organization and this method were *given* and unalterable. He therefore only partly overcame the dogmatism of his predecessors. He duly unmasked the so-called "facts" upon which they based their philosophy, and he also restricted cosmology to the domain of experience. But he still maintained that certain features of this domain were unalterable, though not for the reasons that had been given before.

162. *Knowledge Without Foundations, op. cit.*

163. That methodological, or normative, considerations may play an important role in philosophical matters seems to be neglected by linguistic philosophers. All they perceive is the *result* of having adopted a certain philosophical doctrine, and this result, it is quite true, is sometimes a purely linguistic change. (In the present case the change would be from a "safe" concept of pain to a concept that is richer in content, and therefore not applicable on the basis of acquaintance.) What is overlooked is that such changes may be desirable for *methodological* reasons. Traditional philosophers may prefer the safe concept because its application leads to statements that are *certain*. In the present paper, safe concepts are rejected because of the paucity of their *content*. And the argument (which is due to Popper) is really very simple: we want our statements to be interesting and informative; we want to convey as much as possible in a single statement; we cannot be content with statements that are almost empty. In addition to this simple argument, it is necessary to realize that we are here confronted with a genuine choice: it is completely up to us either to have knowledge by acquaintance and the poverty of content that goes with it or to have hypothetical knowledge, which is corrigible, which can be improved, and which is informative. (The fact—if it is a fact—that the language which we have learned from our

ancestors seems to favor the one alternative rather than the other one is quite ir-relevant here. Why should we adopt the attitude of our forefathers without further examination?)

164. See our discussion of Nagel's theory of reduction in Secs. III and IV above.

165. See Sec. IV above.

166. This objection was raised in private discussions by Roger Buck.

167. *Concepts, Theories and the Mind-Body Problem,* Herbert Feigl, *et al.,* eds., Minnesota Studies in the Philosophy of Science, Vol. II (Minneapolis: University of Minnesota Press, 1958), esp. pp. 387-397. If the conditions discussed are combined with the demand for meaning invariance, then the solution of the mind-body problem becomes a very difficult matter indeed, which illustrates the remark made in the introduction about the character of many philosophical problems.

168. See fn. 150.

169. For details, see fn. 151.

170. The requirement also overlooks the pragmatic character of the notion of observability. For this, see Sec. XV.

171. It was such a reinterpretation that played a most important part in Galileo's defense of the motion of the earth. See also Hall, *op. cit.*

172. See Sec. XIII.

173. The term "secondary meaning" and the example are both due to Witt-genstein, *Philosophical Investigations* (Oxford: Basil Blackwell & Mott, Ltd., 1953), p. 216.

174. "Perspectoid Distances," *Acta Psychologica,* Vol. XI (1955), 297 ff. See also E. Rubin, "Visual Figures Apparently Incompatible with Geometry," *ibid.,* Vol. VII (1950), 365 ff. These two papers deserve more attention than they have so far received from philosophers.

175. This distinction has been emphasized with great clarity by E. Kaila. See his article "Det fraemmande sjaelslivets kunskapsteoretiska problem," *Theoria,* Vol. II (1933), 144 ff., as well as his essay "Über das System de Wirklichkeitsbe-griffe," *Act Philosophica Fennica,* Vol. II (1936), 17 ff.

176. See also the discussion of the double language model in Sec. IV above.

177. Combining the empirical criterion of meaning that is characteristic for the subjective point of view with the objective point of view, we arrive at the following classification of sentences:

	possessing factual content	factually empty
meaningful	empirical theories	formal mathematics
meaningless	metaphysical theories	intuitive mathematics

178. For details, see "Explanation, Reduction, and Empiricism," *op. cit.*

179. In my paper, "An Attempt at a Realistic Interpretation of Experience," I have tried to give a more detailed account.

180. For Brecht's idea of a critical theatre making use of alternative inter-pretations of the action on stage, see his *Schriften zum Theater* (Berlin: Suhr-kamp Verlag, 1957).

NICHOLAS RESCHER
University of Pittsburgh

The Ethical Dimension of Scientific Research

It seems to me better for people to build a house, to plow a field, or at least to plant a fruit to an ordinary tree, rather than merely to gather in a few flowers or fruits. Such diversions are to be praised rather than forbidden, but we must not neglect the more important things. We are responsible for our talents to God and to the community.

—G. W. LEIBNIZ, *Precepts for Advancing the Sciences and Arts*

The Ethical Dimension of Scientific Research

It has been frequently asserted that the creative scientist is distinguished by his objectivity. The scientist—so it is said—goes about his work in a rigidly impersonal and unfeeling way, unmoved by any emotion other than the love of knowledge and the delights of discovering the secrets of nature.

This widely accepted image of scientific inquiry as a cold, detached, and unhumane affair is by no means confined to the scientifically uninformed and to scientific outsiders, but finds many of its most eloquent spokesmen within the scientific community itself. Social scientists in particular tend to be outspoken supporters of the view that the scientist does not engage in making value judgments, and that science, real science, deals only with what is, and has no concern with what ought to be. Any recitation of concrete instances in which the attitudes, values, and temperaments of scientists have influenced their work or affected their findings is dismissed with the scornful dichotomy that such matters may bear upon the psychology or sociology of scientific inquiry, but have no relevance whatever to the *logic* of science.

This point of view that science is "value free" has such wide acceptance as to have gained for itself the distinctive, if somewhat awesome, label
as the thesis of the *value neutrality of science.*

Now the main thesis that I propose is simply that this supposed division between the evaluative disciplines on the one hand and the nonevalu-
ative sciences on the other is based upon mistaken views regarding the nature of scientific research. In paying too much attention to the abstract
logic of scientific inquiry, many students of scientific method have lost
sight of the fact that science is a human enterprise, carried out by flesh
and blood men, and that scientific research must therefore inevitably exhibit some normative complexion. It is my aim to examine the proposition that evaluative, and more specifically *ethical,* problems crop up at
numerous points within the framework of scientific research. I shall attempt to argue that the scientist does not, and cannot put aside his common humanity and his evaluative capabilities when he puts on his laboratory coat.

Ethical Issues and the Collectivization of Scientific Research

Before embarking on a consideration of the ethical dimension of scientific research, a number of preliminary points are in order.

In considering ethical issues within the sciences, I do not propose to
take any notice at all of the various moral problems that arise in relation
to what is *done with* scientific discoveries once they have been achieved. I
want to concern myself with scientific work as such, and insofar as possible
to ignore the various technological and economic applications of science.
We shall not be concerned with the very obviously ethical issues that have
to do with the use of scientific findings for the production of the instrumentalities of good or evil. The various questions about the morality of
the *uses* to which scientific discoveries are put by men other than the scientists themselves—questions of the sort that greatly exercise such organizations as, for example, the Society for Social Responsibility in Science—
are substantially beside the point. We all know that the findings of science
can be used to manufacture wonder drugs to promote man's welfare, or
bacteriological weapons to promote his extermination. Such questions of
what is done with the fruits of the tree of science, both bitter and sweet, are
not problems that arise *within* science, and are not ethical choices that
confront the scientist himself. This fact puts them outside of my limited
area of concern. They relate to the exploitation of scientific research, not
to its pursuit, and thus they do not arise *within* science in the way that
concerns us here.

Before turning to a description of some of the ethical issues that affect the conduct of research in the sciences, I should like to say a word
about their reason for being. Ethical questions—that is, issues regarding

the rightness and wrongness of conduct—arise out of people's dealings with each other, and pertain necessarily to the duties, rights, and obligations that exist in every kind of interpersonal relationship. For a Robinson Crusoe, few, if any, ethical problems present themselves. One of the most remarkable features of the science of our time is its joint tendency toward collectivization of effort and dispersion of social involvement.

The solitary scientist laboring in isolation in his study or laboratory has given way to the institutionalized laboratory, just as the scientific paper has become a thing of almost inevitably multiple authorship, and the scientific calculation has shifted from the back of an envelope to the electronic computer. Francis Bacon's vision of scientific research as a group effort has come to realization. The scientist nowadays usually functions not as a detached individual unit, but as part of a group, as a "member of the team."

This phenomenon of the collectivization of scientific research leads increasingly to more prominent emphasis upon ethical considerations within science itself. As the room gets more crowded, if I may use a simile, the more acute becomes the need for etiquette and manners; the more people involved in a given corner of scientific research, the more likely ethical issues are apt to arise. It seems that these phenomena of the collectivization and increasing social diffusion of modern science are the main forces that have resulted in making a good deal of room for ethical considerations within the operational framework of modern science.

Ethical Problems Regarding Research Goals

Perhaps the most basic and pervasive way in which ethical problems arise in connection with the prosecution of scientific research is in regard to the choice of research problems, the setting of research goals, and the allocation of resources (both human and material) to the prosecution of research efforts. This ethical problem of choices relating to research goals arises at all levels of aggregation—the national, the institutional, and the individual. I should like to touch upon each of these in turn.

The National Level. As regards the national level, it is a commonplace that the United States government is heavily involved in the sponsorship of research. The current level of federal expenditure on research and development is 8.4 billion dollars, which is around 10 per cent of the federal budget, and 1.6 per cent of the gross national product. If this seems like a modest figure, one must consider the historical perspective. The rate of increase of this budget item over the past ten years has been 10 per cent per annum, which represents a doubling time of seven years. Since the doubling time of our GNP is around twenty years, at these present rates our government will be spending all of our money on science and technology in about sixty-five years. But even today, long before this awkward juncture of affairs is reached, our government, that is to say our collective selves, is heavily involved in the sponsorship of scientific work. And since the man who pays the piper inevitably gets to call at least some of the tune, our society is confronted with difficult choices of a squarely ethical nature regarding the direction of these research efforts. Let me cite a few instances.

In the Soviet Union, 35 per cent of all research and academic trained personnel is engaged in the engineering disciplines, compared with 10 per cent in medicine and pharmaceutical science. Does this 3.5 to 1 ratio of technology to medicine set a pattern to be adopted by the United States? Just how are we to "divide the pie" in allocating federal support funds among the various areas of scientific work?

In our country, the responsibility for such choices is, of course, localized. The President's Science Advisory Committee and the Federal Council for Science and Technology give a mechanism for establishing an overall science budget and thereby for making the difficult decisions regarding resource allocation. These decisions, which require weighing space probes against biological experimentation and atomic energy against oceanography are among the most difficult choices that have to be made by, or on behalf of, the scientific community. The entrance of political considerations may complicate, but cannot remove, the ethical issues that are involved in such choices.

What is unquestionably the largest ethical problem of scientific public policy today is a question of exactly this type. I refer to the difficult choices posed by the fantastic costs of the gadgetry of space exploration. The costs entailed by a systematic program of manned space travel are such as to necessitate major sacrifices in the resources our society can commit both to the advancement of knowledge as such in areas other than space and to medicine, agriculture, and other fields of technology bearing directly upon human welfare. Given the fact, now a matter of common knowledge, that modern science affords the means for effecting an almost infinite improvement in the material conditions of life for at least half the population of our planet, are we morally justified in sacrificing this opportunity to the supposed necessity of producing cold war spectaculars? No other question could more clearly illustrate the ethical character of the problem of research goals at the national level.

The Institutional Level. Let me now turn to the institutional level—that of the laboratory or department or research institute. Here again the ethical issue regarding research goals arises in various ways connected with the investment of effort, or, to put this same matter the other way around, with the selection of research projects.

One very pervasive problem at this institutional level is the classical issue of pure, or basic, versus applied, or practical, research. This problem is always with us and is always difficult, for the more "applied" the research contribution, the more it can yield immediate benefits to man; the more "fundamental," the deeper is its scientific significance and the more can it contribute to the development of science itself. No doubt it is often the case unfortunately that the issue is not dealt with on this somewhat elevated plane, but is resolved in favor of the applied end of the spectrum by the mundane, but inescapable, fact that this is the easier to finance.

I need scarcely add that this ethical issue can also arise at the institutional level in far more subtle forms. For instance, the directorship of a virology laboratory may have to choose whether to commit its limited resources to developing a vaccine which protects against a type of virus that is harmless as a rule but deadly to a few people, as contrasted with a variant type of virus that, while deadly to none, is very bothersome to many.

The Individual Level. The most painful and keenly felt problems are often not the greatest in themselves, but those that touch closest to home. At the level of the individual, too, the ethical question of research goals and the allocation of effort—namely that of the individual himself—can arise and present difficulties of the most painful kind. To cite one example, a young scientist may well ponder the question of whether to devote himself to pure or to applied work. Either option may present its difficulties for him, and these can, although they need not necessarily, be of an ethical nature.

Speaking now just of applied science, it is perfectly clear that characteristically ethical problems can arise for the applied scientist in regard to the nature of the application in question. This is at its most obvious in the choice of a military over against a nonmilitary problem context— A-bombs versus X-rays, poison gas versus pain killers. On this matter of the pressure of ethical considerations upon the conscience of an individual, I cannot forbear giving a brief, but eloquent, autobiographical quotation from C. P. Snow:

> I was an official for twenty years. I went into official life at the beginning of the war, for the reason that prompted my scientific friends to begin to make weapons. I stayed in that life until a year ago, for the same reason that made my scientific friends turn into civilian soldiers. The official's life in England is not quite so disciplined as a soldier's, but it is very nearly so. I think I know the virtues, which are very great, of the men who live that disciplined life. I also know what for me was the moral trap. I, too, had got onto an escalator. I can put the result in a sentence: I was coming to hide behind the institution; I was losing the power to say no.[1]

How many scientists in our day are passengers riding along on Snow's escalator and have dulled their moral sensitivities to this question of personal goals? I have myself known more than one scientist who has forgone the chance of being a public benefactor in favor of the more immediate opportunity to be a public servant.

Ethical Problems Regarding the Staffing of Research Activities

The recruitment and assignment of research personnel to particular projects and activities poses a whole gamut of problems of an ethical nature. I will confine myself to two illustrations.

It is no doubt a truism that scientists become scientists because of their interest in science. Devotion to a scientific career means involvement with scientific work: *doing* science rather than *watching* science done. The collectivization of science creates a new species—the science administrator whose very existence poses both practical and ethical problems. Alvin Weinberg, director of the Oak Ridge National Laboratory, has put it this way:

> Where large sums of public money are being spent there must be many administrators who see to it that the money is spent wisely. Just as it is easier to spend money than to spend thought, so it is easier to tell other scientists how and what to do than to do it oneself. The big scientific community tends to acquire more and more bosses. The Indians with bellies to the bench are hard to discern for all the chiefs with bellies to the mahogany desks. Unfortunately, science dominated by administrators is science understood by administrators, and such science quickly becomes attenuated, if not meaningless.[2]

The facts adduced by Weinberg have several ethical aspects. For one thing there is Weinberg's concern with what administrationitis may be doing to science. And this is surely a problem with ethical implications derived from the fact that scientists have a certain obligation to the promotion of science itself as an ongoing human enterprise. On the other hand, there is the ethical problem of the scientist himself, for a scientist turned administrator is frequently a scientist lost to his first love.

My second example relates to the use of graduate students in university research. There seems to me to be a very real problem in the use of students in the staffing of research projects. We hear a great many pious platitudes about the value of such work for the training of students. The plain fact is that the kind of work needed to get the project done is simply not always the kind of work that is of optimum value for the basic training of a research scientist in a given field. Sometimes instead of doing the student a favor by awarding him a remunerative research fellowship, we may be doing him more harm than good. In some instances known to me, the project work that was supposedly the training ground of a graduate student in actuality derailed or stunted the development of a research scientist.

Ethical Problems Regarding Research Methods

Let me now take up a third set of ethical problems arising in scientific research—those having to do with the *methods* of the research itself.

Problems of this kind arise perhaps most acutely in biological or medical or psychological experiments involving the use of experimental animals. They have to do with the measures of omission and commission for keeping experimental animals from needless pain and discomfort. In this connection, let me quote Margaret Mead:

> The growth of importance of the study of human behavior raises a host of new ethical problems, at the head of which I would place the need for consent to the research by both observer and subject. Studies of the behavior of animals other than man introduced a double set of problems: how to control the tendency of the human observer to anthropomorphize, and so distort his observations, and how to protect both the animal and the experimenter from the effects of cruelty. In debates on the issue of cruelty it is usually recognized that callousness toward a living thing may produce suffering in the experimental subject, but it is less often recognized that it may produce moral deterioration in the experimenter.[3]

It goes without saying that problems of this sort arise in their most acute form in experiments that risk human life, limb, well-being, or comfort.

Problems of a somewhat similar character come up in psychological or social science experiments in which the possibility of a compromise of human dignity or integrity is present, so that due measures are needed to assure treatment based on justice and fair play.

Ethical Problems Regarding Standards of Proof

I turn now to a further set of ethical problems relating to scientific research—those that are bound up with what we may call the standards of proof. These have to do with the amount of evidence that a scientist accumulates before he deems it appropriate to announce his findings and put forward the claim that such-and-so may be regarded as an established fact. At what juncture should scientific evidence be reasonably regarded as strong enough to give warrant for a conclusion, and how should the uncertainties of this conclusion be presented?

This problem of standards of proof is ethical, and not merely theoretical or methodological in nature, because it bridges the gap between scientific understanding and action, between thinking and doing. The scientist cannot conveniently sidestep the whole of the ethical impact of such questions by saying to the layman, "I'll tell you the scientific facts and then *you* decide on the proper mode of action." These issues are usually so closely interconnected that it is the scientific expert alone who can properly adjudge the bearing of the general scientific considerations upon the particular case in hand.

Every trained scientist knows, of course, that "scientific knowledge" is a body of statements of varying degrees of certainty—including a great deal that is quite unsure as well as much that is reasonably certain. But in presenting particular scientific results, and especially in presenting his own results, a researcher may be under a strong temptation to fail to do justice to the precise degree of certainty and uncertainty involved.

On the one hand, there may be some room for play given to a natural human tendency to exaggerate the assurance of one's own findings. Moreover, when much money and effort have been expended, it can be embarrassing—especially when talking with the nonscientific sponsors who have footed the bill—to derogate from the significance or suggestiveness of one's results by dwelling on the insecurities in their basis. The multiple studies and restudies made over the past ten years in order to assess the pathological and genetic effects of radioactive fallout afford an illustration of a struggle to pinpoint the extent of our knowledge and our ignorance in this area.

On the other hand, it may in some instances be tempting for a researcher to underplay the certainty of his findings by adopting an unreasonably high standard of proof. This is especially possible in medical research, where life-risking actions may be based upon a research result. In this domain, a researcher may be tempted to "cover" himself by hedging his findings more elaborately than the realities of the situation may warrant.

Especially when communicating with the laity, this matter of indicating in a convincing way the exact degree of assurance that attaches to a scientific opinion may be a task of great complexity and difficulty. Let me illustrate this by a quotation from W. O. Baker, Vice President for Research at Bell Laboratories:

> I happened to be one of a task force that was gathered officially, with State Department sanction, at the very beginning of 1946, to prepare a detailed scientific estimate . . . about the probable duration of the United States nuclear monopoly. We found, of course, the engineering truth that another country, explicitly the Soviet Union, would have nuclear weapons in a certain number of years after 1946—a number which we carefully estimated. Our estimate, which is a matter of record, was off by little more than a year, and it was, indeed, too conservative an estimate. But it was by no means trusted, and—an equally sorry circumstance—we lacked the skill to make people believe and heed it.[4]

Ethical Problems Regarding the Dissemination of Research Findings

A surprising variety of ethical problems revolve around the general topic of the dissemination of research findings. It is so basic a truth as to be almost axiomatic that, with the possible exception of a handful of unusual cases in the area of national security classification, a scientist has not only the right, but even the duty, to communicate his findings to the community of fellow scientists, so that his results may stand or fall in the play of the open market place of ideas. Modern science differs sharply in this respect from science in Renaissance times, when a scientist shared his discoveries only with trusted disciples, and announced his findings to the general public only in cryptogram form, if at all.

This ethical problem of favoritism in the sharing of scientific information has come to prominence again in our day. Although scientists do generally publish their findings, the processes of publication consume time, so that anything between six months and three years may elapse between a scientific discovery and its publication in the professional literature. It has become a widespread practice to make prepublication announcements of findings, or even pre-prepublication announcements. The ethical problem is posed by the extent and direction of such exchanges, for there is no doubt that in many cases favoritism comes into the picture, and that some workers and laboratories exchange findings in a preferential way that amounts to a conspiracy to maintain themselves ahead of the state of the art in the world at large. There is, of course, nothing reprehensible in the natural wish to overcome publication lags or in the normal desire for exchanges of ideas with fellow workers. But when such practices tend to become systematized in a prejudicial way, a plainly ethical problem comes into being.

Let us consider yet another ethical problem regarding the dissemination of research findings. The extensive dependence of science upon educated public opinion, in connection with its support both by the government and the foundations, has already been touched upon. This factor has a tendency to turn the reporting of scientific findings and the discussion of issues relating to scientific research into a kind of journalism. There is a strong incentive to create a favorable climate of public opinion for certain pet projects or concepts. Questions regarding scientific or technical merits thus tend to get treated not only in the proper forum of the science journals, but also in the public press and in Congressional or foundation committee rooms. Not only does this create the danger of scientific pressure groups devoted to preconceived ideas and endowed with the power of retarding other lines of thought, but it also makes for an unhealthy emphasis on the spectacular and the novel, unhealthy, that is, from the standpoint of the development of science itself. For such factors create a type of control over the direction of scientific research that is disastrously unrelated to the proper issue of strictly scientific merits.

The fact is that science has itself become vulnerable in this regard through its increasing sensitivity to public relations matters. Let me cite just one illustration—that of the issue of the fluoridation of municipal water supplies. Some scientists appear to have chosen this issue as a barricade at which to fight for what (to use a political analogue) might be called the "grandeur" of science.

Not long ago, local referenda in the state of Massachusetts gave serious defeats to the proponents of fluoridation. Not only were proposals to introduce this practice defeated in Wellesley and Brookline, but

Andover, where fluoridation had been in effect for five years, voted discontinuance of the program. These defeats in towns of the highest educational and socioeconomic levels caused considerable malaise in the scientific community, and wails of anguish found their way even into *Science,* official journal of the American Association for the Advancement of Science. This annoyance over what is clearly not a *scientific* setback, but merely a failure in public relations or political effectiveness, sharply illustrates the sensitivity that scientists have developed in this area.

Ethical Problems Regarding the Control of Scientific "Misinformation"

Closely bound up with the ethical problems regarding the dissemination of scientific information are what might be thought of as the other side of the coin—the control, censorship, and suppression of scientific misinformation. Scientists clearly have a duty to protect both their own colleagues in other specialties and the lay public against the dangers of supposed research findings that are strictly erroneous, particularly in regard to areas such as medicine and nutrition, where the public health and welfare are concerned. And quite generally, of course, a scientist has an obligation to maintain the professional literature of his field at a high level of content and quality. The editors and editorial reviewers in whose hands rests access to the media of scientific publication clearly have a duty to preserve their readership from errors of fact and trivia of thought. But these protective functions must always be balanced by respect for the free play of ideas and by a real sensitivity to the possible value of the unfamiliar.

To give just one illustration of the importance of such considerations, I will cite the example of the nineteenth-century English chemist J. J. Waterson. His groundbreaking papers on physical chemistry, anticipating the development of thermodynamics by more than a generation, were rejected by the referees of the Royal Society for publication in its *Proceedings,* with the comment (among others) that "the paper is nothing but of nonsense." As a result, Waterson's work lay forgotten in the archives of the Royal Society until rescued from oblivion by Rayleigh some forty-five years later. Let me quote J. S. Haldane, whose edition of Waterson's works in 1928 decisively rehabilitated this important researcher:

> It is probable that, in the long and honorable history of the Royal Society, no mistake more disastrous in its actual consequences for the progress of science and the reputation of British science than the rejection of Waterson's papers was ever made. The papers were foundation stones of a new branch of scientific knowledge, molecular physics, as Waterson called it, or physical chemistry and thermodynamics as it is now called. There is every reason for believing that, had the papers been published, physical chemistry and thermodynamics would have developed mainly in this country [i.e., England], and along much simpler, more correct, and more intelligible lines than those of their actual development.[5]

Many other examples could be cited to show that it is vitally important that the gatekeepers of our scientific publications be keenly alive to the possible but unobvious value of unfamiliar and strange seeming conceptions.

It is worth emphasizing that this matter of "controlling" the dissemination of scientific ideas poses special difficulties due to an important, but much underrated, phenomenon: *the resistivity to novelty and innovation by the scientific community itself.* No feature of the historical course of development of the sciences is more damaging to the theoreticians' idealized conception of science as perfectly objective—the work of almost disembodied intellects governed by purely rational considerations and actuated solely by an abstract love of truth. The mere assertion that scientists can resist, and indeed frequently have resisted, acceptance of scientific discoveries clashes sharply with the stereotyped concept of the scientist as the purely objective, wholly rational, and entirely open-minded man. Although opposition to scientific findings by social groups other than scientists has been examined by various investigators, the resistance to scientific discoveries by scientists themselves is just beginning to attract the attention of sociologists.[6]

The history of science is, in fact, littered with examples of this phenomenon. Lister, in a graduation address to medical students, bluntly warned against blindness to new ideas such as he had himself encountered in advancing his theory of antisepsis. Pasteur's discovery of the biological character of fermentation was long opposed by chemists, including the eminent Liebig, and his germ theory met with sharp resistance from the medical fraternity of his day. No doubt due in part to the very peculiar character of Mesmer himself, the phenomenon of hypnosis, or mesmerism, was rejected by the scientifically orthodox of his time as so much charlatanism. At the summit of the Age of Reason, the French Academy dismissed the numerous and well-attested reports of stones falling from the sky (meteorites, that is to say) as mere folk stories. And this list could be prolonged *ad nauseam.*

Lord Rayleigh, the rediscoverer of J. J. Waterson, who had also himself been burned by scientific opposition to his research findings, became so pessimistic about the difficulties that new conceptions encounter before becoming established in science that he wrote:

> Perhaps one may . . . say that a young author who believes himself capable of great things would usually do well to secure the favorable recognition of the scientific world by work whose scope is limited, and whose value is easily judged, before embarking on greater flights.[7]

(The value of Rayleigh's advice is, of course, very questionable, in view of the fact that it is more than likely that any young scientist of promise who fritters away the maximally creative years of youthful freshness and enthusiasm by doing work of routine drudgery will almost inevitably blunt the keen edge of his productive capacities to a point where "great things" are simply no longer within his grasp.)

Those scientists who have themselves fallen victim to the resistance to new ideas on the part of their colleagues have invariably felt this keenly, and have given eloquent testimony to the existence of this phenomenon. Oliver Heaviside, whose important contributions to mathematical physics were slighted for over twenty-five years, is reported to have exclaimed bitterly that "even men who are not Cambridge mathematicians deserve justice." And Max Planck, after encountering analogous difficulties, wrote:

> This experience gave me also an opportunity to learn a new fact—a remarkable one in my opinion: A new scientific truth does not triumph by convincing its opponents and making them see the light, but rather because its opponents eventually die, and a new generation grows up that is familiar with it.[8]

In summary, the prominence, even in scientific work, of the human psychological tendency to resist new ideas must temper the perspective of every scientist when enforcing what he conceives to be his duty to safeguard others against misinformation and error.

At no point, however, does the ethical problem of information control in science grow more difficult and vexatious than in respect to the boundary line between proper science on the one hand and pseudo-science on the other. The plain fact is that truth is to be found in odd places, and that scientifically valuable materials turn up in unexpected spots.

No one, of course, would for a moment deny the abstract thesis that there is such a thing as pseudo-science, and that it must be contested and controlled. The headache begins with the question of just what is pseudo-science and what is not. We can all readily agree on some of the absurd cases so interestingly described in Martin Gardner's wonderful book *Fads and Fallacies in the Name of Science* (New York: Dover Publications, Inc., 1957). But the question of exactly where science ends and where pseudo-science begins is at once important and far from simple. There is little difficulty indeed with Wilbur Glenn Voliva, Gardner's Exhibit No. 1, who during the first third of this century thundered out of Zion (Illinois) that "the earth is flat as a pancake." But parapsychology, for example, is another study and a much more complicated one. And the handful of United States geneticists who, working primarily with yeasts, feel that they have experimental warrant for Lamarckian conclusions, much to the discomfort of the great majority of their professional colleagues, exemplify the difficulties of a hard and fast compartmentalization of pseudo-science in a much more drastic way. Nobody in the scientific community wants to let pseudo-science make headway. But the trouble is that one man's interesting possibility may be another man's pseudo-science.

On the one hand, reputable scientists have often opposed genuine scientific findings as being pseudo-scientific. Lord Bacon, the high priest of early modern science, denounced Gilbert's treatise *On the Magnet* as "a work of inconclusive writing," and he spoke disparagingly of the "electric energy concerning which Gilbert told so many fables." A more recent, if less clearcut, example is the extensive opposition encountered by psycho-analysis, particularly in its early years.

But on the other hand, we have the equally disconcerting fact that reputable scientists have advanced, and their fellow scientists accepted, findings that were strictly fraudulent. One instructive case is that of the French physicist René Blondlot, which is interestingly described in Derek Price's book *Science Since Babylon* (New Haven: Yale University Press, 1961). Blondlot allegedly discovered "N-rays," which were supposed to be something like X-rays. His curious findings attracted a great deal of attention, and earned for Blondlot himself a prize from the French government. But the American physicist Robert W. Wood was able to show by careful experimental work that Blondlot and all who concurred in his findings were deluded. It is thus to be recognized not only that pseudo-science exists, but that it sometimes even makes its way into the sacred precincts of highly orthodox science. This, of course, does not help to simplify the task of discriminating between *real* and *pseudo*-science.

But let us return to the ethical issues involved. These have to do not with the uncontroversial thesis that pseudo-science must be controlled, but with procedural question of the *means* to be used for the achievement of this worthy purpose. It is with this problem of the means for its control that pseudo-science poses real ethical difficulties for the scientific community.

The handiest instrumentalities to this end and the most temptingly simple to use are the old standbys of thought control—censorship and suppression. But these are surely dire and desperate remedies. It is no doubt highly unpleasant for a scientist to see views that he regards as "preposterous" and "crackpot" to be disseminated and even to gain a considerable public following. But surely we should never lose sensitivity to the moral worth of the methods for achieving our ends or forget that good ends do not justify questionable means. It is undeniably true that scientists have the duty to prevent the propagation of error and misinformation. But this duty has to be acted on with thoughtful caution. It cannot be construed to fit the conveniences of the moment. And it surely cannot be stretched to give warrant to the suppression of views that might prove damaging to the public "image" of science or to justify the protection of one school of thought against its critics. Those scientists who pressured the publisher of Immanuel Velikovsky's fanciful *Worlds in Col-*

lision by threatening to boycott the firm's textbooks unless this work were dropped from its list resorted to measures that I should not care to be called on to defend, but the case is doubtless an extreme one. However, the control exercised by editors and guardians of foundation purse-strings is more subtle, but no less effective and no less problematic.

The main point in this regard is one that needs little defense or argument in its support. Surely scientists, of all people, should have sufficient confidence in the ability of truth to win out over error in the market place of freely interchanged ideas as to be unwilling to forgo the techniques of rational persuasion in favor of the unsavory instrumentalities of pressure, censorship, and suppression.

Ethical Problems Regarding the Allocation of Credit for Scientific Research Achievements

The final set of ethical problems arising in relation to scientific research that I propose to mention relate to the allocation of credit for the achievements of research work. Moral philosophers as well as students of jurisprudence have long been aware of the difficulties in assigning to individuals the responsibility for corporate acts, and thus to allocate to individual wrongdoers the blame for group misdeeds. This problem now faces the scientific community in its inverse form—the allocation to individuals of credit for the research accomplishments resulting from conjoint, corporate, or combined effort. Particularly in this day of collectivized research, this problem is apt to arise often and in serious forms.

Let no one be put off by stories about scientific detachment and disinterestedness. The issue of credit for their findings has for many centuries been of the greatest importance to scientists. Doubts on this head are readily dispelled by the prominence of priority disputes in the history of science. Their significance is illustrated by such notorious episodes as the bitter and long-continuing dispute between Newton and Leibniz and their followers regarding priority in the invention of the calculus—a dispute that made for an estrangement between English and continental mathematics which lasted through much of the eighteenth century, considerably to the detriment of the quality of British mathematics during that era.

But to return to the present, the problem of credit allocation can come up nowadays in forms so complex and intricate as to be almost inconceivable to any mind not trained in the law. For instance, following out the implications of an idea put forward as an idle guess by X, Y, working under W's direction in Z's laboratory, comes up with an important result. How is the total credit to be divided? It requires no great imagination to think up some of the kinds of problems and difficulties that can come about in saying who is to be credited with what in this day of corporate and collective research. This venture lends itself to clever literary exploitation in the hands of a master like C. P. Snow.

Retrospect on the Ethical Dimension of Scientific Research

Let us now pause for a moment to survey the road that we have traveled thus far. The discussion to this point has made a guided tour of a major part of the terrain constituting the ethical dimension of scientific research. In particular, we have seen that questions of a strictly ethical nature arise in connection with scientific research at the following crucial junctures:

1. the choice of research goals
2. the staffing of research activities
3. the selection of research methods
4. the specification of standards of proof
5. the dissemination of research findings
6. the control of scientific misinformation
7. the allocation of credit for research accomplishments

In short, it seems warranted to assert that, at virtually every juncture of scientific research work, from initial inception of the work to the ultimate reporting of its completed findings, issues of a distinctively ethical character may present themselves for resolution.

It is a regrettable fact that too many persons, both scientists and students of scientific method, have had their attention focused so sharply upon the abstracted "logic" of an idealized "scientific method" that this ethical dimension of science has completely escaped their notice. This circumstance seems to me to be particularly regrettable because it has tended to foster a harmful myth that finds strong support in both the scientific and the humanistic camps—namely, the view that science is antiseptically devoid of any involvement with human values. Science, on this way of looking at the matter, is so purely objective and narrowly factual in its concerns that it can, and indeed should, be wholly insensitive to the emotional, artistic, and ethical values of human life.

I hope that my analysis of the role of ethical considerations within the framework of science has been sufficiently convincing to show that this dichotomy, with its resultant divorce between the sciences and the humanities, is based on a wholly untenable conception of the actual division of labor between these two areas of intellectual endeavor. It is my strong conviction that both parties to this unasked for divorce must recognize the spuriousness of its alleged reasons for being, if the interests of a wholesome unity of human understanding are to be served properly.

The humanist, for his part, must not be allowed to forget that in the whole course of the intellectual history of the West, from Aristotle and his predecessors to Descartes, Newton, Kant, James, and Einstein, science has been a part of the cultural tradition in its larger sense. Throughout the whole course of the development of our civilization, science has always merited the historic epithet of "natural philosophy." No matter how much our way of describing the facts may change, there is little doubt that this basic circumstance of the formative role of science in molding the *Weltanschauung* basic to all of our areas of thought will remain invariant.

On the other hand, the scientist, for his part, should realize that science has worth and status enough in its own right that its devotees can dispense with claiming that, although the handiwork of imperfect humans, it is somehow mysteriously endowed with virtually superhuman powers, such as are implied by claims of actual achievement of what in fact are remote ideals, and in some instances unworkable idealizations, like pure open-mindedness, complete objectivity, and perfect rationality.

From the standpoint of a realistic appreciation of the nature of science as a human creation and activity, it seems to me that a heightened awareness of the humanistic dimension of science—which I have tried to illustrate on the ethical side—can serve the best interests of both of these two important working areas of the human intellect. Instead of being nearly separable, these domains are interpenetrating and interdependent. Rather than being strange bedfellows, the sciences and the humanities are ancient and mutually beneficial partners in that pre-eminently humane enterprise of leading man to a better understanding both of himself and of the world in which he lives.

Notes

1. *Science,* Vol. CXXXIII (1961), 258-259.

2. *Ibid.,* Vol. CXXXIV (1961), 162.

3. *Ibid.,* Vol. CXXXIII (1961), 164.

4. *Ibid.,* Vol. CXXXIII (1961), 261.

5. Quoted by Stephen G. Brush in *American Scientist,* Vol. XLIX (1961), 211-212.

6. To anyone interested in this curious topic, I refer the eye-opening article by Bernard Barber, "Resistance by Scientists to Scientific Discovery," *Science,* Vol. CXXXIV (1961).

7. Quoted by Brush, *op. cit.,* 210.

8. Quoted by Barber, *op. cit.*

Indexes

Index of Names

Abro, A. d', 5
Agassi, Joseph, 226, 231, 244
Albert of Saxony, 8, 22
Archimedes, 7, 11, 32, 160
Aristotle, 7, 12, 22, 40, 41, 43, 52, 102, 103, 104, 105, 108, 115, 275
Austin, J. L., 187

Bacon, Francis, 156, 181, 219, 226, 232, 253, 273
Barber, Bernard, 276
Barnes, Harry E., 259
Becker, Howard, 259
Bedau, Hugo, 100
Bergmann, Peter G., 26
Berkeley, George, 147, 148, 211, 229, 234, 240
Bernoulli, Daniel, 25
Bohm, David, 84, 87, 115, 116, 153, 226, 233
Bohr, Niels, 83, 84, 91, 93, 147, 161, 162, 233, 241, 256
Boltzmann, Ludwig, 172, 239
Bondi, Hermann, 144
Born, Max, 4, 5, 79, 81, 83, 86, 88, 90, 99, 155, 166, 235
Boyle, Robert, 20, 234
Brahé, Tycho, 40, 52
Braithwaite, Richard B., 76
Breasted, James H., 258
Brecht, Bertolt, 218, 231
Bridgman, Percy W., 241
Broad, C. D., 27
Broglie, Louis de, 78, 80, 86, 87
Brush, Stephen B., 276
Buck, Roger, 260
Buridan, Jean, 7, 8, 22, 35, 73
Butterfield, Herbert, 31, 40, 66, 228

Cajori, Floriam, 36, 234
Carnap, Rudolf, 76, 77, 152, 171, 253, 254
Caspar, Max, 22
Caspari, Ernst, 116
Cassirer, Ernst, 27, 225
Chandrasekhar, Subrahmanyan, 123
Clagett, Marshall, 73, 243
Collingwood, Robin G., 252, 254
Colodny, Robert G., 1, 115, 116, 226
Copernicus, Nicolaus, 7, 17, 52, 134, 139, 141

Darwin, Charles, 110
De Broglie, Louis (see Broglie, Louis de)
Democritus, 233
Descartes, René, 9, 17, 22, 31, 33, 34, 35, 36, 40, 41, 72, 73, 147, 160, 181, 233, 255, 275
Dirac, Paul, 4, 5, 128
Dodds, Eric R., 221, 258
Duhem, Pierre, 22, 155, 168, 236, 246
Dürer, Albrecht, 221, 222

Eddington, Arthur, 240
Ehrenfest, Paul, 243, 245
Einstein, Albert, 5, 25, 26, 101, 105, 135, 140, 142, 154, 171, 176, 234, 236, 240, 241, 245, 249, 275
Ellis, Brian, 29, 69, 70, 72, 74
Euclid, 33
Euler, Leonard, 16

Feigl, Herbert, 154, 196, 219, 227, 230, 236, 242, 245
Fermat, Pierre de, 3
Feyerabend, Paul K., 115, 116, 145
Fischer, Kuno, 259

Fischl, Johann, 254
Frankfort, Henri, 225, 258

Galen, 233
Galilei, Galileo, 7, 9, 10, 17, 20, 22, 24, 25,
    30, 31, 32, 33, 39, 67, 72, 115, 153,
    160, 162, 168, 218, 227, 228, 229, 231
Gardner, Martin, 272
Gasking, D. A. T., 68
Gassendi, Pierre, 9
Gauss, Karl F., 16, 118
Geikie, Archibald, 219
Gilbert, William, 273
Goethe, Johann W., 158, 237, 238, 248,
    253
Gombrich, Ernst, 221
Grünbaum, Adolf, 15, 16, 116, 240, 241

Hamilton, William, 4
Hanson, Norwood R., 6, 27, 69, 248, 252
Hartmann, Nicolai, 240, 244, 249
Hawkins, David, 102, 116
Heaviside, Oliver, 272
Hegel, Georg W. F., 155, 193, 224, 235, 253,
    254
Heisenberg, Werner, 83, 84, 93, 97, 98, 154,
    163, 224, 231, 233, 244, 248
Helmholtz, Hermann, 237
Hempel, Carl, 242
Herschel, John F. W., 218
Hertz, Heinrich, 3, 11, 12, 16, 21, 25
Hilbert, David, 73
Hill, Edward L., 242, 245
Hobbes, Thomas, 114, 115
Hooke, Robert, 156, 157, 234, 235, 248
Hubble, Edwin P., 135
Hume, David, 147, 233
Huyghens, Christian, 25, 31, 36, 40, 62,
    248

Jaspers, Karl, 220, 252
Jordan, Pascual, 225

Kant, Immanuel, 104, 119, 238, 239, 275
Kepler, Johannes, 9, 17, 40, 45, 155, 229
Keynes, John M., 24, 235
Körner, Stephan, 247
Koyré, Alexandre, 231, 233
Kramers, Hendrik A., 163
Kuhn, Thomas S., 5, 153, 158, 232, 239,
    244, 247, 250

Lagrange, Joseph, 11, 12
Landau, Lev D., 244
Laplace, Pierre S., 3, 16, 73, 230
Leibniz, Wilhelm, 16, 25, 105, 261

Liebig, Justus von, 271
Locke, John, 147, 233
Lorentz, Hendrik A., 171, 241, 245
Ludwig, Günther, 95, 97
Lukács, Georg, 224

McCrea, William H., 135, 140, 144
McCulloch, Warren S., 117
Mach, Ernst, 3, 12, 16, 22, 24, 25, 27, 131,
    143, 166, 169, 182, 239, 240, 246
McLaurin, Colin, 243
McVittie, George C., 135
Mann, Golo, 243
Margenau, Henry, 101
Maupertuis, Pierre, 3
Maxwell, Grover, 236, 245
Maxwell, James, 110, 116, 171, 172
Mead, Margaret, 267
Mesmer, Franz, 271
Michelet, Jules, 220
Milne, Edward A., 135, 140, 144
Minkowski, Hermann, 3
Morrison, Philip, 118

Naess, Arne, 219
Nagel, Ernest, 169, 242, 245, 249, 253, 254,
    260
Neumann, John von, 16, 17, 79, 91, 93
Newton, Isaac, 6, 8, 9, 11, 12, 13, 14, 16,
    18, 19, 20, 21, 24, 25, 29, 30, 33, 36,
    37, 38, 39, 40, 41, 42, 43, 48, 49, 52,
    53, 54, 57, 58, 59, 60, 61, 62, 64, 65,
    75, 103, 128, 131, 155, 156, 157, 160,
    161, 162, 171, 180, 182, 229, 230,
    231, 233, 235, 237, 239, 240, 248,
    275
Nicholas of Cusa, 8, 22

Occam, William of, 73
Oppenheim, Paul, 100
Orêsme, Nicolas, 35, 73

Pasteur, Louis, 271
Pauli, Wolfgang, 122, 225, 233
Pearson, Karl, 25
Perrin, Jean B., 176, 249
Philoponus, 7, 8, 22
Planck, Max, 122, 245, 272
Plato, 113, 165, 232
Poincaré, Henri, 3, 16, 20, 29, 52
Polanyi, Michael, 4, 240
Popper, Karl, 152, 153, 161, 164, 165, 201,
    239, 240, 241, 245, 247
Price, Derek, 273
Ptolemy, 7
Putnam, Hilary, 75, 115

Quine, Willard, 170

Radin, Paul, 224
Rayleigh, Lord (John W. Strutt), 270, 271
Reichenbach, Hans, 229, 230, 246
Rescher, Nicholas, 261
Rosenfeld, L., 244
Russell, Bertrand, 25, 26, 145, 195, 242, 258
Rutherford, Ernest, 240
Rynin, David, 153

St. Augustine, 243
Sambursky, Samuel, 232
Santillana, Giorgio de, 225, 228
Sarton, George, 219
Schrödinger, Erwin, 1, 2, 4, 79, 95, 97, 99, 100
Sciama, Dennis W., 27, 144
Sellars, Wilfrid, 27
Sharp, David, 101
Slipher, Vesto M., 135
Smart, John J., 153, 256, 257
Snow, Charles P., 265, 275
Spencer, Herbert, 116
Spinoza, Benedict, 105, 115

Stallo, John B., 25
Stebbing, Susan, 256
Svedberg, Theodore, 176

Toulmin, Stephen, 76
Tranekjaer-Rasmussen, Edgar, 206
Tyndall, John, 237

Velikovsky, Immanuel, 273
von Neumann, John (*see* Neumann, John von)

Weisskopf, Victor, 116
Weyl, Hermann, 3, 6, 102, 118, 245
Wiener, Philip, 236
Wigner, Eugene, 101
Wittgenstein, Ludwig, 246
Wood, Robert W., 273

Young, Thomas, 172

Zeno, 107

Index of Topics

Ancient science 3, 7, 8, 11-12, 41, 103, 165, 219, 228, 232, 243 (see also Archimedes, Aristotle, Euclid, Galen, Platonism, Pythagoreanism)

Andromeda nebula, 126, 129 (see also Galactic structure, Physics of the large)

Anti-matter, 130

Antiperistasis, 7 (see also Aristotelian mechanics)

Aristotelian mechanics, 8, 43, 155, 228, 232, 246

Astrophysics, 1, 3 (see also Physics of the large)

Atomism, 20, 143, 186

Autonomy principle, 174

Baryon number, 130, 142

Biology and geometrical framework, 112

Biophysics, 1 (see also Thermodynamics of purpose)

Born interpretation of quantum mechanics, 83, 84, 86, 88, 99

Broglie relation, 100

Brownian movement, 107, 175-176, 249

Cartesian physics, 31, 33, 34, 35, 37, 40-41, 65, 67, 72, 181, 233 (see also Scientific revolution of the seventeenth century)

Causal principle, 53, 67, 104, 113, 115, 140, 144, 148, 235, 244, 258 (see also Laws of motion and philosophical analysis)

Chandrasekhar number, 122, 123

Classical mechanics, 3, 12, 15, 16, 75, 80, 82, 87, 105, 107, 115, 128, 131, 161, 233, 240 (see also Classical physics, Newtonian methodology)

Classical physics:
and macro-observables, 93, 232
and teleology, 105

Complementarity principle (see Quantum mechanics)

Conservation laws, 3, 30-31, 33, 36-37, 39, 50, 62, 104, 107, 108, 162, 169 (see also Dynamical theory and mathematics)

Consistency condition, 172-177, 179, 247

Conventionality of dynamic principles, 52 (see also Dynamical laws and semantic problems)

Conventionality of Newton's second law, 58, 61 (see also Newtonian methodology)

Copenhagen interpretation of quantum mechanics, 79-80, 84-85, 88, 89, 90, 91-93, 94, 95, 100, 162-163, 166

Coriolis force, 24, 25, 28, 49

Correspondence principle, 162, 170 (see also Quantum mechanics)

Cosmic time scales, 136-137, 139, 142, 144

Cosmological principle, 134, 137, 139

Cosmological theory, 1, 3, 60, 75, 102, 108-109, 118, 119, 130, 132-133, 134, 135, 140, 141, 143, 144, 214, 215, 219
and geometrical framework, 130, 134, 135, 140, 141

Darwinism, 110

Determinism, 105, 115-116, 166, 171, 231

Determinism (*Cont.*)
(*see also* Causal principle, Laws of motion and philosophical analysis, Materialism)
Dialectical philosophy, 150, 153, 247, 252 (*see also* Materialism)
Dogmatism, 149, 150, 163, 172, 177, 179, 217, 229, 233, 243, 247, 252
Doppler shift, 128, 134, 135, 136, 138, 139, 140, 142, 144 (*see also* Physics of the large)
Dynamical laws:
qua contingent, 104
and semantic problems, 12, 29, 65, 72 (*see also* Laws of motion and philosophical analysis)
Dynamical theory and mathematics, 17, 19, 60, 68, 73

Einsteinian theory of gravitation, 131, 140, 141, 142 (*see also* Relativity theory)
Electrodynamics, 26, 56, 105, 128, 131, 161, 241 (*see also* Classical physics)
Empirical content of Newtonian mechanics (*see* Newtonian mechanics, Newtonian methodology)
Empiricism:
Baconian version, 156
and experimentalism, 146, 156, 158, 159-160
historical background, 145, 147, 152, 154, 160, 161, 171, 217-218, 233, 238-239
logical, 149, 242, 245
and observation, 146, 147, 170, 188-189, 204-205, 207, 212, 214-215, 216-217
and quantum theory, 154, 226, 232-233, 242
radical, 149, 154, 158, 163, 177, 238, 241, 256
and semantic problems, 150, 151, 171, 184-185, 188-189, 194-195, 200, 208, 209, 210, 226, 227, 249, 253, 255, 256, 258
and seventeenth-century science, 30-31, 41, 57, 147, 152, 155, 156, 159, 160, 228-229, 230, 231, 235, 237, 248
twentieth-century, 161, 162, 231, 241
Entropy, 112, 113, 166 (*see also* Thermodynamics, second law)
Epistemology, problems of, 3, 80, 93, 100, 119, 153, 155, 165, 168, 171, 172, 193, 198, 201, 209, 211, 216, 218, 232, 240, 249, 258, 259 (*see also* Laws of motion and philosophical analysis)
Ethical dimensions of scientific research, 2, 261-276
Euclidian geometry, 62, 118, 138, 139
Euler-Lagrange formalism, 3, 71

Evolution, 102, 108, 109, 110, 119 (*see also* Darwinism, Thermodynamics of purpose)
of chemical elements, 121, 125
Expansion of the universe, 133, 134, 135, 139 (*see also* Physics of the large, Relativity theory)

Force concepts, 7, 8, 13, 19, 20-21, 24-25, 26, 27, 28, 31, 32, 35, 37, 39, 42, 44, 46-47, 49, 53, 54, 56-58, 63, 65, 67, 69-70, 72, 132 (*see also* Inertial principles, Laws of motion, Newtonian mechanics)
Foucault's pendulum experiment, 24, 28, 131

Galactic structure, 124, 125, 127 (*see also* Physics of the large)
Galilean mechanics, 9-11, 22-23, 32, 67, 72, 115, 168, 227-229, 231 (*see also* Scientific revolution of the seventeenth century)
Gestalt psychology, 195
God, 8, 14, 33, 40, 115, 124, 165, 231, 233, 242 (*see also* Mechanics and concepts of deity)
Gravitational laws, 19, 26, 27, 41, 48, 49, 68, 70, 119, 125, 128, 129, 130, 131, 132, 134, 140, 142, 155, 234, 236 (*see also* Einsteinian theory of gravitation, Relativity theory)

Hamilton-Jacobi formalism, 3
Hilbert spaces, 100 (*see also* Mechanics and geometrical framework)
History of science, explanatory problems of, 3, 31, 72, 182, 226, 229, 231, 232, 235, 243, 244, 247-248, 250, 270-271
Hooke's law, 56, 207
Hubble's constant, 136-137, 141
Huyghens and clarification of momentum vector, 36 (*see also* Force concepts, Scientific revolution of the seventeenth century)

Impetus theory, 8, 180, 227 (*see also* Medieval mechanics)
Inertial principle, 8, 9, 13, 19, 24, 25, 26, 27, 31, 32, 34, 39, 45, 46, 50, 51, 65, 69, 131, 132, 137, 233 (*see also* Force concepts, Laws of motion and philosophical analysis, Mass concepts)
Information theory, 109, 111-112, 113

Kepler's laws, 45, 71, 120, 235
Kinetic theory, 15, 20, 24, 59, 166, 171, 175,

Kinetic theory (*Cont.*)
186, 230, 249, 256-257 (*see also* Classical physics, Statistical mechanics)

Laws of motion and philosophical analysis, 6, 16-17, 19, 21, 25, 29, 30, 31, 46, 50-52, 57, 59, 61, 63, 65, 69-71, 106, 236 (*see also* Epistemology)
Laws of succession, 45, 46, 49
Logical empiricism (*see* Empiricism, logical)
Logical positivism (*see* Positivism)

Mach's principle, 25, 27, 131, 143, 169
Mass concepts, 44, 46, 58, 60-61, 65, 132, 139, 168-169, 245
Materialism, 103, 104, 181, 185-186, 187, 192-193, 195, 197, 201, 254, 255, 257 (*see also* Determinism, Dialectical philosophy)
Maxwell's demon, 116
Maxwell's equations, 131
Meaning invariance, 151, 164, 167, 168, 170, 172, 179-180, 232, 247
Mechanics:
  and concepts of deity, 33, 41
  and geometrical framework, 13, 15, 17, 18, 19, 33, 71, 100
  and theory of measurement, 19, 24-25, 55, 66, 91, 106
Medieval mechanics, 8, 34-35, 41, 43, 73, 180 (*see also* Aristotelian mechanics)
Metaphysics, 150, 152, 162-163, 172, 181, 182-183, 186, 210, 226, 231, 239, 246 (*see also* Determinism; Dogmatism; Laws of motion and philosophical analysis; Materialism; Monism; Pluralism, theoretical)
Mind-body problem, 186, 189-192, 195-197, 199-203, 205-207, 211, 212-213, 220, 234, 254-256, 257, 258, 260
Mohs-scale, 257-258
Monism, 149, 151, 167-168, 171, 173, 227, 244 (*see also* Metaphysics; Pluralism, theoretical)
Myth, 146, 178-179, 220, 223-225, 227 (*see also* Dogmatism, Witchcraft)

Neurophysiology, 199 (*see also* Information theory, Mind-body problem)
Neutrino, 120, 123, 133, 143 (*see also* Physics of the large, Quantum mechanics)
Newtonian mechanics, 3, 6, 9, 11-12, 15, 18, 20, 27, 36-40, 48, 49, 50, 53, 54, 60-64, 70, 75, 77, 119, 129, 132, 138, 139, 161, 227, 233 (*see also* Force

Newtonian mechanics (*Cont.*)
concepts, Newtonian methodology, *Principia*, Scientific revolution of the seventeenth century)
empirical content of, 59-65 (*see also* Laws of motion and philosophical analysis)
Newtonian methodology, 31, 40-41, 53, 57, 61, 75, 155, 156, 157, 159, 171, 182, 198, 229, 230, 232, 234, 237, 239, 242, 248

Olber's paradox, 139 (*see also* Physics of the large)
Ontology, 150, 170 (*see also* Metaphysics)
Operationalism, 76-77, 241 (*see also* Empiricism)
Ordering procedures, 55-56, 58-59 (*see also* Laws of succession, Scales)
Origin of life, 109 (*see also* Darwinism, Evolution, Physics of the large)
Orion, 124 (*see also* Galactic structure, Physics of the large)

Philosophical analysis (*see* Laws of motion)
Photons, 81, 82, 94, 109 (*see also* Physics of the large, Quantum mechanics)
Physics of the large, 118-144
Planck's quantum of action, 122 (*see also* Physics of the large, Quantum mechanics)
Platonism, 164-165, 233 (*see also* Ancient science, Metaphysics, Pythagoreanism)
Pleiades, 123, 124 (*see also* Galactic structure, Physics of the large)
Pluralism, theoretical, 149, 151, 153, 164, 167, 172, 175, 216 (*see also* Consistency condition, Empiricism, Meaning invariance, Monism)
Positivism, 77, 245, 246, 254
Pragmatic principle of meaning, 203, 214
Pragmatic theory of observation, 152, 212-213, 217, 247
Principia Mathematica Philosophiae Naturalis, 3, 18, 24, 25, 34, 36, 37, 67, 75, 234-235 (*see also* Newtonian mechanics, Newtonian methodology)
Psi (ψ) function, 77, 82 (*see also* Quantum mechanics)
Pythagoreanism, 4, 5 (*see also* Ancient science, Platonism)

Quantum mechanics, 5, 75, 77, 79, 80, 86, 94, 97-98, 100, 107, 122, 143, 162, 251
Born interpretation, 83, 84, 86, 88, 99

Quantum mechanics (*Cont.*)
and classical physics, 162, 163, 166-167, 171-172, 251
and complementarity principle, 85, 100, 163, 166, 178, 231, 239, 250
Copenhagen interpretation, 79-80, 84-85, 88, 89, 90, 91-93, 94, 95, 100, 162-163, 166 (*see also* Dogmatism)
and empiricism, 177-178, 251 (*see also* Empiricism and quantum theory)
and "hidden variable" theory, 84, 85, 86, 88, 89, 90
and problems of epistemology, 80, 84, 93, 100, 251 (*see also* Epistemology)
and problems of mathematics, 77-78, 80-81, 90, 94, 100, 251
and problems of measurement, 76, 79-81, 84-85, 87, 88, 89, 91-92, 95, 98-99, 115, 244, 251
and problems of semantics, 76-77, 256
and statistical interpretation, 90, 91, 99 (*see also* Born interpretation)
and teleology, 107, 108, 109
Quantum pressure, 122 (*see also* Physics of the large)
Quantum theory, history of, 78-79, 86, 90, 150 (*see also* Empiricism of the twentieth century)

Radar, 134, 143 (*see also* Physics of the large)
Radical empiricism (*see* Empiricism, radical)
Reductionism, 169, 245, 252, 260 (*see also* Mind-body problem)
Relativistic mechanics, 3, 26, 44 (*see also* Einsteinian theory of gravitation, Force concepts, Mass concepts, Physics of the large, Relativity theory)
Relativity theory, 26, 71, 119, 130, 134, 139, 140, 141, 155, 168-169, 171, 198, 227, 236, 240-241, 245 (*see also* Einsteinian theory of gravitation, Mass concepts, Mechanics and geometrical framework)
Royal Society, 146, 155, 157-158, 159, 162, 219, 248, 270 (*see also* Newtonian methodology, *Principia*, Scientific revolution of the seventeenth century)

Scales, 55, 57, 58, 61, 62, 64, 66, 108 (*see also* Mass concepts, Mechanics and theory of measurement)
Schrödinger's "cat" case, 94-100 (*see also* Quantum mechanics)
Schrödinger's wave equation, 5, 79 (*see also* Quantum mechanics)

Science:
and government, 264, 269 (*see also* Ethical dimensions of scientific research, Science and orthodoxy)
and orthodoxy, 4, 270-273 (*see also* Science and government)
and philosophy, 1, 2, 3, 30, 231, 232, 238, 243, 246, 254 (*see also* Determinism, Dogmatism, Laws of motion and philosophical analysis, Materialism, Metaphysics)
and values, 262 (*see also* Dogmatism, Ethical dimensions of scientific research, Science and orthodoxy)
and warfare, 265
Scientific revolution of the seventeenth century, 17, 30-31, 52, 65, 72, 104, 105, 152, 153, 168, 180-181, 219, 227, 228, 233, 248, 273, 274 (*see also* Aristotelian mechanics, Cartesian physics, Empiricism, historical background, Galilean mechanics, Hooke's law, Huyghens, Leibniz, Newtonian methodology, *Principia*, Royal Society)
Semantic theory of observation, 203 (*see also* Pragmatic principle of meaning, Pragmatic theory of observation)
Space exploration, 120
Space-time, 130 (*see also* Mechanics and geometrical framework, Physics of the large, Time)
Statistical mechanics, 106, 107, 116, 175-176 (*see also* Kinetic theory, Quantum mechanics)
Steady state theory, 141-142 (*see also* Expansion of the universe, Hubble's constant, Physics of the large)
Stellar thermodynamics, 120-123 (*see also* Physics of the large, Thermodynamics of purpose)
Stress-strain concepts, 43-44, 47, 50, 56 (*see also* Force concepts, Inertial principles)
Supernovae, 128, 130, 218 (*see also* Physics of the large, Stellar thermodynamics)
Superposition of states, 81-82, 90, 96-97 (*see also* Quantum mechanics)
Synonymy, 197, 200, 202, 254 (*see also* Meaning invariance, Mind-body problem)

Teleological theory, 3, 103, 105, 107, 108, 113, 114 (*see also* Causal principles, Evolution, Thermodynamics of purpose)

Theoretical pluralism (*see* Pluralism, theoretical)

Thermodynamics:
and biology, 2, 108, 109, 110-111 (*see also* Origin of life, Thermodynamics of purpose)
and cosmology, 108-109 (*see also* Cosmological theory, Physics of the large)
of purpose, 102-117
second law, 108, 166, 168, 171, 172, 175, 230, 239, 243 (*see also* Physics of the large)

Thermonuclear reactions, 120, 122-123, 124, 128 (*see also* Physics of the large, Stellar evolution)

Time, 103, 104, 105, 110, 119, 130, 131, 136, 138, 139, 141, 142, 144 (*see also* Cosmic time scales, Relativity theory, Space-time)

Uncertainty principle, 80, 83, 94, 107 (*see also* Quantum mechanics)

Virgo, 126, 135 (*see also* Galactic structure, Physics of the large)

Von Neumann axiomatization of quantum theory, 91 (*see also* Quantum mechanics)

Von Neumann's Body Alpha, 17-18, 25 (*see also* Laws of motion and philosophical analysis)

Wave mechanics (*see* Quantum mechanics)

Wave packets, 79-80, 86, 97-98 (*see also* Quantum mechanics)

Weyl postulate, 136, 141 (*see also* Physics of the large)

Witchcraft, 181, 220, 225, 251, 255 (*see also* Dogmatism, Myth, Scientific revolution of the seventeenth century)